# Sanitary Landfilling:
# Process, Technology and Environmental Impact

# RELATED SOLID WASTE TITLES

W.S. FORESTER and J.H. SKINNER: International Perspectives on Hazardous Waste Management, 1987

L. BONOMO and A.E. HIGGINSON: International Overview on Solid Waste Management, 1988

J.Aa. HANSEN and K. HENRIKSEN: Nitrogen in Organic Wastes Applied to Soils, 1989

J.S. CARRA and R. COSSU: International Perspectives on Municipal Solid Wastes and Sanitary Landfilling, 1990

# Sanitary Landfilling:
# Process, Technology and Environmental Impact

Edited by

## THOMAS H. CHRISTENSEN
*Department of Environmental Engineering*
*Technical University of Denmark*
*Lyngby, Denmark*

## RAFFAELLO COSSU
*Instituto di Idraulica, Università di Cagliari*
*Cagliari, Sardinia, Italy*

## RAINER STEGMANN
*Technische Universität Hamburg-Harburg*
*Hamburg, FRG*

## ACADEMIC PRESS
*Harcourt Brace Jovanovich, Publishers*
London   San Diego   New York   Berkeley
Boston   Sydney   Tokyo   Toronto

This book is printed on acid free paper ∞

ACADEMIC PRESS LIMITED
24–28 Oval Road
London NW1 7DX

United States Edition published by
ACADEMIC PRESS INC.
San Diego, CA 92101

British Library Cataloguing in Publication Data is available

ISBN 0-12-174255-5

Typeset by EJS Chemical Composition, Bath
Printed in Great Britain by TJ Press (Padstow) Ltd, Padstow, Cornwall

# Contents

v

# 6   ENVIRONMENTAL  IMPACTS

# 7   DESIGN

# Preface

Sanitary landfilling constitutes a major component of the solid waste disposal system in almost every country. During the last two decades the practice of sanitary landfilling has developed into fully engineered facilities subject to extensive environmental regulations. Currently a variety of new developments are under study or being implemented in various countries. Regrettably often landfills do not meet the desired standard partly for lack of knowledge about relevant information and experience. There is therefore the manifest need for an international reference book or compilation on sanitary landfilling, but to write a comprehensive reference book would be a major undertaking. However, the success of the First International Symposium on Sanitary Landfilling held in Cagliari (Sardinia, Italy) in October 1987 gave rise to the idea of editing the appropriate papers from this symposium, supplemented with a few new contributions, into such a reference book. This present work is the outcome of this endeavour.

The science, technology and management of sanitary landfilling have become a truly multidisciplinary and international activity as is apparent from the spread of contributors to this volume. The many authors participating in the book have added to its diversity but, we hope, not at the expense of its homogeneity.

We hope that the readers will appreciate the book in their professional work and research, we urge you to make reference directly to the author and title of chapters. The responsibility for the technical content of the book primarily rests with the individual authors and we cannot take any credit for their work. Our role as editors has been one of reviewing and of making constructive suggestions during the preparation of the final manuscripts. Should any errors have remained undetected at the proof reading stage then the responsibility for these rests with us.

Finally we would like to thank all our contributors for allowing us to edit their manuscripts and ask for everybody's forbearance with our unintended mistreatment of the English language. For this reason we would like to give credit to Nicki Dennis of Academic Press for correcting the English where necessary and giving shape to the book.

We would like to thank Giorgio Carta and Franco Mannoni, Assessors for the Sardinian Regional Government which strongly supported and made possible the Symposium, and Paola Cannas and Silvaldo Gadoni from the Sardinian Department of the Environment for their invaluable help in organizing the Symposium. We furthermore would also like to thank Cinzia Acaia, Anna Farmer and Francesca Carnevali for their patient assistance with the editing and the many drafts of the book.

*Thomas H. Christensen*
*Raffaello Cossu*
*Rainer Stegmann*

# 1. INTRODUCTION

# 1.1 Role of Landfilling in Solid Waste Management

R. COSSU

*Institute of Hydraulics, University of Cagliari, Piazza d'Armi, 09100 Cagliari, Italy*

## HISTORY

Since wastes were produced the routes for its disposal have been landfilling, recycling and combustion. From ancient times to the present, waste has been burned, reused or dumped with a qualitative level of processing which depends on the economical, cultural, social and political developments of organized people in their own particular situations (geography, climate, resources, etc.).

Increasing urbanization of society, which forced people to live nearer the wastes, increasing levels of awareness, with higher perception of nuisances caused by waste disposal and environmental risks, increasing democracy in political administration, problems for economic activities caused by obstacles which originate from poor waste management, are some of the factors which in the last century and particularly in recent years have necessitated an upgrading in the quality of waste disposal.

The history of waste management, with peculiar reference to landfill, is largely *reactive* history in the sense that concepts of design were developed as a reaction to problems and to fluctuations of the above mentioned factors (Ham, 1989).

It is time to take a proactive approach to consider what the different waste disposals methods should be doing, to define the aims of each individual method and to design concepts to meet these objectives (Ham, 1989).

The role of sanitary landfilling is changing in the light of the incoming waste disposal concept, but it must be considered carefully according to the present and the desired future situation.

3

In fact the present situation reflects the levels in different countries and the gap with the future has to be gradually bridged, taking into account the starting step.

Each country has to develop its own disposal strategy, where landfilling, as discussed in the following paragraphs, invariably plays the most important role.

Experts and researchers have the task of providing the best possible landfill (BPL) for each stage from the present to the future.

## CURRENT SITUATION

Although in many countries there is increasing concern regarding the environmental risks of landfilling (and incineration) and many people believe that separate collection and recycling represents a correct solution, the current situation of waste disposal does not allow dreaming.

In Fig. 1 the results of a recent survey on municipal solid waste (MSW) disposal methods in 15 industrialized countries is reported (Carra and Cossu, 1989). The countries were Canada, Denmark, Finland, France, West Germany, Italy, Japan, Netherlands, Poland, South Africa, Sweden, Switzerland, United Kingdom, and the United States. The population in these countries amounts to 730 million.

The average MSW production per-capita is 0.94 Kg/day and the total quantity of MSW in these countries is 250 Mt (million tons).

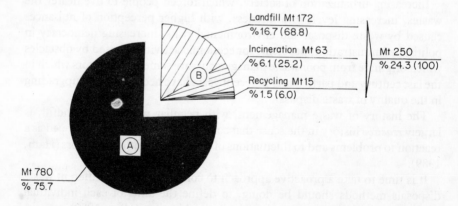

**Figure 1.**World MSW production (A + B) and disposal routes according to a recent investigation in 15 industrialised countries (B). Data from Carra and Cossu (1989).

Landfill in these countries represents the prevailing system for waste disposal (68.8%) while incineration is the most-widely used second choice (25.2%). Recycling, formed almost exclusively of composting and RDF production, represents 6% of MSW disposal routes.

The rate of the national production of MSW which is directly landfilled varies from country to country, ranging from 30–50% (Switzerland, Japan, France, Denmark, Netherlands) to 70–95% of the most industrialized countries (Italy, Germany, Canada, United Kingdom, Austria) to 95–100% in the rest of the world.

It seems important to point out that not necessarily does landfill in the different countries mean sanitary landfill and even in well developed countries open dumps or poorly managed landfills exist and are a large part of the waste disposal system. This leads to the initial statement that there is still an important role for traditional sanitary landfilling in industrialized countries.

Secondly, we must consider those countries with low landfill utilization. Here landfill plays the important role of receiving ashes and slag from the combustion processes which represent the leading method of waste disposal.

Thirdly, the role of landfilling in tourism areas where the growth of the population in summertime requires flexibility in waste disposal should be considered. This could be offered by landfilling alone.

A fourth remark refers to the overall world MSW production. Assuming 5000 million people as the world population and estimating a MSW generation of 0.5 Kg/capita/day for the population not included in the above mentioned survey (4270 million inhabitants) the amount of MSW of this population could be calculated as 780 Mt/year. With few exceptions this production is entirely disposed of in the land, mainly in the form of uncontrolled tipping or poorly managed landfilling.

Taking into account therefore the overall MSW world production (1030 Mt/year) a resulting 92–93% of the same is landfilled, and at least 60–70% is managed in an uncontrolled way. It follows therefore that there is a great potential need of sanitary landfill in the world.

## FUTURE TRENDS

As previously mentioned many industrialized countries are approaching the problem of waste disposal with a new strategy. The main aim of this strategy is to ensure the minimal environmental impact, saving raw materials by promoting resource recovery and reducing the amount of refuse to be disposed. The general scheme of an Integrated Solid Waste Management (ISWM) is shown in Fig. 2.

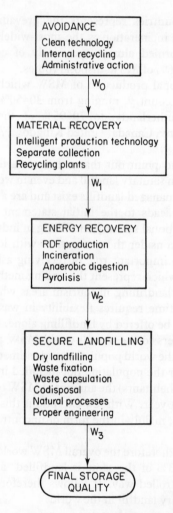

**Figure 2.** Scheme of ISWM (Integrated Solid Waste Management).

The first step of ISWM is *Avoidance*. This concept involves Clean Technology and Internal Recycling in order to minimize waste production at the source. Administrative and legal actions are necessary to promote more durable products, to prolong the circle of use and re-usage of the same good, etc.

The second step of ISWM is *Material Recovery*. A stream $(W_0)$ (Fig. 2) of waste containing many valuable materials (putrescible organics, glass, aluminium, paper, plastics) and hazardous fractions (batteries, old medicines, pesticides, PCB'S, street cleansing dust) is involved in this step.

The hazardous fraction should be separated and valuable material should be recovered according to technical and economical feasibility. This could be done by means of separate collection and automatic or manual sorting in recycling plants.

Some recovered materials could require further processing to assure recycling as in the case of putrescible organics for composting.

Naturally the success of separate collection and recycling systems is strongly related to applied technology, which must be as simple as possible and should take into account the number of people to be served and availability of a market for the material collected and the recycled products.

Furthermore an intelligent production technology is necessary to assure products which render recycling easier: i.e. avoidance of composite products (plastic and paper, glass and metal and other combinations) and marking of plastics according to their quality.

After recovering all possible materials, a waste stream ($W_1$) which contains organic and inorganic substances, of a quality and proportion which depend from the efficiency and strategy of the previous stage. In each case the waste stream ($W_1$) appears to be suitable to enter an energy recovery stage.

Where organic putrescible fraction has not been previously separated it can be at this stage by means of automatic selection, with one or more of the following goals:

1. Achieving anaerobic digestion of the separated organics, with biogas production and energy recovery.
2. Increasing the heating value of the stream ($W_1$) and assuring a better homogeneity to the same stream. This effect is positive both for RDF production and incineration.
3. Reducing the occurence of organic micropollutants in the gas effluent from the combustion process.

With reference to the abovementioned points 2. and 3. separation of inorganics such as glass, stones, ceramics and fractions containing heavy metals or heavily chlorinated plastics could be separated by automatic sorting.

This concept of assuring high quality fuel for combustion process has been applied for instance in the design of the energy recover plant in Venice.

Among the different options for energy recovery from wastes the most experienced is the incineration process which is only currently emerging from the 'dioxine tunnel'. In the near future we will assist in an increasing utilization of this system which surely assures a drastic reduction of the MSW volume.

Of course all new incineration plants will be equipped with proper facilities for gas cleaning. Both these aspects lead to transfer to landfill of high amounts of combustion and gas cleaning residues, with related environmental problems. The best marketable form of energy from MSW incineration is electricity and this is the overall most marketable resource which can be recovered from MSW.

With reference to the process of pyrolisis a lot of experience has still to be gathered in order to confirm the validity of this system. Similar argument could be considered in the full-scale application of anaerobic digestion.

Subsequent to all efforts to avoid MSW production and to recover materials and energy, 'there will always be a residue which is non-avoidable, non-recoverable, non-recyclable, non-burnable which should be disposed to land' (Ham, 1989).

At this stage the residual waste stream ($W_2$) (Fig. 2) must be disposed with minimal environmental impact. This means assuring an acceptable leaching potential on long-term basis. In order to reach this goal many technical options are available such as dry landfilling, waste fixation and waste capsulation, together of course with the occurrence of natural attenuation processes and proper engineering.

Codisposal of MSW and industrial waste, as discussed in Chapter 2.6, is a controversial method. According to some researchers it is environmentally sound, whilst others claim it is detrimental. The illustrated scheme of ISWM meets different application concepts in various countries.

For instance in West Germany separate collection for paper recovery is strongly promoted. On the contrary in Denmark paper combustion with energy recovery is considered as a better recovery option.

In Switzerland the main goal for waste treatment is to turn garbage into rocks, as safe material for final storage quality (Baccini, 1989). This is done by complete incineration and subsequent treatment of the incineration products (bottom ashes and gas cleaning residues).

## CONCLUDING REMARKS

Sanitary landfill remains an integral part of existing as well as new strategies for solid waste disposal.

At the moment landfill is applied widely with big differences in the quality (design, management) in the different countries, reflecting the economic and social development.

The quality range in the waste disposal systems varies between uncontrolled landfill, which is prevailing in most countries, and the first application of the ISWM strategies.

The main role of sanitary landfill in the current situation is to ensure a waste disposal with minimal environmental impact. This, coupled with a simple

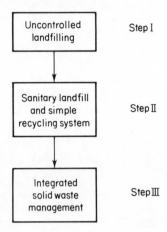

**Figure 3.** Transition steps for Solid Waste Management, from present to future.


recycling system like non-sophisticated composting, would represent the solution for third world and underdeveloped countries (Fig. 3).

The technical level in landfill realization (impermeabilization, leachate disposal, biogas management) should be compatible with the existing situation and upgraded according to further general development in the countries.

Moreover in already developed countries where waste management is still poor, sanitary landfilling is guaranteeing a transaction stage towards a correct management. As illustrated in Fig. 3 the higher level of waste management is reached in a third step, when ISWM strategy is applied.

At the moment few countries are in this situation. The ISWM System requires lower volume for landfilling but higher quality. The separation from MSW of hazardous components, material and energy recovery and the generated residues lead to final waste streams which concentrate elements potentially hazardous to the environment. For this reason it is essential to use pre-treatment steps before landfilling and adopt high quality standards for landfill design and construction in order to achieve the final storage quality.


## REFERENCES

Ham, R.K. (1989). Sanitary landfill, state of the art. Sardinia '89, Second Landfill Symposium, Porto Conte, 9–13 October, 1989.

Carra, J. and Cossu, R. (1989). 'International Perspectives in Waste Management and Landfilling'. Academic Press Ltd., London (in press).

Baccini, P. (ed) (1989). 'The Landfill Reactor and Final Storage'. Springer Verlag.

# 1.2 Principles of Landfilling—the Current Approach

R. STEGMANN

*Technical University of Hamburg-Harburg, Eissendorfer Str. 40,
D-2100 Hamburg 90, West Germany*

## PROBLEMATICS

During the realization of remedial actions for old landfill sites it became obvious that demands concerning construction and operation of landfills are increasing. Among other things, we have learned that:

1. natural soils, even if they consist of clayey or clay-like materials, may not have the expected low permeability, due to enclosed sand lenses or sand layers, as well as shrinking phenomena and clefts;
2. the liners that were installed in the past on the bases of the landfills could not meet present standards;
3. the problems associated with the co-disposal of organic industrial waste had not been realized (e.g. volatility, solubility of toxic organic components in oils, high mobility);
4. drain systems installed to collect leachate often clogged due to precipitation of different components;
5. the amount of pollutants that can be immobilized in the soil had been overestimated in several cases;
6. the loading of landfill gas and leachate with aromatic and halogenated hydrocarbons in several cases resulted in the need to increase treatment;
7. the remediation of contamined sites is very costly, and some contamination always remains. This is especially true if old landfills are encapsulated.

In addition the indefinite period of existence of a landfill has not been sufficiently taken into consideration.

11

## CONSEQUENCES OF CONTAMINATED SITES AND
## PROBLEMATICS OF SITING NEW LANDFILLS

The amount of waste to be landfilled has first to be reduced to an acceptable
level. In this regard the production of waste has to be decreased (e.g. by using
less packing material, re-using used bottles).

The amount of waste can also be reduced through using recycling centres,
and by incineration as well as composting. In order to produce a compost with
a low pollutant content, only the organic kitchen waste (bio-waste) collected
separately in the household should be used (Krogmann, 1988). In addition the
hazardous waste fraction collected in the households such as batteries,
medicine, paint and detergents, has to be treated separately. In doing so,
leachate and gas from municipal solid waste landfills will contain lower
concentrations of toxic and biologically non-degradable substances such as
halogenated hydrocarbons, mercury and heavy metals. This is also the case
with the emissions of incinerators (Stegmann, 1986).

Concerning the siting of landfills several points must be made. Firstly,
the criteria for the siting of a landfill should not only include the existence
of natural clayey soils. In many cases those areas are only available if areas
covered with forest or meadows are used. Secondly, natural barriers are still
the main criterion for the siting of a landfill but, in general, additional
measures for ground-water control are necessary. Thirdly, plants for
treatment and landfilling of waste should be sited in industrial or polluted
areas. If this request is met, together with the construction of the plant
remedial actions are carried out; in addition, an intact biotop has been saved.
In those areas more efforts may be necessary to install an adequate lining
system. Finally, the area for new waste management plants should be prepared
beforehand by planting trees and bushes.

Concerning the time dimension of a landfill several consequences have to
be borne in mind. The emission potential of landfilled material (e.g. from
municipal solid waste) has such a dimension that it cannot be assumed that the
significant amounts of soluble components are leached out only during the
period while the constructed installations for emissions control are working.
Baccini *et al.* (1987) estimated that components like carbon, copper and
sulphur will leach out of landfills for hundreds of years.

Since landfills will surely last forever, the installations for emissions control
should also last forever. Since no material and no technical installation can
meet these criteria we have to aim at controlling and, if necessary, repairing
those installations.

Investigations have to be made if waste materials can be deposited in landfills
that would not need technical installations for emissions control. In this

context salt mines should be investigated for potential use. Today we cannot say which salt mine can be used and for which kind of waste. Phenomena like the movement of salt mines are still being investigated (Hermann *et al.*, 1985). For example, salt resulting from a waste treatment process might be feasible for deposition in salt mines.

Waste could be deposited anywhere without any precautions if it is possible to modify the waste qualities so that no components will be emitted through the dust, gas or water phase. Until now this aim has not been reached even by means of solidification (Wiedemann, 1982; Wiles, 1987) but, as a result of further investigations, this may change. Investigations tend to find a way of leaching out the soluble components of specific wastes (e.g. ashes or slag) under controlled conditions.

## TYPES OF WASTE TO BE LANDFILLED

Depending on how much waste has been minimized and the extent of recycling activities there will still be waste to be landfilled. The landfill technique might depend on the kind of wastes involved. Although the separate collection of biowaste will be practised, there will still be organic degradable components in the remaining waste. In addition, landfills where municipal solid waste or municipal solid waste-like materials have to be landfilled will be necessary still for quite a while. Wastes may be classified as:

1. municipal solid waste that may change in the future due to separate collection activities;
2. mass waste (e.g. residues from gas treatment activities, mine waste and industrial sludges that are produced in huge amounts);
3. demolition waste or residues from demolition waste recycling plants);
4. soil;
5. waste water sludges;
6. bulky waste.

## LANDFILL STRATEGIES DEPENDENT UPON THE KIND OF WASTE

### Landfills Containing Significant Proportions of Biodegradable Components

This kind of landfill is operated in such a way as to optimize the biochemical degradation processes in the landfill (Stegmann *et al.*, 1986). This could mean:

1. installation of a layer of composted or compost-like municipal solid waste as

the first lift of a landfill in which the organic high-polluted leachate produced in the high density upper layers is anaerobically pretreated.

2. collected leachate on the landfill base and practising a controlled leachate recirculation (this method is very much dependent on the climatic conditions, it is feasible for example at annual precipitation rates of about 750 mm and in the Middle European climate; Stegmann, 1979);

3. no use of cover material containing high amounts of low permeability materials during the operation phase of the landfill;

4. high compaction of the waste in thin layers over large areas;

5. reaching a good shredding and mixing of the waste by means of compactors;

6. gas extraction during landfill operation.

After the main biological processes are terminated the landfill surface should be sealed in order to minimize the infiltration of precipitation. The assumption is that the landfill liner and the leachate collection system do not work forever and cannot be repaired. On a long-term basis the top cover overtakes the function of leachate minimization, and this results also in decreasing amounts of leachate to be treated. On the other hand the top seal must be monitored forever and, if necessary, repaired.

## Landfilling of Inorganic Waste (Mass Waste, Demolition Material)

The aim of this landfill conservation method should be the minimization of infiltration of water in order to avoid, as far as possible, leaching of salts and metals. These kinds of waste include fly ash, gipsum from gas treatment and slag. The waste should be landfilled in such a way that relatively small areas are actually in operation and each landfill portion is soon brought to final height and then immediately sealed. If this is not possible, intermediate cover using low-permeability material should be used. The top seal has to satisfy the same criteria. This landfill type needs also a bottom liner system including leachate collection. Those kinds of waste that might be re-used later should not be mixed with others.

The demolition waste landfill as presently in operation in West Germany without bottom liner and sufficient control of the incoming waste should not be practised in the future.

## Landfilling of Soil

In many cases soil is deposited on municipal solid waste landfills. This should no longer happen. Excavated soil should be stored and re-used.

## CRITERIA FOR LANDFILL CONSTRUCTION

The following chapters present basic recommendations for the design and construction of landfills. There is a variety of possible approaches.

Landfills should for several reasons be constructed above ground with a safe distance to the natural ground-water surface. Installation in pits has enormous disadvantages especially because of the possibility of uncontrolled water inflow from the sides as well as from working in deep leachate collection manholes often constructed in the landfill itself. The inspection of those manholes is dangerous. If the drains are to be inspected and cleaned by means of high-pressure water, manholes with large diameters are necessary.

The design of basic liner systems are discussed in several chapters and will not be discussed here. There is no best system that can be recommended. It is important to describe how a liner system should be constructed. For several reasons the tendency is to double or triple liner systems. In West Germany a combination liner consisting of 60 cm of clay directly on top of which a HDPE liner of about 2 mm thickness is placed (August, 1986).

The construction methods of liners have to be improved. It is important to minimize the influence of weather on the quality of the liner system (avoiding shrinking and so on). Since natural material is not always homogeneous, it might be necessary to upgrade the clay to a predetermined quality by specific pretreatment (e.g. by mixing, addition of additives as bentonite, installation of the optimum water contact). By those measures it is possible to reach permanently an equal quality of the natural lining material. The shape of the surface should also meet optimum construction criteria for the installation of the surface cap. The development of long-term resistant lining materials should be enforced.

During the design of the basic liner and the dewatering system at the base of the landfill in many cases no slopes have been constructed. In order to collect the leachate at the bottom of the landfill and discharge it as far as possible, a roof profile with slopes greater than or equal to 2% should be constructed; the leachate collected at the lowest point should be transported out of the landfills by means of drain pipes that at least should have a slope of 1%. In addition the dewatering system should consist of non-desolvable coarse gravel built in layers about 30 cm thick over the entire landfill base. As already mentioned, drain pipes have to be inspected by camera and cleaned by high-pressure water at least twice a year (see also Chapter 5.4). At refuse mounds in the area of the side slopes of the landfill underneath the cover a drain system should also be installed; by this means leachate penetration through the slopes can expect to be avoided.

The above liner systems are appropriate for municipal solid waste landfills, as well as for landfills for mass wastes, and partly for industrial waste landfills.

These landfills should have a final cap that is constructed similarly to the bottom liner system. This cap should be built in when the main biological processes and the associated settling have taken place. Doing so, the municipal solid waste will have enough moisture content for the biological processes to continue after the final cap is installed. It is obvious that landfill gas extraction has to be guaranteed if a landfill is capped. Otherwise gas pressure would rise in the landfills and the gas will find its way through the surface cap or through weak parts of the liner system.

Concerning the final shape of the landfill certain aesthetic aspects have to be considered. In addition, optimum surface water run-off should be possible, but erosion should be avoided. The final soil cover above the liner system should encourage plant growth; with increasing depth this cover soil will have its own water budget. Underneath the soil cover and on top of the liner a drain system should be installed, so that there will be no water build-up in the root zone. In addition the penetration of roots into the liner system should be avoided. The surface run-off system could be supported by installing a ditch system.

With regard to landfill operation the advantage of the growth of trees before the landfill is constructed has already been mentioned. In addition, the dams of the landfill should always be higher than the working area. By these measures, emissions of dust, paper, noise and so on can be minimized. Leachate recirculation should only be practised with pretreated leachate and should take place over the year, depending on the specific potential evaporation rate. For the average climate in West Germany, about 10 mm of leachate could be recirculated per week. Leachate recirculation should only take place during the operation phase of the landfill. It should not be practised on the landfill cap. The distribution of the leachate over the landfill surface should not produce any sprays of possibly harmful substances in the leachate.

The aftercare of a closed landfill has to be improved. In addition to ground-water observation leachate quality and quantity should be monitored and registered continuously; this should also be done for the extracted gas. The observation of the surface cap is a major task for the future. These are no optimum ways of finding out if the liner system is still functioning. There are some proposals for the control of surface caps, but improvements are still necessary. So the long-term monitoring of leachate production rate dependent upon the segments of the landfill could indicate the functioning of the surface cap. In addition, such control systems as double-liners on the surface are to be discussed. Gas composition could be measured in the drainage system installed above the surface liner system. Measuring landfill gas in this area could indicate leakage. Lysimeters could be installed in the surface cap from which the liner material (e.g. plastic liners samples) could be inspected. Measuring the high settling rates of the surface could indicate stress on the liner system.

The surface liner system could also be inspected by digging holes into the surface of the liner and taking samples. As an additional measure, root control is necessary.

## SUMMARY

The experiences with remedial actions of old landfill sites made obvious that current landfill practices depend on the different kinds of waste type to be improved. Apart from the need for waste minimization and recycling, landfills still have to be operated in the future. All the above-mentioned measures lead to a decrease of waste to be landfilled and also may change the quality of those wastes; but the landfill problem cannot ultimately be solved.

Landfills have to be operated and constructed depending on what kinds of waste are deposited:

1. municipal solid waste and municipal solid waste after separate collection;
2. inorganic mass waste;
3. industrial waste.

Landfills for municipal solid waste should be operated in such a way that the biochemical degradation processes in the landfill body are optimized. When the main processes are terminated, the landfill should be capped. After this period the same principle is valid as for inorganic waste: minimization of water penetration into the landfill (principle of conservation).

With regard to the constructed installations of the landfill, the following demands should be made, highly resistant materials with a long lifetime should be used and installations should be both controllable and repairable.

Depending on the different landfill types, a major task for the future is the aftercare of the landfills. This means mainly the control of the different installations for emissions control. If failures are detected, repair should be possible. The effects of the new technical developments should be monitored extensively to find out the weak points so that it will be possible to improve emissions control in landfills in the future.

## REFERENCES

August, H. (1986). 'Untersuchungen zur Wirksamkeit von Kombinationsdichtungen', Abfallwirtschaft in Forschung und Praxis', Heft 16. Erich Schmidt Verlag, Berlin.
Baccini, P., Henseler, G., Figi, R. and Belevi, H. (1987). 'Water and element balances of municipal solid waste landfills', *Waste Management and Research*, **4**.

Hermann, A.G., Brumsack, H.J. and Heinrichs, H. (1985). 'Notwendigkeit, Möglichkeiten und Grenzen der Untergrund-Deponie anthropogener Abfälle', *Naturwissenschaften* 72.

Krogmann, U. (1988). 'Separate Collection and Composting of Putrescrible Municipal Solid Waste in West Germany'. Andersen, Müller, ISUA 88 Proceedings, Academic Press, London.

Stegmann, R. (1979). 'Reinigung und Verregnen von Müllsickerwasser unter Betriebsbedingungen'. *Veröffentlichungen des Instituts für Stadtbauwesen TU Braunschweig* 27, Eigenverlag.

Stegmann, R. (1986). 'Altlastenproblematik und Konsequenzen für den Bau und Betrieb zukünftiger Abfalldeponien', 10 Internationale Fachmesse, PRO AQUA—PRO VITA 1986, Basel, 7–10 October 1986.

Stegmann, R. and Spendlin, H.H. (1986). 'Research activities on enhancement of biochemical processes in sanitary landfills'. *Water Pollution Research Journal Canada* 21 (4).

Wiedemann, H.U. (1982). 'Verfahren zur Verfestigung von Sonderabfällen und Stabilisierung von verunreinigten Böden. UBA-Berichte 1/82, Erich Schmidt Verlag, Berlin.

Wiles, C.C. (1987). 'A review of solidification/stabilization technology, *Journal of Hazardous Materials* 14, 5–21.

# 1.3 Environmental Aspects of Sanitary Landfilling

THOMAS H. CHRISTENSEN

*Department of Environmental Engineering, Building 115, Technical University of Denmark, DK-2800 Lyngby, Denmark*

## INTRODUCTION

The potential environmental impacts arising from landfilling of waste are the main reasons for the many regulations enforced, and the technical innovations seen, during recent years concerning landfill siting, design and operation.

The actual impact which a landfill has on its surrounding environment highly depends on the practice at the actual landfill and the quality, or quality expectations, of the surroundings. For example, the impact of the noise at a high capacity landfill located close to a residential area is supposedly much more severe than of the noise at a small, low-technology landfill neighbouring an existing industrial facility. While the impacts thus depend on the specific surroundings, the emissions from the landfill, causing the impacts, are primarily related to the landfill itself and thus easier to describe in more general terms. In this context emissions must be seen in their broadest sense—as any kind of release of matter to the area surrounding the landfill.

Fundamentally, the major environmental impacts from a sanitary landfill originate from the fact that the waste, in terms of composition, significantly differs from the surrounding land. As a part of the ecological stabilization of the landfill, fluxes of matter are released from the landfill, differing in composition and/or concentration from the fluxes of the surroundings. Depending on the route of these landfill generated fluxes they may cause unacceptable changes in the quality of the surrounding environment.

In this chapter a short introduction is given to the environmental aspects of sanitary landfilling focusing on the potential emissions from a landfill.

## LIFE CYCLE OF A LANDFILL

With respect to the environmental aspects of sanitary landfilling it is important to realize that a landfill has different phases in its life cycle. These are, related to the actual site:

1. Planning phase, typically involving preliminary hydrogeological and geotechnical site investigation as a basis for actual design.
2. Construction phase, typically involving earthwork, road and facility construction and preparation (eventually liners and drains) of the fill area.
3. Operation phase from the filling of the first waste load to the filling of the last waste load years later. This phase has a high intensity of traffic, work at the front of the fill, operation of environmental installations and completion of finished sections.
4. Completed phase from the termination of the actual filling to the day where the environmental installations do not need to be operated anymore, because the emissions have decreased to a level where they do not need to be treated, but may discharge into the surroundings.
5. Final storage phase. Now the emissions are at acceptable levels, which often means at the same order of magnitude as the fluxes found in the surrounding environment. The composition of the landfilled area may still differ from the composition of the surroundings, but the landfill area does not impact the surroundings. The landfill is now integrated into the surroundings and does not need special attention. The fill area may be used for many purposes, but like many other land sites, it may not be suitable for all kinds of use.

Of these five phases introduced, the planning phase and the construction phase do not need to be mentioned further, because the environmental aspects of these phases, although they may be important, do not differ from the environmental aspects of planning and construction of any physical facility. The final storage phase, as it is defined here, does not need any further discussion. In this phase the landfill is integrated into the environment.

The environmental concern related to landfills are connected with the operation phase and the completed phase. [The operation phase may typically cover 5 to 20 years and paying attention to the intensity of this phase, neighbourhood nuisances become an important part of the environmental aspects related to landfilling.] Since the sanitary landfill, as an engineered facility, is a rather recent creation (emerging around 1970) there is no direct field experience on how long the completed phase may last. From current experience, the completed phase will extend for several decades, but it is uncertain if it may last several hundred years. The completed phase, of course, depends on the size of the fill and the type of wastes in the fill. The uncertainty about the length of the completed phase should cause caution about introducing energy and maintenance intensive environmental installations to abate the landfill emissions of the completed phase.

## POTENTIAL ENVIRONMENTAL EMISSIONS

The potential environmental emissions from a sanitary landfill are summarized in Fig. 1 for the operation phase and the completed phase. In addition to this, incidental events such as flooding, fires, landslides and earthquakes, which

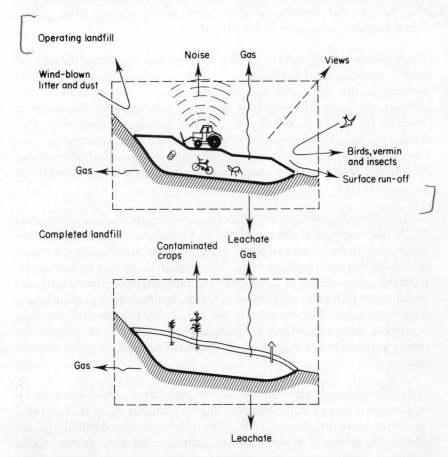

**Figure 1.** Illustration of the major environmental aspects related to sanitary landfilling.

could result in severe environmental impacts, must be included in the environmental impact assessment and eventually demand preventive measures with respect to landfill siting, design and operation.


## Operating Landfills

The environmental aspects of the operating landfill is dominated by the nuisance imposed on the neighbourhood: Wind-blown litter and dust, noise, odorous gases, birds, vermin and insects attracted by the waste, surface run-off and the physiological disturbance of the view to the landfilled waste. But also the gas and leachate problems arise during the operating phase, demanding significant environmental controls.

*Wind-blown litter and dust.* This is a continuous reminder of the ongoing landfill operation and a significant nuisance to the neighbourhood. By careful covering of the waste cells with soil, the problem may be reduced, but complete avoidance is impossible at the tipping front of the landfill. Spraying water on the dirt roads and the waste in dry periods, in combination with fencing and movable screens at the tipping front, and frequent cleansing of the fencing plantations, may minimize the problem. Further improvements may be obtained paying attention to dominant wind directions and seasonal use of the adjacent areas.

*Noise.* This is caused by the traffic of waste collection trucks arriving at the fill, by the emptying of the trucks and by the compactors and earthmoving equipment. In some cases large gatherings of birds, attracted by the waste, may in itself create a noise problem. The noise problem may be reduced by technical improvements of compactors and earthmoving equipment, by surrounding the fill area by soil embankments and by limiting the working hours. Plantations may reduce the noise level some, if they provide a tall and tight vegetation. Also with respect to noise, improvements may be obtained by paying attention to dominant wind directions and seasonal use of the adjacent areas.

*Gas.* Gas released from the waste, by degradation of the waste or by volatization of waste components, will migrate vertically out of the filled area or into the soil of the adjacent fields. The problems associated with the gas are odours, the release of explosive and/or flammable methane, health aspects related to specific compounds released with the gas and damage of the vegetation primarily due to depletion of oxygen from the root zone. The methane generation may start a few months after the disposal of waste at the site and may continue for several decades. The effect of methane gas is usually restricted to a zone of a couple of hundred meters from the fill area. Explosions and fires in sewers and basements in the vicinity of landfills have been reported, in some cases with fatal injuries. However, the most common effect of methane gas is the kill or phytotoxic effects on the vegetation on the landfill area itself. Most plants require aerobic root zones, but migration of landfill gas may displace the oxygen or methane oxidation bacteria may use all the available oxygen in the root zone. In the recent years the presence of specific volatile compounds in the gas (specific organics, mercury) has gained some focus and seems to constitute a problem that needs further evaluation in specific cases. The gas control measures introduced are liners, soil covers, passive venting or active extraction of gas and use or treatment before discharge to the atmosphere. The interest in utilization of the landfill gas for energy production has reduced the problem of gas migration significantly at many landfills. However, the odour problems associated with the tipping front of the landfill still exists, but fortunately this is only a local nuisance.

*Birds, vermin, insects.*   These and other animals are attracted by the landfill for feeding or breeding. Since many of these animals may act as vectors (disease transmitters), their presence may constitute a potential health problem (for example seagulls carrying salmonella). Because of this, rat reduction campaigns are often performed at landfills and compaction and daily covering of the waste cells is prescribed by the authorities. Besides the potential health problem, birds and insects may be a nuisance because of noise and unaesthetic droppings and excreta, in particular if residential areas are located in the vicinity. Birds, in particular seagulls, have been reported to constitute a problem for neighbouring airports during take-off and landing of aircraft. The aggressive and effective feeding patterns of seagulls right at the tip front, make it very difficult to effectively reduce their presence at landfills.

*Surface run-off.*   Run-off which has been in contact with the landfilled waste may be a problem in areas with intensive rainfall or snow melt. If not controlled, heavily polluted run-off may enter directly into creeks and streams, causing, for example, oxygen depletion and fish kill. However, careful design and maintenance of surface drains and ditches, together with final soil covers on completed landfill sections, may eliminate the problem of surface run-off.

*Leachate.*   The polluted leachate from the landfilled waste appears usually shortly after disposal of the waste. In most cases the leachate is heavily polluted, containing several grammes of dissolved matter per litre and may, through subsurface migration, cause extensive pollution of streams, creeks and water wells. Groundwater pollution is the major environmental concern at sanitary landfills demanding extensive controlling measures, such as liners, drainage collection, treatment of leachates and ground water quality control monitoring downstream of the landfill. Since the subsurface migration of leachate may be slow, the consequences of improper leachate controls may not emerge until decades later, where the leachate finally appears in a water well or a stream. Many countries today spend vast resources on reclaiming aquifers polluted by leachate from earlier waste disposal sites with or without improper leachate controls. In many cases, remedial schemes must be operated for decades in order to clean up a polluted aquifer.

*Views.* These are often important elements of the quality of residential and recreational areas. Thus an operating landfill where waste, waste trucks, compactors and alike are exposed, may psychologically affect the appreciation of an attractive area. Careful planning of the site with respect to screening soil embankments, extensive plantation, rapid covering and revegetation of filled sections combined with tidy operation and maintenance schemes, may reduce this problem significantly.

## Completed Landfill

At the completed landfill the local nuisances are negligible. If properly landscaped and thoroughly completed with appropriate vegetation, the completed landfill may even become an asset for the local community.

The environmental aspects related to gas and leachate still persist. Generation rates and composition of gas and leachate may have changed, but environmental controls still need to be operated.

If insufficiently covered an additional environmental aspect may emerge at the completed landfill. Vegetation and crops grown on the completed site may have been in contact through their roots with the waste material and may have become contaminated. For example, crops grown on top of a landfill where heavy metal containing sludges have been disposed of may be inappropriate for human consumption. However, at properly completed landfills—with no contaminated waste materials contributing to the final cover—this risk of contaminated vegetation and wildlife at the site seems avoidable.

## MINIMIZING ENVIRONMENTAL IMPACTS

As can be seen from the previous discussion, the environmental aspects related to sanitary landfilling are versatile, ranging from local nuisance to be abated by a tidy operation of the site, to the potential contamination of regional ground water resources by migrating leachate. While many of the environmental aspects may be minimized by current technology, the long term aspects of gas and leachate still raise questions about the appropriation of the current technology. Significant developments in the concepts and technology of landfilling, with a view to the long term aspects, are expected in the decades to come.

Meanwhile, meeting today's need for landfill capacity, acceptable environmental impacts of a sanitary landfill can be obtained only if the environmental aspects are paid proper attention at all stages and phases of a landfill: siting, design, construction, operation and maintenance. And, of course, the low level of impact on the environment from the landfill must be carefully documented by quality control measurements in order to improve our experience and hence to improve future sanitary landfilling.

# 2.   LANDFILL DEGRADATION PROCESSES

## 2.1.   Basic Biochemical Processes in Landfills

THOMAS H. CHRISTENSEN AND PETER KJELDSEN

*Department of Environmental Engineering, Bldg. 115, Technical University of Denmark, 2800-Lyngby, Denmark*

### INTRODUCTION

The environmental impacts of landfilling are today well recognized and at most sanitary landfills measures are taken to control these. The degradation processes inside the landfill are the key to understanding and controlling the environmental impacts. Physical, chemical and microbial processes are taking place in the waste and result in the release of gaseous and dissolved compounds in terms of landfill gas and leachate. In most landfills, assuming that they receive some organic wastes, the microbial processes will dominate the stabilization of the waste and hence govern the generation of landfill gas and the composition of the leachate.

This chapter briefly describes the basic biochemical processes taking place in a landfill in terms of the active microbial consortium undertaking the degradation and of the abiotic factors such as oxygen, hydrogen, pH and alkalinity, sulphate, nutrients, inhibitors, temperature and moisture/water content. Finally, a typical waste degradation sequence is described.

### ECOLOGY OF METHANE FORMATION

The predominant part of the landfill waste will soon after disposal become anaerobic, and a consortium of bacteria will start degrading the solid organic carbon, eventually to carbon dioxide ($CO_2$) and methane ($CH_4$). However, the microbial processes converting the organic carbon in the waste are rather complex and a short presentation of the consortium is needed to understand the overall process.

29

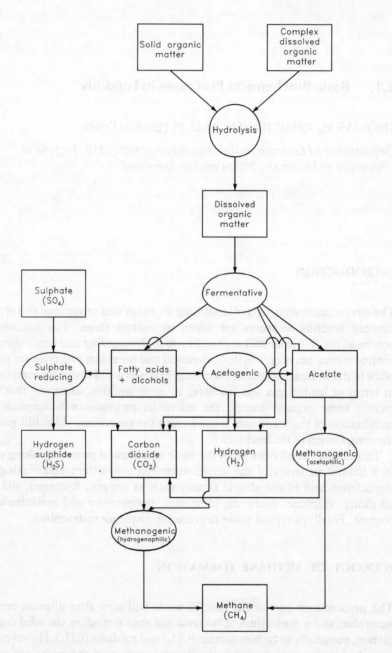

**Figure 1.** Substrates and major bacterial groups involved in the methane generating ecosystem.

Figure 1 illustrates the most important interactions between the involved bacterial groups, the involved substrates and intermediate products. The anaerobic degradation can be viewed as consisting of three stages. In the first stage solid and complex, dissolved organic compounds are hydrolysed and fermented by the fermenters to primarily volatile fatty acids, alcohols, hydrogen and carbon dioxide. In the second stage, an acetogenic group of bacteria converts the products from the first stage to acetic acid, hydrogen and carbon dioxide. In the final stage methane is produced by the methanogenic bacteria. This may be done by acetophilic bacteria converting acetic acid to methane and carbon dioxide or by hydrogenophilic bacteria converting hydrogen and carbon dioxide to methane. The overall process of converting organic compounds to methane and carbon dioxide may stoichiometrically be expressed by (Buswell and Mueller, 1952):

$$C_n H_a O_b + \left( n - \frac{a}{4} - \frac{b}{2} \right) H_2O \rightarrow$$

$$\rightarrow \left( \frac{n}{2} - \frac{a}{8} + \frac{b}{4} \right) CO_2 + \left( \frac{n}{2} + \frac{a}{8} - \frac{b}{4} \right) CH_4$$

But, as indicated by Fig. 1, the actual process is not at all a simple reaction.

The hydrolysis process is a very important process in the landfill environment since the present solid organic waste must be solubilized before the microorganisms can convert it. After the smaller, easily soluble part of the organic matter has been converted, the hydrolysis may prove to be the overall rate-limiting process in the landfill environment (Leuschner, 1983; McInerney and Bryant, 1983). The hydrolysis is caused by extracellular enzymes produced by the fermenting bacteria (Jones *et al.*, 1983).

The fermenters are a large, heterogeneous group of anaerobic and facultatively anaerobic bacteria. Some of the important reactions are shown in Table 1. The acetogenic bacteria are also a large heterogenic group. The acetogenic bacteria produce acetic acid, hydrogen and also carbon dioxide (McInerney and Bryant, 1983), if the volatile fatty acid being converted contains an odd number of carbon atoms. The acetogenic bacteria may also convert aromatic compounds containing oxygen (e.g. benzoic acid and phenols), while aromatic hydrocarbons (e.g. benzene and toluene) apparently are not degraded. Some of the important acetogenic processes are also shown in Table 1.

The methanogenic bacteria are obligate anaerobic and require very low redox potentials. One group, the hydrogenophilic, converts hydrogen and carbon dioxide to methane, while another group, the acetophilic, converts primarily acetic acid to methane and carbon dioxide. The methanogenic bacteria may also convert formic acid and methanol. Some of the important

**Table 1.** Examples of important reactions for four groups of bacteria involved in anaerobic waste degradation, based on Hansson and Molin (1981b), McInerney and Bryant (1983), Zehnder (1978) and Postgate (1979).

| Reactants converted to products |
| --- |

Fermentative processes

$$C_6H_{12}O_6 + 2H_2O \qquad\qquad 2CH_3COOH + H_2 + 2CO_2$$
$$C_6H_{12}O_6 \qquad\qquad CH_3C_2H_4COOH + 2H_2 + 2CO_2$$
$$C_6H_{12}O_6 \qquad\qquad 2CH_3CH_2OH + 2CO_2$$

Acetogenic processes

$$CH_3CH_2COOH + 2H_2O \qquad\qquad CH_3COOH + CO_2 + 3H_2$$
$$CH_3C_2H_4COOH + 2H_2O \qquad\qquad 2CH_3COOH + 2H_2$$
$$CH_3CH_2OH + H_2O \qquad\qquad CH_3COOH + 2H_2$$
$$C_6H_5COOH + 4H_2O \qquad\qquad 3CH_3COOH + H_2$$

Methanogenic processes

$$4H_2 + CO_2 \qquad\qquad CH_4 + 2H_2O$$
$$CH_3COOH \qquad\qquad CH_4 + CO_2$$
$$HCOOH + 3H_2 \qquad\qquad CH_4 + 2H_2O$$
$$CH_3OH + H_2 \qquad\qquad CH_4 + H_2O$$

Sulphate reducing processes

$$4H_2 + SO_4^{2-} + H^+ \qquad\qquad HS^- + 4H_2O$$
$$CH_3COOH + SO_4^{2-} \qquad\qquad CO_2 + HS^- + HCO_3^- + H_2O$$
$$2CH_3C_2H_4COOH + SO_4^{2-} + H^+ \qquad\qquad 4CH_3COOH + HS^-$$

HCOOH: formic acid, $CH_3COOH$: acetic acid, $CH_3CH_2COOH$: propionic acid, $CH_3C_2H_4COOH$: butyric acid, $C_6H_{12}O_6$: glucose, $CH_3OH$: methanol, $CH_3CH_2OH$: ethanol, $C_6H_5COOH$: benzoic acid, $CH_4$: methane, $CO_2$: carbon dioxide, $H_2$: hydrogen, $SO_4^{2-}$: sulphate, $HS^-$: hydrogen sulphide, $HCO_3^-$: hydrogencarbonate, $H^+$: protone, $H_2O$: water.

reactions are shown in Table 1. The conversion of acetic acid to methane is by far the most important part of the methane-forming process (about 70%).

Finally, the sulphate-reducing bacteria, dominated by *Desulfovibrio* and *Desulfotomaculum* (Postgate, 1979), is mentioned, since this group of bacteria in many ways resembles the methanogenic group and since sulphate is a major compound of many waste types (demolition waste, incinerator slag, fly ashes). The sulphate-reducing bacteria are obligate anaerobic and may convert hydrogen, acetic acid and higher volatile fatty acids during sulphate reduction. However, the organic carbon is always oxidized to carbon dioxide as opposed to the convertion by the methanogenic group of bacteria. A high activity of sulphate reducers hence may decrease the amount of organics available for methane production. Some of the sulphate reducing reactions are shown in Table 1.

## GOVERNING ABIOTIC FACTORS

The sketched consortium of bacterial groups participating in the methane-forming ecosystem is exposed to a variety of highly variable abiotic factors in the landfilled waste. The heterogeneity of the landfilled waste may make the landfill a highly diverse, but rather inefficient ecosystem. Figure 2 illustrates the major abiotic factors affecting the methane formation in the landfill: oxygen, hydrogen, pH/alkalinity, sulphate, nutrients, inhibitors, temperature and water content. These factors are discussed below.

The organic fraction of the waste making up the substrate for the microbial consortium is of highly varying nature, ranging from easily degradable organics such as food wastes to hardly degradable organics such as lignin and polymers. The degradability of the waste highly effects the degradation rates but is not discussed in the following paragraphs which focus on the basic abiotic factors.

### Oxygen

The absence of free oxygen is a must for the anaerobic bacteria to grow and perform the above-mentioned processes. The methanogenic bacteria are the most sensitive to oxygen; they require a redox potential below $-330$ mV.

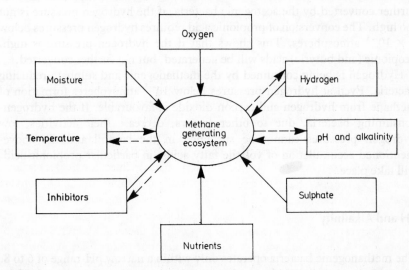

**Figure 2.** Major abiotic factors affecting the methane-generating ecosystem.

Oxygen will always diffuse from the atmosphere into the landfill waste. However, aerobic bacteria in the top of the landfill waste will readily consume the oxygen and limit the aerobic zone to less than 1 m of compacted waste. Extensive gas-recovery pumping may create a substantial vacuum in the landfill, forcing atomospheric air to enter the landfill. This may extend the aerobic zone in the waste and eventually prevent formation of methane in these layers.

Although no spore-forming methanogenic bacteria are known (Zehnder, 1978), the methanogenic community apparently is not completely wiped out by the introduction of oxygen. Anaerobic incubation of aerobic sewage and aerobic soil fairly rapidly result in methane production (Zehnder *et al.*, 1982). The existence of anaerobic microenvironments in the overall aerobic environment may be one of the main reasons for the survival of the methanogenic bacteria in aerobic environments.

## Hydrogen

Hydrogen is produced by both the fermentative and the acetogenic bacteria and the generated hydrogen pressure affects the biochemical pathways. The fermentative bacteria yield hydrogen, carbon dioxide and acetic acid at low hydrogen pressures, while hydrogen, carbon dioxide and ethanol, butyric acid and propionic acid are generated at higher hydrogen pressures, according to McInerney and Bryant (1983). The last three organic compounds may be further converted by the acetogenic bacteria, if the hydrogen pressure is not too high. The conversion of propionic acid requires hydrogen pressures below $9 \times 10^{-5}$ atmospheres. This shows that if the hydrogen pressure is high, propionic (and butyric) acids will be generated, but not further converted.

Hydrogen is being consumed by the methanogenic and sulphate-reducing bacteria. Even at hydrogen pressures below $10^{-5}$ atmospheres formation of methane from hydrogen and carbon dioxide is favourable. If the hydrogen-consuming bacteria, due to other factors, decrease their activities, low hydrogen pressures cannot be sustained in the landfill and the above-mentioned accumulation of volatile fatty acids, in particular propionic acid, will take place.

## pH and Alkalinity

The methanogenic bacteria operates only within a narrow pH-range of 6 to 8. Figure 3 presents the relative methane production rate as a function of pH for

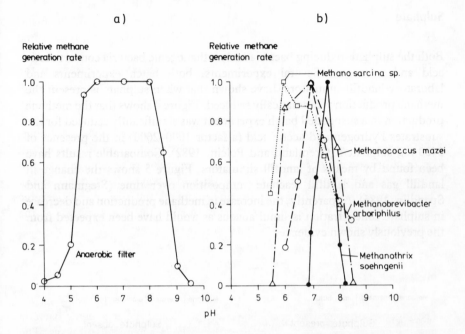

**Figure 3.** Relative methane generation rate as a function of pH for a mixed anaerobic filter (a) and for individual bacterial strains (b) (after Zehnder *et al.*, 1982).

a mixed culture of methanogens (Fig. 3a) and for selected strains (Fig. 3b). The pH range for the fermentative and acetogenic bacteria is much wider than for the methanogenic bacteria. If the methanogens are stressed by other factors their conversion of hydrogen and acetic acid is decreased, the hydrogen pressure builds up and pH decreases, leading to a build-up of propionic and butyric acids and a further lowering of pH. Eventually, the methane production stops. The methanogenic ecosystem in the landfill is rather delicate and balanced relations between the various bacteria groups are crucial for a good methane production rate. The presence of buffering material in the landfill (e.g. demolition waste, soil) will significantly improve the ability of the landfill environment to maintain a reasonable pH range.

The sulphate-reducing bacteria has a somewhat wider pH range than the methanogenic bacteria: less than pH 5 to pH 9 according to Postgate (1979). If sulphate is present in the waste, the sulphate-reducing bacteria may at low pH values dominate the methanogenic bacteria and convert the organics to carbon dioxide.

## Sulphate

Both the sulphate-reducing bacteria and methanogenic bacteria convert acetic acid and hydrogen. Several experiments, both batch experiments and laboratory landfill simulators, have shown that when sulphate is present the methane production is dramatically reduced. Figure 4 shows that the methane production in a suspended batch experiment was significantly reduced for the substrates hydrogen and acetic acid (a factor 1000–2000) in the presence of 2400 mg sulphate/l (Oremland and Polcin, 1982). Comparable results have been found by means of landfill simulators. Figure 5 shows the change in landfill gas and landfill leachate composition over time (Stegmann and Spendlin, 1985). Apparently, the increase in methane production and decrease in sulphate concentration is simultaneous as would have been expected from the previously shown evidence.

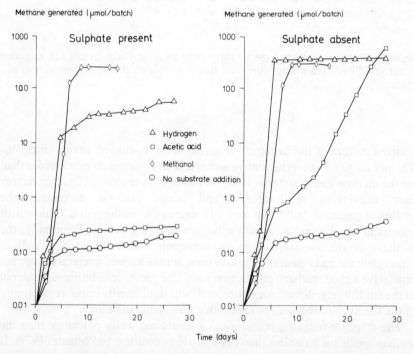

**Figure 4.** Accumulated methane generation in sediment (batch tests) for various substrates in the presence (initial concentration: 2400 mg/l) and absence of sulphate. Based on Oremland and Polcin (1982).

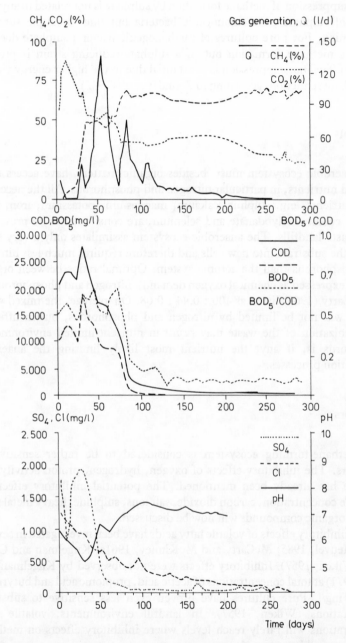

**Figure 5.** Observed gas and leachate composition in laboratory landfill simulators as a function of time. After Stegmann and Spendlin (1985).

The suppression of methane formation by sulphate is not related to any toxic effects of sulphate on methanogenic bacteria but due to simple substrate competition. For pure cultures of methanogenic bacteria sulphate does not suppress methane formation but, if a sulphate-reducing strain is present, e.g. *Desulfovibrio*, suppression is substantial due to the higher energy yielded by sulphate reduction, according to Zehnder *et al.* (1982).

## Nutrients

The anaerobic ecosystem must, besides organic matters, have access to all required nutrients, in particular nitrogen and phosphorous. All the necessary micronutrients, e.g. sulphur, calcium, magnesium, potassium, iron, zinc, copper, cobalt, molybdanate and selenium, are considered to be present in most waste landfills. The anaerobic ecosystem assimilates only a very small part of the substrate into new cells and therefore requires much less nitrogen and phosphorous than the aerobic system. Optimal ratios between organic matter (expressed as chemical oxygen demand), nitrogen and phosphorous are by McCarty (1964) listed as 100 : 0.44 : 0.08. On average, the mixed waste landfill will not be limited by nitrogen and phosphorous, but insufficient homogenization of the waste may result in nutrient-limited environments. Phosphorus is, if any, the nutrient most likely limiting the anaerobic degradation processes.

## Inhibitors

The methane-forming ecosystem is considered to be rather sensitive to inhibitors. The inhibitory effects of oxygen, hydrogen, proton activity and sulphate has already been mentioned. The potential inhibitory effects of substrate concentration, carbon dioxide, salt ions, sulphide, heavy metals and specific organic compounds will now be discussed.

The inhibitory effects of volatile fatty acids have been investigated in several cases (Heuvel, 1985; McCarty and McKinney, 1961a; Kugelman and Chin, 1971; Wiken, 1957). Inhibitory effects were not observed by Kugelman and Chin (1971) at total concentrations of acetic acid, propionic acid and butyric up to 6000 mg/l. Pure cultures have proven even less sensitive to substrate concentrations (Wiken, 1957). In landfill environments, volatile acid concentrations will rarely reach levels where inhibitory effects on methane production will be expected.

Carbon dioxide is produced in many of the degradation processes previously

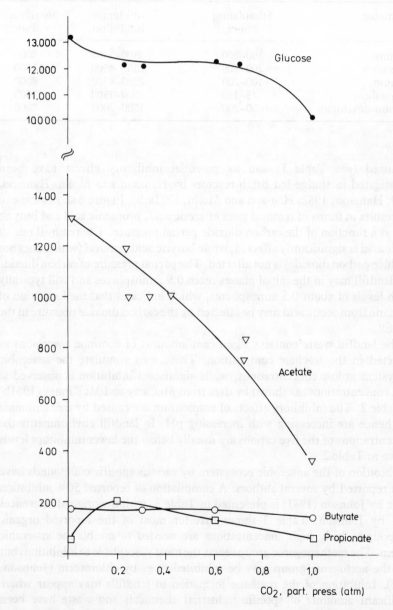

**Figure 6.** Degradation rates of various substrates as a function of $CO_2$ partial pressure. After Hansson and Molin (1981a).

**Table 2.** Effect of cations on methane generation (mg/l), after McCarty and McKinney (1961b).

| Parameter | Stimulating effect | Moderate inhibition | Significant inhibition |
|---|---|---|---|
| Sodium | 100–200 | 3500–5500 | 8000 |
| Potassium | 200–400 | 2500–4500 | 12 000 |
| Calcium | 100–200 | 2500–4500 | 8000 |
| Magnesium | 75–150 | 1000–1500 | 3000 |
| Ammonium (total) | 50–200 | 1500–3000 | 3000 |

discussed (see Table 1) and its potential inhibitory effects have been investigated in sludge-fed batch reactors by Hansson and Molin (Hansson, 1979; Hansson, 1982; Hansson and Molin, 1981a,b). Figure 6 shows some of the results in terms of removal rates of acetic acid, propionic acid and butyric acid as a function of the carbon dioxide partial pressure. The removal rate of acetic acid is significantly affected, while butyric acid removal (which does not produce carbon dioxide) is not affected. The partial pressure of carbon dioxide in a landfill may in the initial phases reach 0.9 atmospheres and will typically reach levels of about 0.5 atmospheres, which indicates that the formation of methane from acetic acid may be affected by the carbon dioxide pressure in the landfill.

The landfill waste contains significant amounts of common macroions as reflected in the leachate composition. These ions stimulate the anaerobic ecosystem at low concentrations, while significant inhibition is observed at high concentrations, as shown by data from McCarty and McKinney (1961b) in Table 2. The inhibitory effects of ammonium are caused by free ammonia and hence are increasing with increasing pH. In landfill environments the concentrations of the five cations are usually below the lower inhibitory levels shown in Table 2.

Inhibition of the anaerobic ecosystem by various specific compounds have been reported by several authors. A compilation of reported 50% inhibition levels by Johnson (1981) is presented in Table 3, with the original references used by Johnson. Table 3 shows that for most of the reported organic compounds fairly high concentrations are needed to inhibit the anaerobic system. The methanogenic group seems the most susceptible to inhibition, but also the acetogenic group may be inhibited, e.g. by chloroform (Johnson, 1981). Inhibition of the methane formation in landfills may appear where significant amounts of specific industrial chemicals and waste have been disposed of in the landfill or in local microenvironments with leaking chemical containers.

**Table 3.** Inhibition levels of specific organic compounds on methane generation (Johnson, 1981).

| Compound | Concentration[a] (mg/l) | Reference |
|---|---|---|
| Acetaldehyde | 440 | Chou *et al.* (1978) |
| Acrolein | 20–50 | Hovious *et al.* (1973) |
| | 11 | Chou *et al.* (1978) |
| Acrylic acid | 864 | Chou *et al.* (1978) |
| Acrylonitrile | 100 | Hovious *et al.* (1973) |
| | 212 | Chou *et al.* (1978) |
| Analine | 2418 | Chou *et al.* (1978) |
| Catechol | 2640 | Chou *et al.* (1978) |
| Crotonaldehyde | 455 | Chou *et al.* (1978) |
| Diethylamine | 300–1000 | Hovious *et al.* (1973) |
| Ethylacetate | 968 | Chou *et al.* (1978) |
| Ethylacrylate | 300–600 | Hovious *et al.* (1973) |
| Ethylbenzene | 339 | Chou *et al.* (1978) |
| Ethyldichloride | 150–500 | Hovious *et al.* (1973) |
| | 2.5–7.5 | Stuckey *et al.* (1978) |
| 2-Ethyl-1-hexanol | 500–1000 | Hovious *et al.* (1973) |
| Formaldehyde | 50–100 | Hovious *et al.* (1973) |
| | 72 | Chou *et al.* (1978) |
| | 200 | Pearson *et al.* (1980) |
| Chloroform | 20 | Stickley (1970) |
| 3-Chloro-1,2-propandiole | 663 | Chou *et al.* (1978) |
| 1-Chloropropane | 149 | Chou *et al.* (1978) |
| 1-Chloropropene | 7.6 | Chou *et al.* (1978) |
| 2-Chloropropionic acid | 868 | Chou *et al.* (1978) |
| Lauric acid | 593 | Chou *et al.* (1978) |
| Methylene chloride | 100 | Thiel (1969) |
| | 1.8–2.2 | Stuckey *et al.* (1978) |
| 2-Methyl-5-ethylpyridine | 100 | Hovious *et al.* (1973) |
| Methyl-isobutyl ketone | 100–300 | Hovious *et al.* (1973) |
| Nitrobenzene | 12.3 | Chou *et al.* (1978) |
| Phenol | 2444 | Chou *et al.* (1978) |
| | 300–1000 | Hovious *et al.* (1973) |
| | 500 | Pearson *et al.* (1980) |
| Propanol | 5200 | Chou *et al.* (1978) |
| Resorcinol | 3190 | Chou *et al.* (1978) |
| Carbontetrachloride | 2.2 | Thiel (1969) |
| Vinyl acetate | 592 | Chou *et al.* (1978) |
| | 200–400 | Stuckey *et al.* (1978) |
| Vinyl chloride | 5–10 | Stuckey *et al.* (1978) |

[a] Represent the concentration causing a 50% reduction in methane generation rate.

## Temperature

Like all other microbic processes, the anaerobic waste degradation rate is highly affected by temperature. The methanogenic bacteria contains a mesophilic group with a rate maximum around 40°C and a thermophilic group with a maximum around 70°C. Only the former group is relevant in landfills. In laboratory simulations of landfill processes, the methane production rate has been proven to increase significantly (up to 100 times), when the temperature is raised from 20 to 30 and 40°C, e.g. Buivid (1980), Ehrig (1984), Scharf (1982).

Both the aerobic and the anaerobic degradation of waste yields heat, although the anaerobic heat generation often is neglected, because its low generation rate seldom results in a temperature rise. Comparing the aerobic and the anaerobic conversion of, e.g. glucose (as suggested by Pirt (1978) and modified by Rees (1980b) and Lagerkvist (1986)), shows that the anaerobic decomposition only yields 7% of the heat generated by aerobic decomposition.

*Aerobic decomposition of glucose:*

$$C_6H_{12}O_6 + O_2 \rightarrow CO_2 + H_2O + \text{biomass} + \text{heat}$$

$$\begin{array}{llllll} 1\,\text{kg} & 0.64\,\text{kg} & 0.88\,\text{kg} & 0.34\,\text{kg} & (\text{dry weight}) & 9300\,\text{kJ} \\ & & & & 0.40\,\text{kg} & \end{array}$$

*Anaerobic decomposition of glucose:*

$$C_6H_{12}O_6 \rightarrow CH_4 + CO_2 + \text{biomass} + \text{heat}$$

$$\begin{array}{lllll} 1\,\text{kg} & 0.25\,\text{kg} & 0.69\,\text{kg} & (\text{dry weight}) & 632\,\text{kJ} \\ & & & 0.056\,\text{kg} & \end{array}$$

In a deep landfill with a moderate water flux, the flux of heat from the landfill to the surroundings is small due to the insulating capacities of the waste, and the heat generated by the anaerobic decomposition process may cause a temperature rise in the landfill (Rees, 1980b). Landfill temperatures of 30–45°C should be possible even in temperate climates. Figure 7 shows ambient and waste temperatures at a UK landfill for a two-year period. It should be pointed out that such elevated temperatures are not found in all waste landfills. It does require substantial waste thickness, a high methane production rate and a low water flux through the landfill.

## Moisture/Water Content

Several laboratory investigations have shown that the methane production rate increases for increasing moisture content of the waste, e.g. Buivid (1980) and

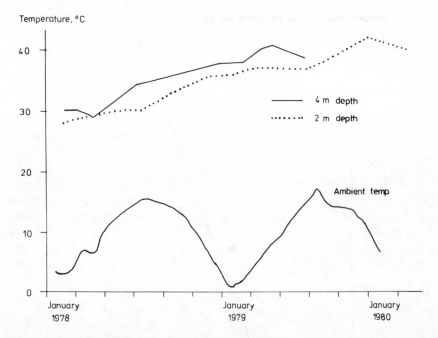

**Figure 7.** Landfill and ambient temperatures in a British landfill over a two-year period. After Rees (1980b).

Rees (1980a). Figure 8 summarizes findings from the literature suggesting an exponential increase in gas production rates between 25 and 60% water content.

The main effect of the increased water content, besides limiting oxygen transport from the atmosphere, is probably the facilitated exchange of substrate, nutrients, buffer, and dilution of inhibitors and spreading of microorganisms between the waste micro environments.

## WASTE DEGRADATION SEQUENCE

The combination of the information previously presented on the abiotic factors and experiences from full-scale landfills have led to the speculation on a theoretical or idealized sequence of the involved anaerobic degradation processes and their consequences as to gas and leachate composition. Figure 9 illustrates such an idealized sequence for a homogeneous volume of waste,

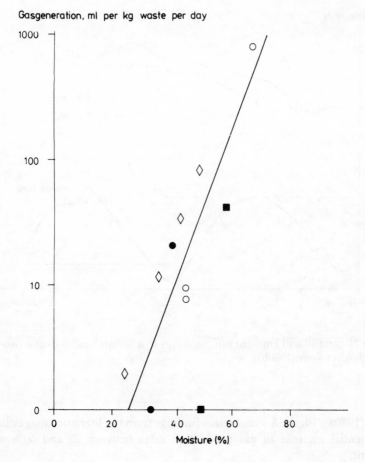

Gasgeneration, ml per kg waste per day

Moisture (%)

**Figure 8.** Gas generation rates as a function of moisture content. After Rees (1980a).

based on Farquhar and Rovers (1973) and Ehrig (1984), involving five distinct phases:

*Phase I* This is a short aerobic phase immediately after landfilling the waste, where easily degradable organic matter is aerobically decomposed during carbon dioxide generation.

*Phase II* A first intermedial anaerobic phase develops immediately after the aerobic phase. The activity of the fermentative and also the acetogenic bacteria results in a rapid generation of volatile fatty acids, carbon dioxide and some hydrogen. The acidic leachate may contain high concentrations of fatty acids,

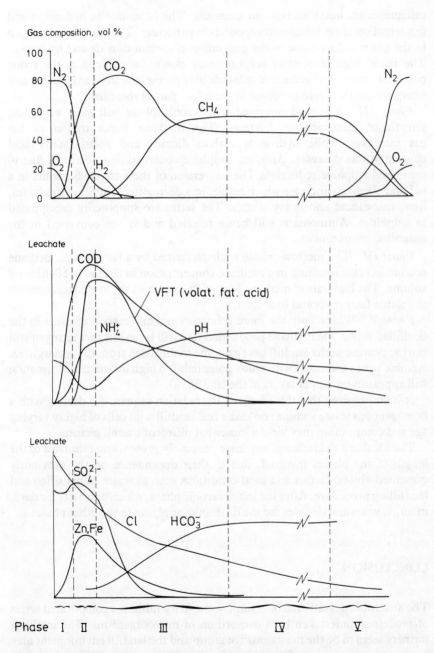

**Figure 9.** Illustration of developments in gas and leachate composition in a landfill cell (partly based on Farquhar and Rovers, 1973).

calcium, iron, heavy metals and ammonia. The latter due to hydrolysis and fermentation of proteineous compounds in particular. The content of nitrogen in the gas is reduced due to the generation of carbon dioxide and hydrogen. The initial high content of sulphate may slowly be reduced as the redox potential drops. The generated sulphide may precipitate iron, manganese and heavy metals that were dissolved in the initial part of this phase.

*Phase III*   A second intermedial anaerobic phase will start with slow growth of methanogenic bacteria. The methane concentration in the gas increases, while hydrogen, carbon dioxide and volatile fatty acid concentrations decrease. Also, the sulphate concentration decreases due to continued sulphate reduction. The conversion of the fatty acids results in a pH and alkalinity increase which results in a decreasing solubility of calcium, iron, manganese and heavy metals. The latter are supposedly precipitated as sulphides. Ammonia is still being released and is not converted in the anaerobic environment.

*Phase IV*   The methane phase is characterized by a fairly stable methane production rate resulting in a methane concentration in the gas of 50–65% by volume. The high rate of methane formation maintains the low concentrations of volatile fatty acids and hydrogen.

*Phase V*   Where only the more refractory organic carbon remains in the landfilled waste, the methane production rate will be so low that nitrogen will start appearing in the landfill gas again due to diffusion from the atmosphere. Aerobic zones and zones with redox potentials too high for methane formation will appear in the upper layers of the landfill.

It is emphasized that this idealized degradation sequence is dealing with a homogeneous waste volume and that a real landfill with cells of highly varying age and composition may yield a somewhat different overall picture.

The idealized degradation sequence purposely presents no estimates of the length of the phases involved, due to their dependence on the previously presented abiotic factors and local conditions such as waste composition and landfilling procedure. After the initial aerobic phase, which only lasts for days, months, years and decades are the time units applying to the other phases.

## CONCLUSION

The anaerobic degradation of organic waste in a landfill is a complicated series of processes undertaken by a consortium of microorganisms. The methane formers seem to be the most sensitive group and the landfill environment may in some cases prevent formation of substantial amounts of methane. In particular low pH values may be detrimental to methane formers, and proper control of the landfill pH is the main factor in insuring degradation of the organic waste.

Abiotic factors as oxygen, hydrogen, sulphate, nutrients, inhibitors and water content may in some cases limit the anaerobic degradation processes, but will in most cases be second to proper pH control. Increased temperatures will significantly increase the turnover rate or organic matter and at proper design of the landfill, the energy release by the anaerobic processes may support a significant and beneficial temperature rise in the landfill.

# REFERENCES

Buivid, M.G. (1980). 'Laboratory simulation of fuel gas production enhancement from municipal solid waste landfills'. Dynatech R & D Co., Cambridge, MA, USA.

Buswell, A.M. and Mueller, H.F. (1952). 'Mechanisms of methane fermentations', *Industrial and Engineering Chemistry* **44**, 550.

Chou, W.L., Speece, R.E., Siddiqi, R.H. and McKeon, K. (1978). 'The effects of petrolchemical structure on methane fermentation toxicity', *Progress in Water Technology* **10**(5), 545–558.

Ehrig, H.-J. (1984). 'Laboratory scale tests for anaerobic degradation of municipal solid waste'. Proceedings from the International Solid Wastes and Public Cleansing Association Congress, Philadelphia, September 15–20.

Farquhar, C.J. and Rovers, F.A. (1973). 'Gas production during refuse decomposition', *Water, Air and Soil Pollution* **2**, 483–495.

Hansson, G. (1979). 'Effects of carbon dioxide and methane on methanogenesis', *European Journal of Applied Microbiology and Biotechnology* **6**, 352–359.

Hansson, G. (1982). 'End product inhibition in methane fermentations', *Process Biochemistry* **17**(6), 45–49.

Hansson, G. and Molin, N. (1981a). 'End product inhibition in methane fermentations: effects of carbon dioxide and methane on methanogenic bacteria utilizing acetate', *European Journal of Applied Microbiology and Biotechnology* **13**, 236–241.

Hansson, G. and Molin, N. (1981b). 'End product inhibition in methane fermentations: effects of carbon dioxide on fermentative and acetogenic bacteria', *European Journal of Applied Microbiology and Biotechnology* **13**, 242–247.

Heuvel, J.C. van der (1985). 'The acidogenic dissimilation of glucose: a kinetic study of substrate and product inhibition'. Laboratory of Chemical Engineering, University of Amsterdam, The Netherlands.

Hovious, J.C., Waggy, G.T. and Conway, R.A. (1973). 'Identification and control of petrolchemical pollutants inhibitory to anaerobic processes'. US Environmental Protection Agency (Environmental Protection Technology Series, EPA-R2-73-194).

Johnson, L.D. (1981). 'Inhibition of anaerobic digestion by organic priority pollutants', PhD Thesis, Iowa State University, Iowa, USA.

Jones, K.L., Rees, J.F. and Grainger, J.M. (1983). 'Methane generation and microbial activity in a domestic refuse landfill site', *European Journal of Applied Microbiology and Biotechnology* **18**, 242–245.

Kugelman, I.J. and Chin, K.K. (1971). 'Toxicity, synergism and antagonism in anaerobic waste treatment processes'. Anaerobic biological treatment processes, *Advances in Chemistry Series* 105, 55–90.

Lagerkvist, A. (1986). 'Landfill strategy. Present practices—future options'. Division of Residual Products Technology, Lulea University, Sweden.

Leuschner, A.P. (1983). 'Feasibility study for recovering methane gas from the Greenwood Street sanitary landfill, Worcester, Mass.', Vol. I. Task 1–Laboratory feasibility. Dynatech R & D Co., Cambridge, Mass., USA.

McCarty, P.L. (1964). 'Anaerobic waste treatment fundamentals', Parts one, two, three and four, *Public Works* 95, (9), 107–112, (10), 123–126, (11), 91–94, (12), 95–99.

McCarty, P.L. and McKinney, R.E. (1961a). 'Volatile acid toxicity in anaerobic digestion', *Water Pollution Control Federation Journal* 33, 223–232.

McCarty, P.L. and McKinney, R.E. (1961b). 'Salt toxicity in anaerobic digestion', *Water Pollution Control Federation Journal* 33, 399–415.

McInerney, M.J. and Bryant, M.P. (1983). Review of methane fermentation fundamentals. In 'Fuel Gas Production from Biomass', Wise, D.L. (ed.) ch. 2. CRC Press, Boca Raton, Florida.

Oremland, R.S. and Polcin, S. (1982). 'Methanogenesis and sulphate reducation. Competitive and noncompetitive substrates in estuarine sediments', *Applied and Environmental Microbiology* 44, 1270–1276.

Pearson, F., Shiun-Chung, C. and Gartier, M. (1980), 'Toxic inhibition of anaerobic biodegradation', *Water Pollution Control Federation Journal* 52, 472–482.

Pirt, S.J. (1978). 'Aerobic and anaerobic microbial digestion in waste reclamation', *Journal of Applied Chemistry and Biotechnology* 28, 232–236.

Postgate, J.R. (1979). 'The Sulphate-reducing Bacteria'. Cambridge University Press, Cambridge.

Rees, J.F. (1980a). 'The fate of carbon compounds in the landfill disposal of organic matter', *Journal of Chemical Technology and Biotechnology* 30, 161–175,

Rees, J.F. (1980b) 'Optimisation of methane production and refuse decomposition in landfills by temperature control', *Journal of Chemical Technology and Biotechnology* 30, 458–465.

Scharf, W. (1982). 'Untersuchungen zur gemeinsamen Ablagerung von Müll und Klärschlamm im Labormaßstab', Gas- und Wasserhaushalt von Mülldeponien. Internationale Fachtagung 29.9–1.10.1982, Braunschweig, 83–98. TU Braunschweig Veröffentlichungen des Instituts für Stadtbauwesen, Heft 33.

Stegmann, R. and Spendlin, H.-H. (1985). 'Research activities on enhancement of biochemical processes in sanitary landfills', Paper presented at the Conference 'New Directions and Research on Enchancement of Biochemical Processes in Sanitary Landfills', June 23–28, University of British Columbia, Vancouver, Canada.

Stickley, D.P. (1970). 'The effect of chloroform in sewage on the production of gas from laboratory digesters', *Water Pollution Control* 69, 585–592.

Stuckey, D.C., Parkin, G.F., Owen, W.F. and McCarty, P.L. (1978). 'Comparative evaluation of anaerobic toxicity by batch and semicontinuous assays'. Paper presented at the Water Pollution Control Federation 51st Annual Conference, Anaheim, Calif., USA.

Thiel, P.G. (1969). 'The effect of methane analogues on methanogenesis in an aerobic digestion', *Water Research* 3, 215–223.

Wiken, T.O. (1957). 'Über den Mechanismus des anaeroben bakteriellen Abbaus von Kohlenhydrat, Eiweiss und Fett in Faulräumen', *Schweizerische Zeitschrift für Hydrologie* 19(1), 428.

Zehnder, A.J.B. (1978). Ecology of methane formation. *In* 'Water Pollution Microbiology', 13, Mitchell, R. (ed.) vol. 2, pp. 349–376. John Wiley, New York.

Zehnder, A.J.B., Ingvorsen, K. and Marti, T. (1982). Microbiology of methane bacteria. *In* 'Anaerobic Digestion, 1981', Hughes, D.E. *et al.* (eds) pp. 45–68. Proceedings of the Second International Symposium of Anaerobic Digestion, Travemünde 6–11 September 1981. Elsevier Biomedical Press, B.V., Amsterdam, The Netherlands.

## 2.2 Activities and Distribution of Key Microbial Groups in Landfill

R. SLEAT, C. HARRIES, I. VINEY AND J.F. REES
*Biotal Limited, Cardiff Industrial Park, Cardiff, CF4 5DL, UK*

## INTRODUCTION

The biological stabilization of both industrial and municipal solid wastes (MSW), is an increasingly important element of waste management by landfill techniques. Inevitably the failure or the incomplete nature of the biological stabilization process results in potentially serious and sometimes catastrophic damage to land, water, air or life. Greater control over these biological processes, integrated with improved engineering management, can play an important role in both mitigating environmental damage from landfill and optimizing commercial benefits.

For biological stabilization to work effectively, reliably and predictably in full-scale landfill systems, two key elements must be satisfied:

1. the numbers and activities of all the key groups of microbes must be adequately high and fully integrated;
2. the physical and chemical environment of the landfill must be conducive to their effective growth and activity.

These are fundamental principles which apply to landfill and all other forms of land treatment systems for wastes or contaminated land.

This chapter sets out for the first time measures of the full range of key microbe groups involved in MSW stabilization to methane and carbon dioxide in an environment exhibiting varying rates of refuse stabilization and physicochemical conditions. It reflects a requirement to understand landfill microbiology that is relevant to the waste stabilization process.

51

## THE KEY MICROBIAL GROUPS

The earliest studies of landfill microbiology have emphasized and been motivated by public health considerations (Cromwell, 1965; Peterson, 1971; Engelbrecht and Amirhor, 1975) where attention has been focused on organisms, particularly bacteria and viruses of faecal origin, in refuse and leachate. More recent studies have begun to examine more closely the number and physiological activities of microbes relevant to MSW stabilization (Jones and Grainger, 1983; Rees, 1980a; Filip and Kuster, 1979; Peck, 1986) including evaluation of extracellular enzyme activity and rapid chemical methods for estimation of numbers of methanogenic organisms.

General schemes for landfill reactions and the flow of carbon compounds have been set out in many papers including Rees (1985); and see Fig. 1. Reactions are dominated by an anaerobic fermentation of lignocellulosic material where carbon dioxide and sulphate are the principal terminal electron

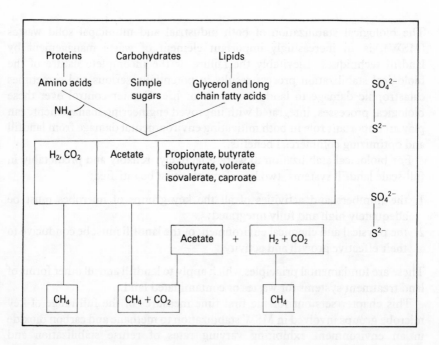

**Figure 1.** General scheme for landfill reactions for carbon and sulphur compounds.

**Table 1.** Key physiological groups of microbes involved in anaerobic landfill degradation processes

| Microbial groups | Chemical substrates |
|---|---|
| Amylolytic bacteria | Starches |
| Proteolytic bacteria | Proteins |
| Cellulolytic bacteria | Cellulose |
| Hemicellulolytic bacteria | Hemicellulose |
| Hydrogen-oxidizing methanogenic bacteria | Hydrogen |
| Acetoclastic methanogenic bacteria | Acetic acid |
| Sulphate-reducing bacteria | Sulphate |

acceptors. As shown in Table 1, seven key physiological groups of microbes have been recognized. They have been selected on the basis that they participate in the often rate-limiting stages of the fermentation, i.e. polymer hydrolysis and methanogenesis.

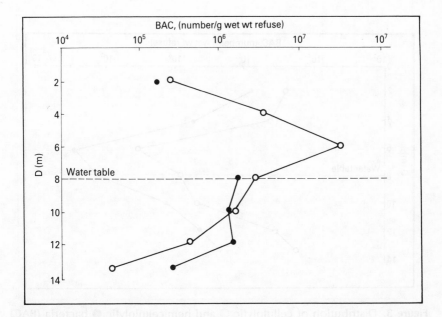

**Figure 2.** Distribution of amylolytic ● and proteolytic ○ bacteria (BAC) with depth (*D*) in borehole BTL 01.

## MICROBIOLOGY AND CHEMISTRY OF THE AVELEY LANDFILL

The Aveley Landfill in Essex, UK, was selected for this study on the basis that the site had already been extensively analysed for its leachate and refuse chemistry. Furthermore, it was known that landfill gas (LFG) production varied substantially with varying physicochemical conditions in the site (Rees, 1980b). Microbiological studies were conducted with strictly anaerobic microbial processing facilities on fresh MSW samples recovered from boreholes which typically penetrated the site to 13 m. Distribution of all seven key microbial groups are shown in Figs 2–4.

Chemical analyses of the landfill have comprised both analysis of refuse solids and leachate pressed from the refuse obtained from various depths. Leachate chemistry data are shown in Fig. 5. The principal species analysed were volatile fatty acids (VFAs), ammonia, sulphate, chloride, potassium and pH. The organic cellulose and hemicellulose content of the refuse solids is shown in Fig. 6.

Rates of gas production for fresh refuse samples recovered from the landfill at various depths are shown in Fig. 7, together with refuse moisture content. Rates were determined on approximately 1 kg refuse samples.

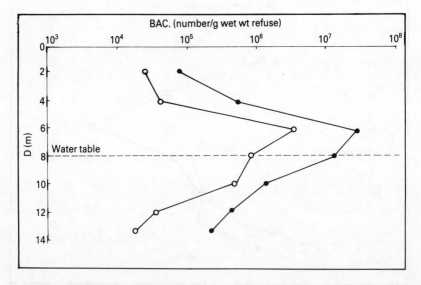

**Figure 3.** Distribution of cellulolytic ○ and hemicellulolytic ● bacteria (BAC) with depth (*D*) in borehole BTL 01.

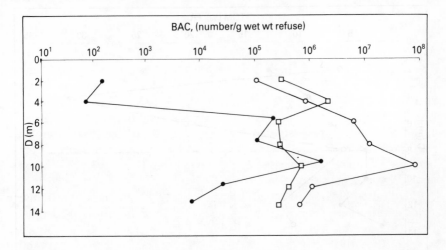

**Figure 4.** Distribution of hydrogen oxidizing methanogens ○, acetoclastic methanogens ● and sulphate-reducing bacteria (BAC) □ with depth (*D*) in borehole BTL 01.

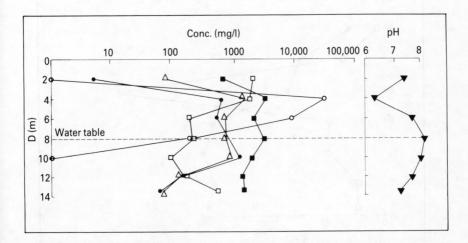

**Figure 5.** Distribution of total VFA ○, ammonia nitrogen ●, $SO_4^{2-}$ □, $Cl^-$ ■, $K^+$ △ and pH ▲ with depth (*D*) in borehole BTL 01.

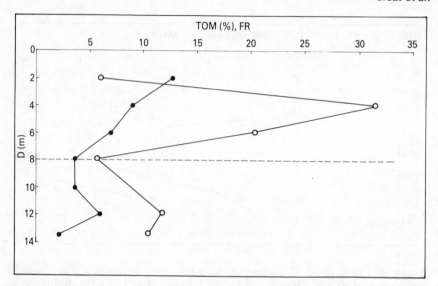

**Figure 6.** Distribution of total organic matter (TOM, in % of dry refuse) ○ and the fibre ratio (FR; hemicellulose + cellulose/lignin) ●, with depth (*D*) in borehole BTL 01.

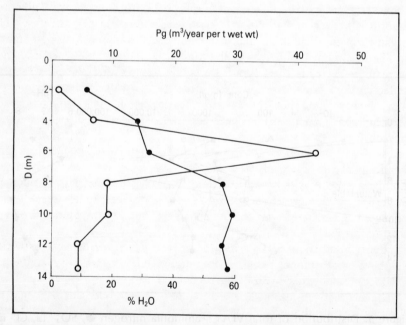

**Figure 7.** Distribution of maximum specific gas production (*P*g) ○ and water content (%H₂O) of refuse ● with depth (*D*) in borehole BTL 01.

## DISCUSSION AND CONCLUSIONS

Even a cursory analysis of the microbial data shows dramatic variations in numbers of all the polymer hydrolysing and methanogenic organisms with depth in the landfill. By comparison, numbers of sulphate-reducing bacteria (SRBs) were relatively constant with depth. These variations in microbial numbers again resemble the large variations in concentration of certain bioreactive carbon species, particularly carboxylic acids (VFAs) that have been observed in a wide range of landfills, including the Aveley Landfill, on previous occasions.

Carboxylic acid concentrations are highest at the 4 m depth (Fig. 5) and diminish rapidly with depth as methane production becomes better established. Similarly, sulphate concentrations are very high at 2 m and 4 m depth and diminish at 6 m and below due to increased activity of SRBs.

The ammonia profiles are typical of what is observed in most British landfills and the chloride and potassium data show the essentially conservative nature of these species, except where there is enhanced dilution of leachate at the 12 m and 13.5 m depths due to ground-water intrusion.

Concentration of cellulose and hemicellulose in the refuse solids (Fig. 6) also show variations with depth and age of refuse. The total organic content of the landfill is highest at 4 m below the surface in recently placed refuse. The organic content then diminishes with depth and the approaching water table.

The maximum number of hydrolytic bacteria of all four types examined were found at 6 m (Figs 2 and 3). Below this depth numbers decreased reflecting depletion of the polymeric substrates. Low numbers of microbes at 4 m, however, do not reflect substrate depletion, but rather that the fermentation is in its early stages and that environmental conditions for microbial growth are suboptimal. These data (Fig. 3) indicate that the higher rates of gas production are associated with numbers of cellulolytic and hemicellulolytic organisms between $10^6$ and $10^8$/g wet wt MSW. These numbers compare with between $10^4$ and $10^5$ present at 2 m, thus the numbers at the 6 m depth represent between a 2 and 3 log increase in numbers. Similar changes are associated with the other polymer hydrolysing groups shown in Fig. 2.

Higher numbers of SRBs are found at 4 m, which corresponds with high sulphate concentration, but there is remarkably less variation in numbers of SRBs than in polymer-hydrolysing bacteria.

The methanogens again show a different profile to both the SRBs and the hydrolytic group. Figure 4 shows that both methanogenic groups show highest numbers at the 10 m depth and not at 6 m—the area showing highest rates of gas production (Fig. 7). However, numbers at the 5 m depth are still substantially higher than they are at 2 m and 4 m. Below 10 m, the methanogens rapidly decline in numbers as the availability of carbon substrates becomes exhausted. The hydrogen-oxidizing methanogens show an increase of some 3 logs between 2 and 10 m, while the acetoclastic methanogens show a 4 log increase over the same depth.

The gas production data (Fig. 7) again show the dramatic variation in specific activity with depth in the landfill with highest rates occurring at 6 m, just above the water table. These high rates of gas production at 6 m certainly correlate well with high numbers of hydrolytic bacteria at that depth and clearly there are high numbers of methanogenic organisms also present. Concentration of organic acids at approximately 10 000 mg/l at this depth nevertheless suggest that, under these conditions, methanogenesis is rate-limiting.

The low gas production rate at 4 m, together with the high organic acid concentration, is also associated with low numbers of methanogenic bacteria. This could be attributed to substrate inhibition of these bacteria at this depth, which is supported by the pH and VFA data.

Recognizing the importance of the prospective control of landfill processes and the acceleration of gas production and refuse stabilization, this chapter describes the measurement of the numbers of key microbes present in landfill systems during various stages of decomposition.

Clearly, numbers of key microbes in fresh refuse are substantially lower than those associated with a high rate of stabilization and gas production. An understanding of why these variations in microbial numbers occur and the factors influencing their growth and activity in landfill are fundamental to the development of better controlled landfill systems.

## ACKNOWLEDGEMENT

This work was undertaken in collaboration with Aveley Methane Limited.

## REFERENCES

Cromwell, D.L. (1965). 'Identification of microbial flora present in landfills'. MSc thesis. West Virginia University, USA.

Engelbrecht, R.S. and Amirhor, P. (1975). Biological impact of sanitary landfill leachate on the environment. *In* Proceedings of the 2nd National Conference on Complete Water Re-use. Chicago, USA.

Filip, Z. and Küster, E. (1979) 'Microbial activity and the turnover of organic matter in municipal refuse disposed to landfill', *Eur. J. Appl. Microbiol.* 7, 371–379.

Jones, K.L. and Grainger, J.M. (1983). 'The application of enzyme activity measurements to a study of protein, starch and cellulose fermentation in domestic refuse', *Eur. J. Appl. Microbiol. Biotechnol.* **18**, 181–185.

Peck, M.W. (1986). Indirect method for the enumeration and identification of methanogenic bacteria in landfill samples. *In* Emberton, J.R. and Emberton, R.F. (eds) 'Energy from Landfill Gas'. Harwell Laboratory, Solihull, UK.

Peterson, M.L. (1971). 'Pathogens associated with solid waste processing'. USEPA SW-49r.

Rees, J.F. (1980a). 'The fate of carbon compounds in landfill disposal of organic matter', *J. Chem. Technol. Biotechnol.* **30**, 161–175.

Rees, J.F. (1980b). 'Optimization of methane production and refuse decomposition in landfills by temperature control', *J. Chem. Technol. Biotechnol.* **30**, 465–485.

Rees, J.F. (1985). Landfills for treatment of solid wastes. *In* 'Comprehensive Biotechnology'. The Principles of Biotechnology, Engineering Considerations **63**, 1071–1076. Pergamon Press, Oxford.

## 2.3 Enhancement of Degradation: German Experiences

RAINER STEGMANN and HANS-HENNING SPENDLIN

*Technical University of Hamburg-Harburg, Eissendorferstrasse 40, D-2100 Hamburg 90, W. Germany*

### INTRODUCTION

Long-term emissions control as well as reclamation of closed-up landfills are the main problems in the field of sanitary landfill operation. Improvement in this area could be achieved if the anaerobic microbiological processes in landfills could be enhanced. This would result in the following practical advantages:

1. Aerobic biological leachate treatment will be less energy intensive, since the organic concentrations of leachate from landfills in the methane phase are up to two orders of magnitude lower (Ehrig, 1980).
2. Gas production will be enhanced; as a result gas recovery rates for utilization are comparatively higher. Gas from leachate treatment inside the landfill is available for recovery, also.
3. Since the main gas production phase is limited to a shorter period, reclamation of landfills should be less problematic.
4. The long-term effectiveness of barrier systems is unknown; enhancement of biochemical processes results in earlier stabilization of municipal solid waste (MSW) and hence later, when some barrier systems may be less effective, in lower emissions of organics.
5. The settling processes will take place earlier; this is advantageous in regard to the design and operation of landfill gas extraction systems as well as to the effectiveness of surface barrier systems, which are being considered more and more as top seals.

Optimal enhancement could be achieved if the processes that take place in a landfill could be described. Steps in this direction were done by studying the

61

literature and anaerobic decomposition processes of MSW in different test scales.

This chapter does not describe the basic degradation processes of MSW since this is already available in the literature (Farquhar and Rovers, 1973; Emcon Ass., 1980; Stegmann and Ehrig, 1980; Mosey, 1981; Sahm, 1981; Rees, 1982; Christensen and Kjeldsen, 1989). Numerous lysimeter and field tests have been performed with different results. Often the methane phase was not reached, which makes the results less meaningful. A summary of the research activities may be found in Emcon Ass. (1980) and Spendlin (1985). New results are presented by Leuschner and Melden (1983), Barlaz et al. (1987), Kinman et al. (1987) and Pacey (1987).

In the following, results from laboratory scale and outdoor lysimeters as well as full-scale landfills will be presented. Preliminary conclusions are drawn from these results but final results cannot be given at this time. These results are discussed together with the findings of other researchers.

## LABORATORY SCALE TESTS

A procedure has been developed at the Technical University of Braunschweig, West Germany, where shredded municipal solid waste is anaerobically degraded; the acid and the methane phase took place during the degradation process (Stegmann, 1981).

The containers (Fig. 1) are made of PVC and are gas-tight sealed on top and bottom (volume 120 l). The leachate is collected in a plastic vessel (volume 2.5 l) and may be pumped back periodically on top of the refuse by means of a small submersible pump. The leachate is analysed periodically and the total gas production is collected in 'gas-tight' aluminium bags laminated with plastic. The gas is analysed and the gas volume is measured by pumping it out of the bag through a flow meter. The containers are placed in a room heated at 30°C.

As a first test, 80 kg of shredded fresh MSW (diameter <5 cm) from a composting facility were placed in the gas-tight test container; the moisture content was adjusted to 65% (based on wet weight) and the temperature was 30°C; no leachate recirculation took place. In a parallel test a similar container has been placed in a room at 11–18°C. During the starting phase the produced gas contained mainly $CO_2$ (30°C: 16.75 l/t dry MSW; 11–18°C: 3.7 l/t dry MSW after about 400 days); the maximum methane concentration was 5% (90% $CO_2$ at the end of the test). This indicates that these containers may never reach the methane phase, since the high production of organic acids results in a low pH value and in a 'conservation process' (analogy silage); under these

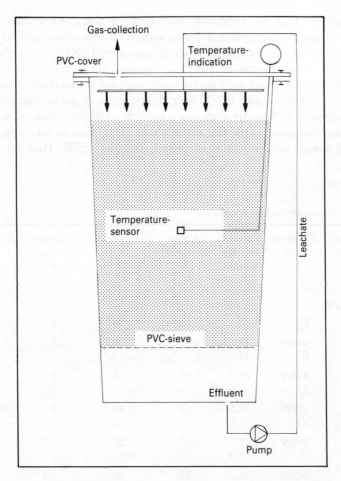

**Figure 1.** Cross-section of a laboratory scale test cell (Stegmann, 1981).

conditions the fermentative (pH <4.5–5.0) and methane- producing bacteria (pH <6.2) are not able to develop. Volatile organic acids are also toxic to methanogenic bacteria at concentrations >6000–16 000 mg/l. This situation could be changed if a leachate flux were provided, so that readily degradable organic acids may be transported out of the MSW with the leachate. Ham (1982) came to the conclusion that the gas production increases faster if there is a leachate flux through the MSW.

From the practice of anaerobic digestion of sewage sludge, it is well known that the mixing of fresh sludge with digested sludge may prove advantageous for starting the anaerobic processes. Bearing this in mind, test series involving various mixtures of compost and sludge with MSW were developed, as shown in Table 1. Figure 2 shows results from the tests series CDO-1,2,4,6, where the influence of the addition of partly composted MSW and sewage sludge becomes obvious. Using only partly composted MSW, methane production starts first, combined with BOD5 degradation and high gas production rates. The addition of 50% partly composted material to MSW enhanced methane production compared to the test where about 15% (based on solid content) of digested sewage sludge had been mixed with the MSW. How far sludge

**Table 1.** Summary of test series.

| Container | Content (−) | Volume (kg dry weight basis) | Water content[a] (%) | Temperature (°C) | Water exchange-rate[b] (l/w) | Remarks |
|---|---|---|---|---|---|---|
| Series CD0: | | | | | | |
| 1 | C | 28 | 65 | 30 | 2 | |
| 2 | MSW C | 14 14 | 65 | 30 | 2 | |
| 4 | MSW SS | 24.4 3.6 | 65 | 30 | 2 | |
| 6 | MSW SS | 23.5 3.5 | 65 | 30 | 2 | pH = 7[c] |
| 9 | MSW C | 18.7 9.3 | 65 | 30 | 2.5 | |
| 10 | MSW C | 18.7 9.3 | 65 | 30 | 2.5 | [d] |
| 11 | MSW C | 18.7 9.3 | 65 | 30 | 2.5 | [e] |
| Series CD2: | | | | | | |
| 1–7 | MSW C | 18.7 9.3 | 65 | 30 | 3.5 | |

C = Partly composted shredded MSW.
SS = Anaerobical digested sewage sludge.
[a] At the beginning of the test.
[b] l/w = 1 per week.
[c] Addition of NaOH, addition of P.
[d] No leachate recirculation.
[e] Compost only in bottom layer.

addition to MSW may enhance methane production cannot be concluded; the addition of sewage sludge (low solid content) is probably not high enough to act as an inoculum. Kinman *et al.* (1987) found the beneficial influence of the addition of sewage sludge in order to enhance biochemical processes; similar conclusions were drawn by Leuschner and Melden (1983) and Gandolla *et al.* (1982). Barlaz *et al.* (1987) did not observe any positive effect of sludge addition. However, these results cannot be compared directly since different amounts of different kinds of sewage sludge have been added at different times. The authors would therefore not support a general statement that sludge addition leads to enhancement.

The effect of the addition of partly composted waste may have different reasons:

1. During composting also anaerobic processes take place up to a certain degree, so that methanogenic bacteria may develop.
2. The readily degradable organics have already been decomposed aerobically so the rapid acid production phase—as can be found in fresh refuse—is overcome and the balance of acid and methane production bacteria can develop earlier. Barlaz *et al.* (1987) came to the same conclusion.
3. There might be also a 'dilution' effect that lowers the organic acids concentrations in the system, and that might be advantageous, particularly in the initial phase of methane production.

The addition of a base (NaOH) to keep the initial pH-value at pH=7 and the addition of phosphorus did not enhance the methane production; in this test container, the lowest gas production was measured (CDO-6; Fig. 2).

Other researchers (Kinman *et al.*, 1987; Leuschner and Melden, 1983; Barlaz *et al.*, 1987) came to the conclusion, that the addition of buffer and nutrients may be helpful. New experiments by the authors (not yet published) indicate that buffer addition may especially be helpful to overcome the inhibition of the acid-producing bacteria. On the other hand, it must be pointed out that, when interpreting and comparing test results, different researchers may come to different conclusions. One reason for this might be that the test conditions are different, although it is difficult to tell how. One possibility might be the different composition of waste and the different periods of time of its exposition to air, before it is placed in the test containers.

Experiments were set up to obtain information concerning the following questions (Series CDO-9,10,11):

1. Do high water addition rates, in an equivalent of 1200–1400 mm/year, make leachate recirculation ineffective?
2. Will a layer of partly composted MSW placed at the bottom of the container result in a degradation of the organic acids produced in the above-placed MSW?

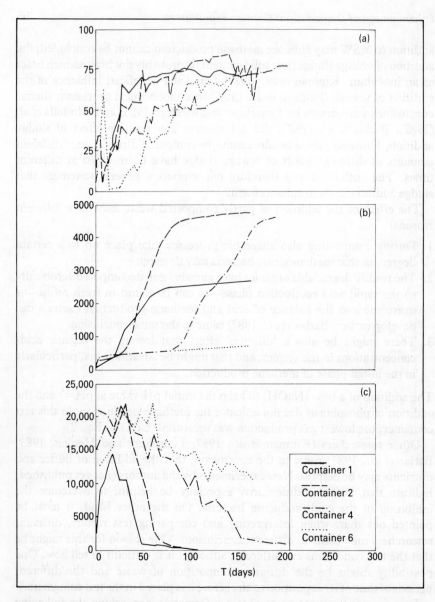

**Figure 2.** Results from test series CDO (Table 1) showing methane concentration (%, a), accumulated gas production (l, b) and $BOD_5$ (mg/l, c) as functions of time (T, days).

Container 1: C
Container 2: MSW, C
Container 4: MSW, SS
Container 6: MSW, SS and addition of NaOH
C = partly composted shredded MSW
SS = anaerob digested sewage sludge

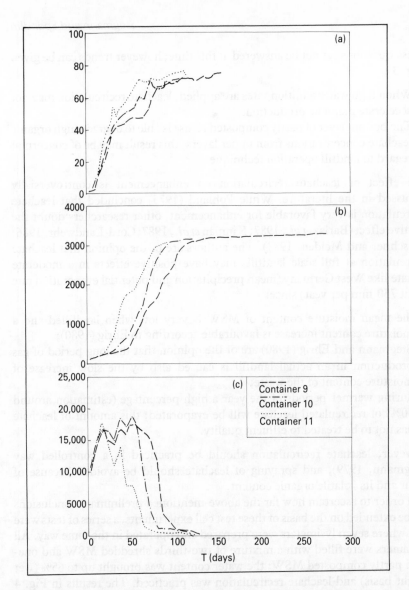

**Figure 3.** Results from test series CDO (Table 1) showing methane concentration (%, a), accumulated gas production (l, b) and BOD₅ (mg/l, c) as functions of time (T, days).

Container  9: MSW, C, leachate recirculation
Container 10: MSW, C, no leachate recirculation
Container 11: MSW, C, compost only in the bottom layer.

These questions cannot be answered at this time; however trends can be given (Fig. 3):

1. When high water addition rates are applied, leachate recirculation may not accelerate methane production.
2. The bottom layer of partly composted refuse is able to degrade high organic leachate concentrations from upper layers; this result may be of concern in regard to landfill operation technique.

The effect of leachate recirculation on enhancement is controversially discussed in the literature. While Pohland (1975) concluded that leachate recirculation is very favorable for enhancement, other researchers doubt the positive effect (Barlaz *et al.*, 1987; Kinman *et al.*, 1987; Cord-Landwehr, 1986; Leuschner and Melden, 1983). The authors are of the opinion that leachate recirculation at full-scale landfills may have positive effects in a moderate climate like West Germany (mean precipitation and potential evaporation rate about 750 mm per year) since:

1. the mean moisture content of MSW is very low when landfilled and a moisture content increase is favourable according to Ehrig (1980);
2. Stegmann and Ehrig (1980) are of the opinion that the long period of gas production in an actual landfill is caused also by the slow increase of moisture content of the waste;
3. during warmer periods of the year a high percentage (estimation around 50%) of recirculated leachate will be evaporated; this amount of leachate has not to be treated to effluent quality.

However, leachate recirculation should be practiced in a controlled way (Stegmann, 1979), and spraying of leachate should be avoided because of odour and its volatile organic content.

In order to ascertain how far the above-mentioned preliminary conclusions can be extended on the basis of these test cell experiments, a series of tests were run, where seven lysimeters were prepared and operated in the same way. All containers were filled with a mixture of two-thirds shredded MSW and one-third partly composted MSW; the water content was brought up to 65% (wet weight basis) and leachate recirculation was practiced. The results in Fig. 4 show that the tests are, to a high degree, reproducible (the general problem of biological tests should be encountered). Barlaz *et al.* (1987) came to the conclusion that, 'it is difficult to duplicate methane production in replicate containers'. The different results may originate in the utilization of different waste materials for the tests.

Figure 5 represents further results (CD2-3) concerning biochemical, physical and chemical processes during anaerobic degradation of MSW. The measured gas compositions in the test container were very similar to the

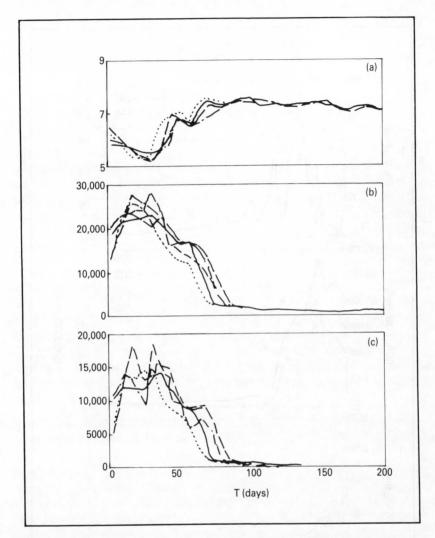

**Figure 4.** Leachate concentrations from test series CD2 (Table 1) showing pH (a), COD (mg/l, b) and $BOD_5$ (mg/l, c) as functions of time ($T$, days).

theoretical curves presented by Farquhar and Rovers (1973). Note the plateau at 50% methane; this kind of plateau was observed in most of our tests; it is likely that acetic acid is to be converted to $CO_2$ and methane, since the anaerobic fermentation of acetic acid results in the above-mentioned gas composition as shown by Gujer and Zehnder (1983) in Fig. 6.

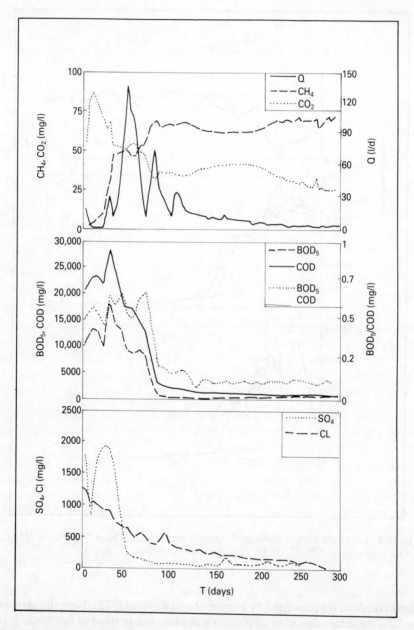

**Figure 5.** Results from test series CD2-3 (Table 1) showing concentrations of $CO_2$, $CH_4$, $BOD_5$, COD, $SO_4$, Cl and gas production rate as functions of time ($T$, days).

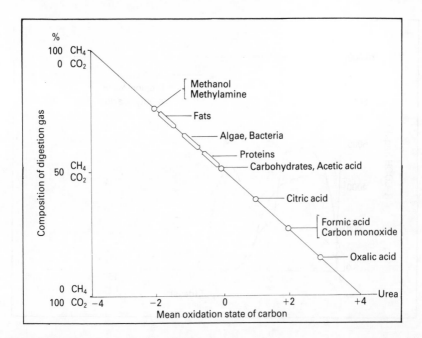

**Figure 6.** Relationship between different kinds of organic acids and the methane concentration (Gujer and Zehnder, 1983).

Hydrogen has been found several times during different tests in the initial phase. Since the period of hydrogen appearance is very short, it might happen that no samples are taken during that period.

The sequence of organic acid concentrations in leachate was also measured during an enhancement project run at the Technical University of Hamburg-Harburg (Fig. 7). In series E10 shredded MSW, which for a short time had been aerobically composted, was anaerobically fermented in the same way mentioned above (Fig. 5). Figure 7 shows that when most of the acetic acid was broken down to $CH_4$ and $CO_2$, high concentrations of propionic acid were measured and these were converted, about 50 days later, to biogas. Often, a second (much smaller) peak at 65–70% methane can be detected (Fig. 5). Further research is needed to identify the succession of the degradation of the organic components in the MSW. Since methane concentration increases as the degradation process proceeds, this might be a way to interpret gas concentrations measured in actual landfills.

In order to extract leachate samples in the test container, 1.5 litres of leachate had to be removed: the same amount of tap water was put back into

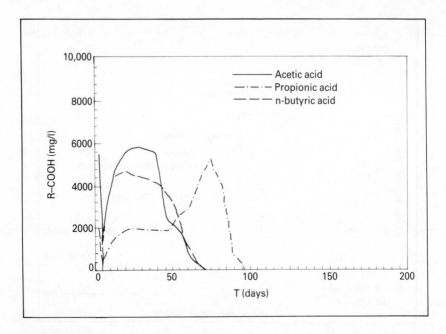

**Figure 7.** Sequence of organic acid concentrations in the leachate as a function of time (Spendlin, 1985).

the container in order to keep the moisture content constant. As a consequence, a leaching effect may be observed if the chloride concentration is taken as an indicator (see Fig. 5). Comparing other components with the chloride concentration during the experiment, it became obvious that the organic concentrations behaved very differently. The $BOD_5$ and COD concentrations decrease dramatically when stable methane production takes place. Note that organic leachate concentrations may still be high, although gas production is at a peak. For this reason organic leachate concentration cannot always be used as an indicator for the gas production and the gas composition. This fits well with observations in the field, where high gas production rates, with methane concentrations around 50%, have been measured in refuse one year after placement (high compaction, $BOD_5$ concentrations in the leachate of the order of 20 000 mg/l).

In addition, the behaviour of different elements, dependent upon the biological situation, can be observed (Fig. 8); while the elements K, Ca, Mg and Na are to a high degree diluted (similar curve as Cl-curve), Fe, Zn and Mn are converted in the landfill to a chemical form that does not dissolve well

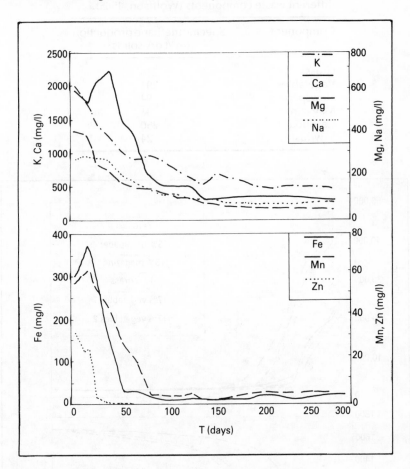

**Figure 8.** Results from test series CD2-3 (Table 1) showing concentrations of various inorganic elements as functions of time (*T*, days).

(sulphides). This reaction is similar to the biochemical sulfate reduction (Fig. 5).

The influence of the waste composition on the kind and sequence of the biochemical processes in MSW landfill is also of interest. This is of concern when comparing results from research activities in different countries, where waste compositions differ (USA waste has higher paper content; West German waste has higher food-waste content). Wolffson (1985) presented results on gas production effects of various waste compounds.

Table 2 presents specific methane yields for various waste compounds.

**Table 2.** Specific methane production rates from different waste components (Wolffson, 1985).

| Component | Specific methane production ($m^3$/t dry solids) |
|---|---|
| Grass | 216 |
| Vegetables | 291 |
| Newspapers | 94 |
| Magazines | 81 |
| Cardboard | 250 |
| Sawdust | 24 |

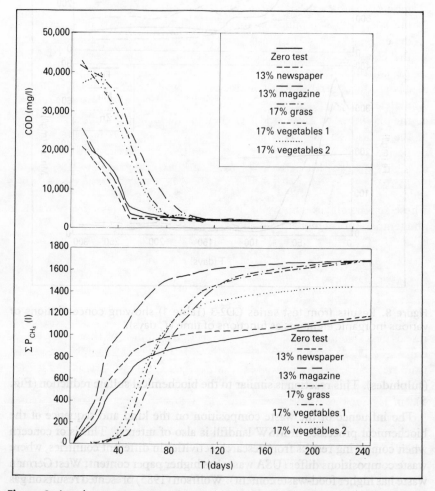

**Figure 9.** Leachate concentrations and accumulated methane production of laboratory scale tests where different amounts of waste components (based on dry solids) had been added (Wolffson, 1985).

These results indicate that addition of vegetable waste and grass may result in a higher production of methane but, as shown in Fig. 9, methane production is retarded. From these experiments it was concluded that waste composition influenced gas enhancement strategies. These should be taken into account when results from enhancement studies in West Germany and the US are compared.

As a result of these laboratory scale tests the following aspects may be important for gas enhancement in full scale landfills:

1. the main gas production phase is terminated after a period of about 6–8 months (lysimeter tests);
2. the specific gas production rate per unit of MSW can be determined, depending on the waste composition. The shredded MSW that was used in the container tests produced about 150–180 l gas/kg dry MSW (2.4–2.9 cu.ft/lb);
3. gas production rates fluctuate due to the degradation of different organic materials (food waste, paper, etc.);
4. the pH increases 'automatically' when the methane production starts;
5. the addition of composted MSW to MSW, or a short aerobic degradation period of MSW, helps to initiate the methane phase relatively early;
6. highly polluted leachate can be degraded in a bottom layer of composted MSW.

These statements are only valid, however, if sufficient moisture is available and temperature is higher than 30°C (86°F).

## LYSIMETER TESTS

Lysimeter tests at pilot plant scale were carried out with a group of other researchers (Collins *et al.*, 1979) at the sanitary landfill in Braunschweig. A cross-section of a lysimeter is presented in Fig. 10. As an example, the results from two lysimeters are presented, where MSW and aerobic stabilized sewage sludge were codisposed. In lysimeter 3 this material was highly compacted in 40 cm layers; these layers were built in with a time lag of about 6 weeks. No cover material was used. The refuse–sludge mixture in lysimeter 4 was highly compacted in 2 m layers. The sludge, which had been added to the MSW, has little or no influence on the mechanism described here. About 2 years after the first 2 m of the MSW mixture had been landfilled in the lysimeter, another 2 m of the refuse–sludge mixture was placed on top, using the same technique for lysimeter 3 and 4. When the new layers were added, the organic leachate concentrations in both lysimeters had already decreased dramatically (Fig. 11).

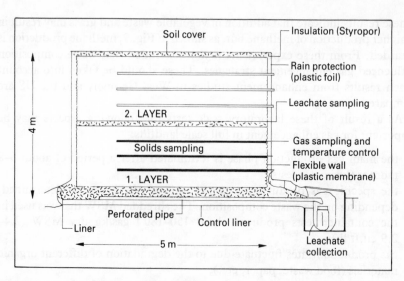

**Figure 10.** Cross-section of the lysimeter.

In order to ascertain how far the highly polluted leachate from the fresh layers of refuse is degraded in the lower layers, small sheets of plastic membrane had been placed between the two 2 m layers. Using this technique, leachate had been collected and analysed with the results presented in Fig. 11:

1. Methane production starts earlier if the refuse is compacted in thin layers using no cover (lysimeter 3). This can be seen from the comparatively lower $BOD_5$ and COD concentrations in the leachate. The difference in organic leachate concentrations between the lysimeters 3 and 4 becomes even more obvious when the leachate concentrations from the upper layer are compared with each other.

2. The degradation of the leachate from the upper layer in the older refuse was more effective in lysimeter 3.

The stimulating effect concerning anaerobic degradation, when the refuse is compacted in thin layers, may be the partly aerobic decomposition of the readily degradable organics. These might produce high organic acid concentrations if the refuse is compacted in 2 m layers, with only one-third of the surface exposed to the air during a short period. It is well known from the literature that high organic concentrations (>10 000 ppm), as well as organic acid–alkinity ratios >0.8, result in inhibition of methane-producing bacteria (Jennett and Dennis, 1975). In addition, the refuse can be more equally wetted by precipitation, if it is disposed of in thin layers.

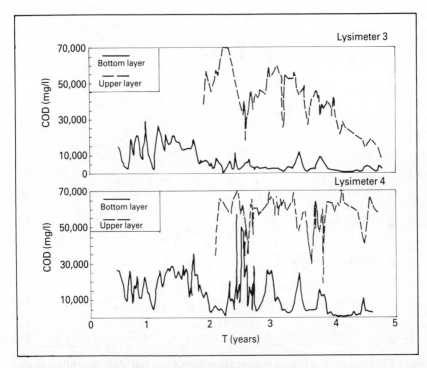

**Figure 11.** Concentrations of leachates collected at the bottom and on top of the first layer of refuse (Fig. 10) as a function of time (years).
   Lysimeter 3: 0.40 m layers, highly compacted, placed every 6 weeks
   Lysimeter 4: 2.00 m layers, highly compacted.

## FULL-SCALE LANDFILLS

The influence of landfill operation on leachate quality is shown in Fig. 12, where results from an investigation of 15 full scale landfill sites are presented (Ehrig, 1980). Enhancement of anaerobic processes can be achieved if landfills are operated with low compaction and the refuse is placed in thin layers without cover (highly compacted). In Fig. 12, enhancement is documented by the early decrease of organic concentrations in the leachate.

As a result of these data, in combination with the lysimeter tests, the following concept for a full-scale landfill has been developed. The first layer of landfill can be prepared in such a way that the polluted leachate from the lifts above can be treated in this area. The first layer of refuse (1.5–2 m height) must not be compacted, so that readily degradable organics can decompose

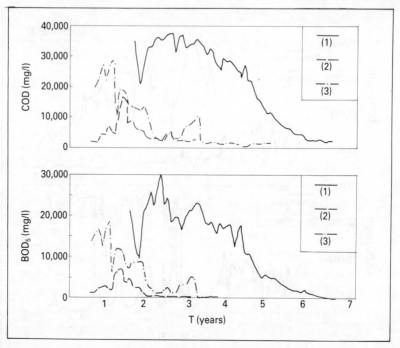

**Figure 12.** Examples of leachate concentrations from full scale landfills showing the effects of various landfill operation schemes as a function of the landfill age (*T*, years).

    (1) 2 m lifts
    (2) thin layers
    (3) 2 m lifts with leachate recirculation.

aerobically. Thinner layers or placement of perforated pipes may increase the air supply. Leachate recirculation should be practised. The rate of recirculation should be moderate in order to avoid anaerobic conditions. After 1 year of placement, the usual landfill operation can be started. The disadvantage of this procedure is that a high leachate production rate will result during the first year.

    This procedure was tested in a research project at the sanitary landfill of Lingen. Two sections of the landfill (each 1 ha size) were filled with MSW. In one section the refuse was highly compacted in 0.5–1 m lifts. In the other section, the refuse was placed 1 m high without any compaction. After 6 months, another lift of one meter was placed on top of the first lift in the same manner. After another 6 months refuse was disposed in highly compacted 2 m lifts. Leachate recirculation has been practised in both fields.

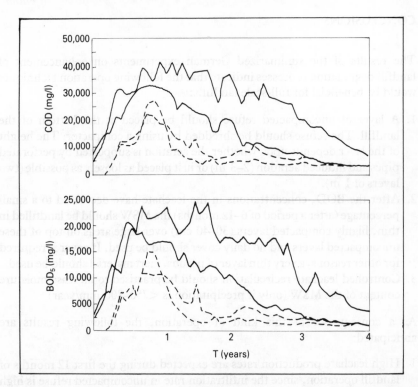

**Figure 13.** Leachate $BOD_5$ concentrations from two test fields (full-scale operation) at the landfill of Lingen, West Germany, as a function of operation and time ($T$, years), solid line: highly compacted refuse in 0.5–1.0 m lifts, leachate recirculation; dashed line: uncompacted layers of refuse covered after 1 year of placement with refuse in 2 m lifts (highly compacted), leachate recirculation.

Figure 13 shows the leachate quality data from both sections, indicating the advantage of the uncompacted layers. Twenty months after placement, the COD concentrations were of the order of 2000 mg/l in the landfill section with the uncompacted bottom layer, while in the parallel section ten times higher COD concentrations were measured. As already mentioned, high leachate production rates have been detected during the first year where the refuse was not compacted. Efforts should be made to reduce the time for the aerobic decomposition of the first layer, e.g. by providing air flow through perforated pipes. Lebsanft (1984) observed at the 'Rottedeponie' (aerobic landfill) of Schwäbisch Hall, that when the landfill was converted to high compacted anaerobic operation, the low $BOD_5$ concentrations of the leachate stayed at the same level.

## CONCLUSIONS

The results of the summarized German experiments on enhancement of landfill degradation processes indicate that the following operation techniques would be beneficial for full-scale landfills:

1. A layer of uncompacted refuse should be placed at the bottom of the landfill. The refuse should be shredded by using a compactor. The height of the layer depends upon whether the aeration is supported by perforated pipes and induced aeration (2–3 m) or just placed as loosely as possible (two layers of 1 m).
2. After the $BOD_5$ concentrations in the leachate have decreased to a small percentage (after a period of 6–12 months) the MSW should be landfilled in thin, highly compacted layers (30–40 cm) over large areas on top of these uncompacted layers and no daily cover should be used. If cover is required for other reasons, a very thin layer of porous, inert material should be used.
3. Controlled leachate recirculation should be practised to increase moisture content of the MSW (only if precipitation is <750–800 mm/year).

As a consequence of this kind of operation, the following results are anticipated:

1. High leachate production rates are expected during the first 12 months of landfill operation, since the infiltration rate in uncompacted refuse is high and large areas are required.
2. After a period of about one year (the climatic situation has to be considered), the concentration of the organics in the leachate is expected to be very low, so that energy requirements for the aerobic leachate treatment are reduced and more gas is produced during anaerobic degradation of the leachate inside the landfill. It can be concluded from column tests that about 0.9–1.3 $m^3$ of gas/kg $BOD_5$ degradation will be produced.
3. Controlled leachate recirculation will result in evaporation loss of leachate (dependent upon climatic conditions and irrigation rates).
4. Leachate treatment is still required but with a minimum energy input (mostly for nitrification processes).
5. Methane production is expected 1–2 years after placement dependent upon climatic conditions. The period of gas production is hard to predict, but it is believed that it will be significantly shorter compared to the present way of operation.

These recommendations have to be verified quantitatively. A first attempt is now being made at the sanitary landfill in Lingen, West Germany, where a system is tested at full scale. Other techniques should be tested at full scale accompanied by a monitoring programme. Sanitary landfill operations in the future will need more sophisticated methods and this is only possible if technical knowhow is available.

## REFERENCES

Barlaz, M.A., Milke, M.W. and Ham, R.K. (1987). 'Gas production parameters in sanitary landfill simulators'. *Waste Management and Research* 5.

Christensen, T.H. and Kjeldsen, P. (1990). Basic biochemical processes in landfills. *In* Christensen, T.H., Cossu, R. and Stegmann, R. (eds) 'Sanitary Landfilling: Process, technology and environmental impact', p. 29. Academic Press, London.

Collins *et al.* (1979) 'Wasser- und Stoffhaushalt in Abfalldeponien und deren Wirkung auf Gewässer' (Water- and compound budgets in sanitary landfill and the influence on natural waters). *Müll und Abfall* 11.

Cord-Landwehr, K. (1986). 'Stabilisierung von Mülldeponien durch eine Sicker-wasserkreislaufführung' (Stabilization of landfills with leachate recirculation), *Veröffentlichungen des Institutes für Siedlungswasserwirtschaft und Abfalltechnik der Universität Hannover* 66.

Ehrig, H.J. (1980). 'Beitrag zum quantitativen und qualitativen Wasserhaushalt von Mülldeponien' (Contribution to the quantitative and qualitative water balance of sanitary landfill), *Veröffentlichungen des Instituts für Stadtbauwesen, Technische Universität Braunschweig* 26, Eigenverlag.

Emcon Ass. (1980). 'Enhancing production of landfill gas'. Report prepared for Argonne National Laboratories, PG and E, Southern California Gas Company.

Farquhar, G.J. and Rovers, F.A. (1973). 'Gas production during refuse decomposition', *Water, Air and Soil Pollution* 2.

Gandolla, M., Grabner, E. and Leoni, R. (1982). 'Ergebnisse von Lysimetern auf der Deponie Croglio, Schweiz' (Results of Lysimeters in the Landfill Croglio, Switzerland), *Veröffentlichungen des Instituts für Stadtbauwesen, Technische Universität Braunschweig* 33, Eigenverlag.

Gujer, W. and Zehnder, A.J.B. (1983). 'Conversion processes in anaerobic digestion', *Water Science and Technology* 15.

Jennett, J.C. and Dennis, N.D. (1975). 'Anaerobic filter treatment of pharmaceutical waste', *Journal, Water Pollution Control Federation* 47(1).

Ham, R.K. (1982). 'Einflußfaktoren auf die Gasproduktionsrate von Proben aus Mülldeponien' (Variables affecting gas generation rates in landfill samples), *Veröffentlichungen des Instituts für Stadtbauwesen, Technische Universität Braunschweig* 33, Eigenverlag.

Kinman, R.N. *et al.* (1987). 'Gas enhancement techniques in landfill simulators', *Waste Management and Research* 5.

Lebsanft, H. (1984). 'Rotte als Unterbau einer hochverdichteten Hausmülldeponie', *Müll und Abfall* 3.

Leuschner, A.P. and Melden, H.A. (1983). 'Landfill enhancement for improving methane production and leachate quality', 56th Annual Conference of the Water Pollution Control Federation, 2–7 October 1983, Atlanta, Georgia, USA.

Mosey, F.E. (1981). 'Anaerobic biological treatment of food industry waste', *Water Pollution Control*.

82 Stegmann and Spendlin

Pacey, J. (1987). Landfill gas production—past and future, a case history. International Symposium Cagliari, Sardinia, 19–23 October, 1987.

Pohland, F.G. (1975). 'Sanitary Landfill Stabilization with Leachate Recycle and Residual Treatment', EPA-Report, EPA-600/2-75-043.

Rees, J. (1982). 'Mikrobiologische Umsetzungsprozesse in Mülldeponien' (Microbiological degradation processes in sanitary landfill), Veröffentlichungen des Instituts für Stadtbauwesen, Technische Universität Braunschweig 33, Eigenverlag.

Sahm, H. (1981). 'Biologie der Methan-Bildung' (Biology of methane formation), Chem.-Ing.-Tech. 53(11).

Spendlin, H.H. (1985). 'Auswirkungen des Deponiebetriebes auf Sickerwasserbelastungen—Messungen im Labormaßstab' (Effect of the landfill operation technique on leachate quality—results from laboratory scale experiments; Sickerwasser aus Mülldeponien, Fachtagung an der T.U. Braunschweig, Inst. f. Stadtbauwesen, 21–22 March 1985, 39, Eigenverlag.

Stegmann, R. (1979). 'Reinigung und Verregnen von Müllsickerwasser unter Betriebsbedingungen' (Treatment and spray irrigation of leachate in praxis), Veröffentlichungen des Instituts für Stadtbauwesen, Technische Universität Braunschweig 27, Eigenverlag

Stegmann, R. and Ehrig, H.J. (1980). 'Entstehung von Gas- und Sickerwasser in geordneten Deponien' (Production of gas and leachate in sanitary landfills), Müll und Abfall 2.

Stegmann, R. (1981). 'Beschreibung eines Verfahrens zur Untersuchung anaerober Umsetzungsprozesse von festen Abfallstoffen im Labormaßstab' (Description of a procedure to investigate on aerobic degradation processes from solid waste materials in laboratory scale), Müll und Abfall 2.

Stegmann, R. and Ehrig, H.J. (1982). 'Enhancement of Gas Production in Sanitary Landfill Sites'. Resource Recovery from Solid Wastes, Proceedings of a Conference held in Miami Beach, Florida, USA, May 10–12, Academic Press, New York.

Stegmann, R. (1983). 'New aspects of enhancing biological processes in sanitary landfill', Waste Management and Research.

Wolffson, C. (1985) 'Auswirkungen des Deponiebetriebes auf Sickerwasserbelastungen—Messungen im Labormaßstab' (Effects of the landfill operation technique on leachate quality—results from lab.-scale experiments); Sickerwasser aus Mülldeponien, Fachtagung an der T.U. Braunschweig, Veroffentlichungen des Instituts für Stadtbauwesen, 39.

# 2.4 Enhancement of Degradation: Laboratory Scale Experiments

ALFRED P. LEUSCHNER

*Dynatech Scientific Inc., Cambridge, Massachusetts 02139, USA*
*(Present address: ReTeC Inc., Concord, Massachusetts 01747, USA)*

## INTRODUCTION

The concept of enhancement, as analysed in this laboratory programme, entails the use of leachate recycle to promote a more conducive environment for anaerobic microorganisms contained in a landfill. Under these conditions they will grow more quickly and thus convert the organic fraction of the municipal solid waste (MSW) to methane gas at a faster rate. It has long been understood that environmental conditions such as the presence of moisture, pH, temperature and nutrients (i.e. nitrogen and phosphorus) have to be regulated to optimize anaerobic biological processes. This can be accomplished in a landfill by adding a variety of material such as anaerobically digested sewage sludge (a source of bacteria), buffer (for pH control), and nitrogen and phosphorus (nutrients) to the leachate and recirculating these through the landfill. The thrust of this study was therefore to simulate a variety of these conditions in a laboratory landfill section and assess the effect each had upon the yield and rate of methane production and upon the quality of the leachate generated.

## EXPERIMENTAL PROGRAMME

Six laboratory scale digesters simulating landfills were built for this programme and operated for one year. Each reactor was made from a modified 55 gallon drum. Figure 1 presents a schematic of the reactors. A cement base

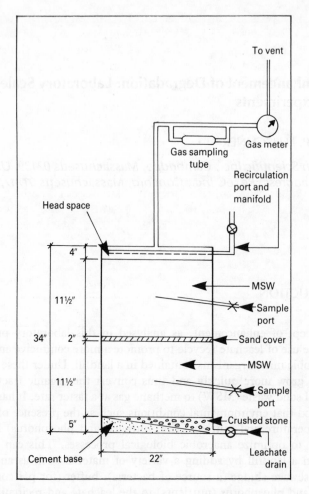

**Figure 1.** Schematic of laboratory reactors.

was installed into the bottom of each reactor, sloped to a leachate drain. Over the cement base, crushed stone was installed to allow the leachate to flow freely out of the reactor. The MSW was placed in each reactor, compacted to landfill densities in two lifts with two inches of sand placed between each lift, simulating a daily cover. The densities achieved ranged from 25.5 to 27.7 lb/ft$^3$.

Reactor 1 was operated as a control, receiving only simulated rainfall. All other reactors were operated with leachate recycle and supplemental addition of water, buffer (in the form of $NaHCO_3$), nitrogen (in the form of urea), phosphorus (in the form of $K_2HPO_4$), anaerobically digested sewage sludge, and septic tank pumpings (septage). Table 1 presents a summary of the additions to each reactor. Water was added to each reactor initially to achieve field capacity; thereafter each reactor received 3.0 l/wk of simulated rainfall. On a once per week basis leachate was drained from each reactor. A portion of the leachate (corresponding to the simulated rainfall) was discarded, with the remainder being recirculated. To this leachate additions of buffer, nutrients, and inoculum were made. The data in Table 1 represent the total quantity of the material added over the length of the study. Unless otherwise noted in Table 1, these additions were made over the first 150 days of operation. From that day until day 365, only leachate recirculation was practised.

On a regular basis solids, leachate and gases were analysed to assess the performance of these digesters. The MSW was analysed prior to placement

**Table 1.** Summary of additions to laboratory reactors.

| Parameter | Reactor | | | | | |
|---|---|---|---|---|---|---|
| | 1 | 2 | 3 | 4 | 5 | 6 |
| MSW added (lb) | 133 | 135 | 134 | 129 | 139 | 140 |
| Water added (l) | 147.0 | 245.5 | 197.2 | 197.5 | 263.3 | 265.9 |
| Water removed (l) | 102.2 | 190.4 | 144.3 | 156.8 | 207.1 | 215.3 |
| Buffer added | | | | | | |
| (g as $CaCO_3$) | | 785.7[a] | 1465.0 | 1324.5 | 957.6 | 1634.0 |
| (% of MSW) | | 1.49 | 2.77 | 2.62 | 1.76 | 2.98 |
| Nitrogen added | | | | | | |
| (g as N) | | | | 457.3 | 427.2 | 437.8 |
| (% of MSW) | | | | 0.90 | 0.78 | 0.80 |
| Phosphorus added | | | | | | |
| (g as P) | | | | 91.7 | 82.8 | 85.0 |
| (% of MSW) | | | | 0.18 | 0.15 | 0.15 |
| Sludge solids added | | | | | | |
| (g) | | 794.1[a] | | | 657.6 | |
| (% of MSW) | | 1.51 | | | 1.21 | |
| Septage solids added | | | | | | |
| (g) | | | | | | 783.3 |
| (% of MSW) | | | | | | 1.43 |

[a] Reactor No. 2 was operated without buffer or sludge addition for the first 188 days; thereafter, it received sludge and buffer.

**Table 2.** Initial analysis of fresh MSW.

| Parameter | Quantity |
|---|---|
| Total solids (TS) | 86.29% |
| Total volatile solids (TVS) | 63.34% of TS |
| COD | 0.70 kg COD/kg MSW |
| COD/volatile solids ratio | 1.28 |
| Total organic carbon (TOC) | 124.30 g TOC/kg MSW |
| Total Kjeldahl nitrogen (TKN) | 3.46 g TKN/kg MSW |
| Total phosphorus (TOT.P) | 0.53 g TOT.P/kg MSW |
| C : N : P | 100 : 2.78 : 0.43 |
| Lignin | 18.5% of TVS |

into reactors. These parameters included total solids (TS), total volatile solids (TVS), chemical oxygen demand (COD), total organic carbon (TOC), total Kjeldahl nitrogen (TKN), total phosphorus (TOT.P) and lignin. Table 2 presents the results of these analyses. The leachate was analysed each time it was drained from the reactors. On a weekly basis, alkalinity, pH, and total volatile acids (TVA) were measured. On a bi-weekly basis, TS, TVS, and COD were measured. The gas evolved was measured daily as well as being analysed for composition.

No attempt was made to control the temperature of these reactors. They were contained in an enclosed room and allowed to operate at ambient temperatures. The basis for this decision were the results of Leckie *et al.* (1979), who showed that moisture movement through test cells at the Sonoma County Landfill exhibited lower temperatures (closer to ambient) then test cells which did not have moisture moving through them. Ambient temperatures for this study ranged from 61°F (16.1°C) to 82°F (27.8°C).

## RESULTS

### Methane Production

Figure 2 presents data for all reactors on methane production, gas composition, and reactor (leachate) pH. The rates and quantity of methane produced varied significantly from reactor to reactor. The following is a summary of each reactor's performance with respect of methane production.

Reactor 1 produced methane intermittently at very low levels throughout this study. As shown in Fig. 2, only slight quantities of methane were produced near days 120, 250, and 280. Starting on day 309, and continuing through day 365, methane was produced on nearly a daily basis, but again at very low levels. The percentage methane in the gas produced started at levels below 10% of the total and rose throughout. By day 318, the percentage methane in the gas produced had climbed to 50%. The percentage methane remained at this level throughout the remainder of the study. The total quantity of methane produced in this reactor was only 2.46 l at standard temperature and pressure (STP) for the entire year.

No methane was produced in Reactor 2. The reason for this becomes evident in examining the data on leachate pH in Fig. 2. For over 300 days, the pH in this unit remained below 6.0. It was not until day 340 that the pH rose above 6.0, and at no time during this study did the pH reach 7.0.

In Reactor 3, from day 0 to day 153, the leachate pH was significantly below 7.0. On day 153, a leachate pH of 7.1 was found and from that point until the end of the study the leachate pH remained above 7.0. On day 159, the first significant methane production began. By day 210, the daily methane production had increased to 16 l/d. From that point, daily methane production exhibited a slow decline. Between day 280 and 365, the daily methane production averaged 5.0 l/d. The total quantity of methane produced in this reactor was 1275 l at STP. The percentage methane in the gas produced was consistently between 55 and 60%.

Methane production first occurred in Reactor 4 on day 75. By day 103, a maximum daily production of 20 l/d was reached. At this point, methane production declined rapidly and between days 120 and 150 the methane production was less than 5.0 l/d. After day 150, methane production increased again and reached a maximum of 20 l/d by day 182. From that point, methane production slowly declined to a level of approximately 2.5 l/d by day 365. The percentage methane in the gas reached 50% shortly after gas production started and remained between 50 and 60% thereafter. The leachate pH was low, approximately 6.0, at the start of this reactor, but rapidly climbed to above 7.0. From day 56 on, the pH was always above 7.0. The total quantity of methane produced in this reactor was 1642 l at STP.

Methane production first started in Reactor 5 on day 47. However, it was not until day 88 that appreciable quantities were produced. Gas production rose to a maximum of 29 l/d by day 182. From that point on, methane production declined to a rate of 5 l/d by day 365. For a 12-day period from day 134 to day 146, a sharp decline in methane production was noted. This was due to a leak in the reactor and does not reflect depressed gas production. The methane concentration in the gas produced initially was below 10%, but climbed to 50% by day 105, and remained between 50 and 60% thereafter. The leachate pH varied significantly (between 5.5 and 7.4) prior to methane production, but remained above 7.0 for the entire time after methane production started (day 88). The total quantity of methane produced by this reactor was 2340 l at STP.

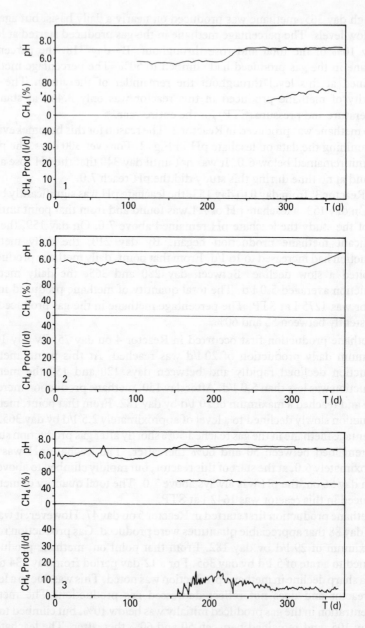

**Figure 2.** Daily methane production, gas composition and pH for reactions 1–6 as a function of time (1–3 this page; 4–6 facing page).

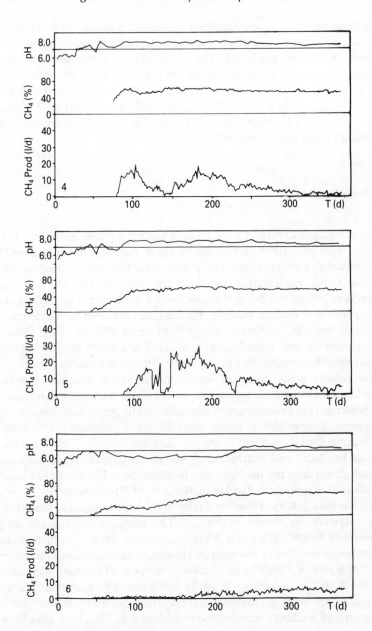

Methane production first occurred in Reactor 6 on day 43 and exhibited a slow, steady increase throughout the study. By day 365 an average of 5.0 l/d of methane was being produced. The percentage of methane in the gas also exhibited a slow, steady increase from less than 10% at the start to 50% by day 190. It remained above 50% thereafter. The leachate pH was consistently below 7.0 (generally 6.0 to 5.2) until day 230, when a pH of 7.0 resulted. Thereafter, the pH remained above 7.0. The total quantity of methane produced by this reactor was 835 l at STP.

## Leachate Quality

A plot of leachate COD and TVA for each reactor is presented in Fig. 3. These analyses were performed on a sample of the discarded leachate removed from each reactor on a weekly basis. The quality of the leachate produced in Reactor 1 is also shown presented in Fig. 3. It was not until day 111 that leachate production began; this being the only reactor not brought to field capacity at the start of the operation. Initially, the leachate contained high concentrations of organic material, a COD of nearly 30 g/l and a TVA of 15 g/l. Over time, these concentrations steadily declined. This is a result of flushing out the volatile acids by continually passing water through the reactor.

The volatile acids produced are an intermediate product of anaerobic degradation. They are an immediate precursor to methane production. The data presented are a summation of five acids (acetic, propionic, butyric, valeric and caproic) expressed as acetic acid. Figure 4 presents the volatile acid distribution for all reactors. Using the data for Reactor 1, the quantities of leachate produced and stoichiometry, the potential quantity of methane lost through discarding the leachate, can be calculated. For the entire period of operation, this represents 410 l of methane at STP for Reactor 1.

The leachate quality of Reactor 2 reflects the sour conditions which resulted in this system, as shown in Fig. 3. The leachate COD never dropped significantly below 10 g/l and TVA consistently ranged from 5 to 10 g/l. Leachate was recycled in this reactor. However, since simulated rainwater was added at a rate of 3.0 l/wk, an equivalent amount of leachate was discarded each week so that consistent moisture levels could be maintained. Figure 4 provides a breakdown on the volatile acids in the leachate. Using this data, data on quantity of leachate discarded and stoichiometry, it has been calculated that the potential methane lost from this reactor was 815 l over this study.

Leachate quality data for Reactor 3 is presented in Fig. 3. Several significant changes occurred in the leachate after methane production started. The COD of the leachate averaged 20 g/l prior to the production of methane, and fell to

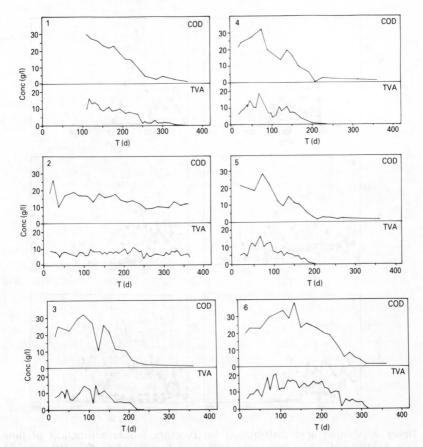

**Figure 3.** Leachate quality in terms of COD and TVA for Reactors 1–6 as a function of time (*T*, days).

below 2.5 g/l within 50 days after methane production started. Similarly, TVA were approximately 10 g/l prior to methane production and fell to less than 100 mg/l within 50 days. Due to the method of operation, volatile acids were lost in discarding some of the leachate (see Fig. 4). The total quantity of potential methane lost through this practice represents 470 l at STP; 88% of these volatile acids were lost prior to the start of methane production.

As was noted in Reactor 3, the leachate produced in Reactor 4 exhibited high concentrations of COD (25–30 g/l) and TVA (10–20 g/l) prior to the start of methane production. After methane production started, the concentrations of these constituents dropped. However, the reductions required a longer period to occur. It was not until day 220 (140 days after the start of methane

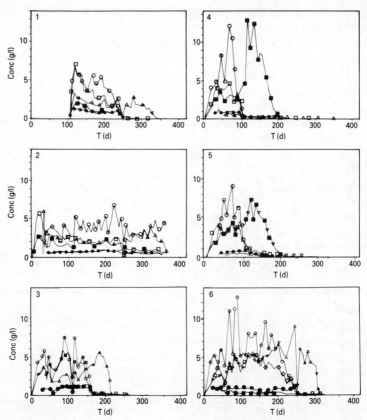

**Figure 4.** Volatile acid distribution for reactors 1–6 as a function of time.
○ = Acetic acid;  △ = Propionic acid;  □ = Butyric acid;  ● = Valeric acid;
▲ = Caproic acid.

production) that the COD fell below 2.5 g/l and TVA fell to 0 g/l. Figure 4 presents a breakdown of TVA components. Using these data, it was determined that 484 l of methane were lost by discarding leachate. Nearly 60% of this total occurred before methane production started.

The leachate quality of Reactor 5 over time is presented in Fig. 3. High concentration of COD (20 to 30 g/l) and TVA (10–15 g/l) existed prior to methane production. These concentrations declined steadily after methane production started. By day 200, the leachate COD had fallen below 2.5 g/l and TVA had fallen to essentially 0 g/l. Removal of leachate from this system occurred on a regular basis. The potential methane production from the TVA lost in the leachate represented 595 l over the length of this study. Approximately 70% of this loss occurred prior to the start of methane production.

Figure 3 presents data on leachate quality for Reactor 6. The leachate COD (20–30 g/l) and TVA (10–20 g/l) remained at this level for the first 200 days of operation. This was 160 days after methane production first occurred. After this time, a decrease in the concentrations of these constituents was noted. It is evident that the performance of this reactor, in terms of methane production, was being inhibited. This most probably resulted from the addition of septage. A significant loss of potential methane production occurred in this reactor from discarding a portion of the leachate. Using the data presented in Fig. 4, it was calculated that a potential methane production of 1304 l was lost.

## DISCUSSION

Analysing the performance of the anaerobic microorganisms responsible for converting the organic material in MSW to methane and improving leachate quality in these reactors is crucial to understanding the effects of enhancement. This performance has been examined on the basis of conversion of the biodegradable volatile solids (BVS) to produce (methane and volatile acids). Two methods were used to determine the biodegradable fraction of the MSW. The first method was to use the lignin content of the fresh MSW as an indicator of biodegradability (Chandler and Jewell, 1980). The second method was to assume that all of the TOC in the fresh MSW was available for conversion. These two methods indicated that the biodegradable fraction of the TVS in the fresh MSW was 39%.

### Decay Rate Constants and Methane Generation

Quantifying reactor performance and enhancement effects can be accomplished by mathematically modelling the results of each reactor. Each reactor was operated as a batch system, thus it is assumed that the rate of substrate (BVS) disappearance and, consequently, product formation (methane gas and volatile acids) will follow the laws of first order batch kinetics. The rate equation for first order batch kinetics is given by:

$$S_{bl} = S_{bo}\, e^{-kt} \tag{1}$$

where: $S_{bl}$ = quantity of substrate (BVS) remaining at any time, $t$:

$S_{bo}$ = initial quantity of substrate (BVS) at time, $t = 0$;

$k$ = substrate decay coefficient, day$^{-1}$; and

$t$ = time, days.

By plotting $S_{bl}/S_{bo}$ versus time on a semilog graph, the rate of substrate degradation, $k$, can be determined as the slope of the resultant straight line.

Measurement of the BVS remaining at any given time ($S_{bl}$) cannot be made directly. However, $S_{bl}$ can be determined indirectly by relating the quality of products formed and leaving the reactor with time (i.e. methane gas and volatile acids) to the quantity of BVS which had to be destroyed to create those products. Thus:

$$\frac{S_{bl}}{S_{bo}} = 1 - \frac{G + A}{\text{Max } G} \qquad (2)$$

where: $G$ = sum of methane produced at any time, $t$;

  $A$ = sum of potential methane produced from volatile acids discarded in the leachate at any time, $t$; and

Max $G$ = ultimate methane production from 100% BVS conversion to methane.

This analysis has been performed for each reactor and the results are presented in Fig. 5. A summary of these data are given in Table 3. From these data several conclusions can be drawn relating to the nature of the anaerobic process that occurs and to the effect of enhancement.

The substrate decay rate coefficient, $k$ (expressed as reciprocal days), is a direct measure of how quickly the MSW is being degraded. The higher the value of $k$, the more rapid the reaction is and, thus, the greater the quantity of methane being produced with time.

Reactors 1 and 2 exhibited a very low decay rate when compared to the other reactors. For Reactor 1, a decay rate of $0.228 \times 10^{-3}$ day$^{-1}$ was determined. This rate is consistent with reported literature values. At this rate, only 7.1% of the BVS were converted in one year. Nearly all of this conversion was to volatile acids which were removed from the reactor with the leachate. Only at the very end of this study was any methane produced in this reactor. At this point, only very small quantities of volatile acids (less than 100 mg/l as acetic acid) were present in the leachate.

Reactor 2, which experienced only leachate recycle for the first 188 days and the additions of sludge and buffer thereafter, exhibited a decay rate of $0.431 \times 10^{-3}$ day$^{-1}$ for the entire study. As with Reactor 1, the entire conversion of BVS resulted in volatile acid production, and no methane was produced. Since the leachate pH was low in this reactor, this result was not unexpected. However, at the end of the study, the leachate pH was rising (due to buffer addition) and it is expected that if this reactor had been operated for a longer time, methane production would have started. This conclusion is based upon what occurred in each of the other five reactors. At the end of one year, 13.9% of the BVS had been converted in this reactor.

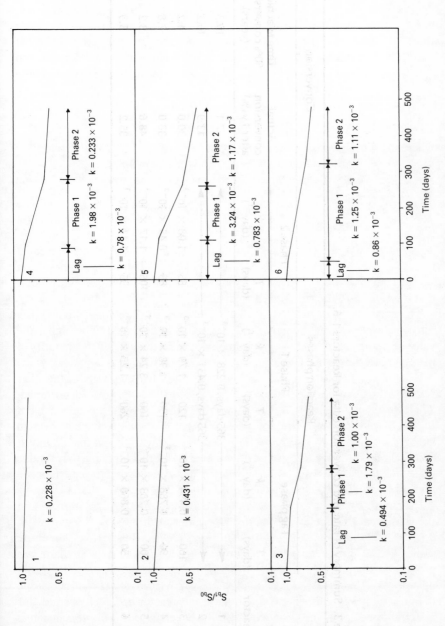

**Figure 5.** Determination of decay rate coefficient $k$ (day$^{-1}$), for Reactors 1–6, by plotting $S_{bt}/S_{bo}$ (see text) as a function of time ($T$, days).

**Table 3.** Summary of rate and conversion data for Reactors 1–6.

| Reactor | Reaction phases | | | | | | Conversion | |
| --- | --- | --- | --- | --- | --- | --- | --- | --- |
| | Lag phase | | Phase 1 | | Phase 2 | | Actual conversion after 1 y (%) | Time to achieve 80% conversion (years) |
| | $T$ (days) | $k$ (day$^{-1}$) | $T$ (days) | $k$ (day$^{-1}$) | $T$ (days) | $k$ (day$^{-1}$) | | |
| 1 | ↓ | 365 days, $0.228 \times 10^{-3}$ | | | ↑ | | 7.1 | 19.3 |
| 2 | ↓ | 365 days, $0.431 \times 10^{-3}$ | | | ↑ | | 13.9 | 10.2 |
| 3 | 160 | $0.494 \times 10^{-3}$ | 120 | $1.79 \times 10^{-3}$ | 85 | $1.08 \times 10^{-3}$ | 30.0 | 4.2 |
| 4 | 90 | $0.788 \times 10^{-3}$ | 190 | $1.98 \times 10^{-3}$ | 85 | $0.45 \times 10^{-3}$ | 38.0 | 7.8 |
| 5 | 100 | $0.783 \times 10^{-3}$ | 160 | $3.24 \times 10^{-3}$ | 105 | $1.17 \times 10^{-3}$ | 48.6 | 3.2 |
| 6 | 50 | $0.860 \times 10^{-3}$ | 260 | $1.25 \times 10^{-3}$ | 55 | $1.11 \times 10^{-3}$ | 35.2 | 3.9 |

Reactors 3 to 6 exhibited a significantly different reaction from Reactors 1 and 2. Three distinct reaction phases occurred in each of these reactors; termed here the lag phase, phase 1 and phase 2. The lag phase is the period of time when the organisms were starting up the process of degradation. In each case the end of the lag phase was marked by the appearance of methane production in the reactor. Phase 1 is the period of time in each reactor when maximum methane production was occurring. As shown in the analysis of the leachate, during phase 1 the leachate pH was at or above 7.0, and there was an excess of volatile acids present. The methane-forming bacteria were increasing in numbers and converting the volatile acids to methane. During phase 1 the maximum rate of degradation was occurring in each reactor. In each case, the end of phase 1 was marked by the disappearance of volatile acids from the leachate. Therefore, in phase 2, a sufficient population of methanogenic bacteria had been established in each reactor to convert any volatile acids being produced to methane. As a result, the controlling factor in BVS destruction in phase 2 is the rate at which the organic material is being solubilized and converted to acids.

From these data, and using Equation 1, the time required for 80% BVS destruction in each reactor has been calculated. These data appear in Table 3. As is clearly shown, enhancement has dramatic effect in decreasing the time required to destroy 80% of the BVS. In the control reactor, a period of nearly 20 years would reduce this time to slightly more than 3 years.

A summary of the quantity of methane produced in each reactor is presented in Table 4. Those reactors which received additions to the leachate that was being recycled (Reactors 3 to 6) all produced appreciable quantities of

**Table 4.** Summary of reactor performance with respect to methane production.

| Reactor | Total methane produced (l) | Total methane produced (ft³/day lb) | Total potential methane lost (l)[a] | Total methane (l) | Percent conversion To methane (%) | Percent conversion Total (%) |
|---|---|---|---|---|---|---|
| 1 | 2.5 | 0 | 410 | 413 | 0 | 7.2 |
| 2 | 0 | 0 | 815 | 815 | 0 | 13.9 |
| 3 | 1275 | 0.39 | 470 | 1745 | 21.9 | 30.0 |
| 4 | 1642 | 0.52 | 484 | 2126 | 29.3 | 38.0 |
| 5 | 2339 | 0.69 | 595 | 2935 | 38.8 | 48.6 |
| 6 | 835 | 0.24 | 1304 | 2140 | 13.7 | 35.2 |

[a] Lost as TVA by leaching.

methane. Reactor 1 (control) produced a very small quantity of methane and Reactor 2 produced no gas at all.

In the operation of each reactor, a quantity of leachate was removed from the system on a regular basis. Associated with that leachate were volatile acids, the immediate precursor to methane production lost through the removal of these acids. In those reactors where methane production became firmly established (Reactors 3 to 5) the quantity of potential methane lost amounted to 25–35% of the amount of methane actually produced. In addition, 60–90% of the potential methane lost occurred before the start of methane production. In Reactor 6, which exhibited methane inhibition due to the addition of septage, the quantity of potential methane lost was greater than the actual methane produced. The final two columns in Table 4 present the percentage of the BVS converted to methane and methane plus volatile acids, respectively.

## Enhancement Factors

Several conclusions on the effect of enhancement on BVS conversion, subsequent methane production, and improved leachate quality can be derived from the data presented. In particular, the effects of leachate recycle, buffer, nutrients and inoculum will be examined.

Comparing the performance of Reactors 1 and 2 over the first 188 days of operation will provide an insight into leachate recycle effects. From these data, it was shown that recycle of leachate did not enhance methane production as no methane was produced. However, an increase in the rate of substrate conversion from $0.228 \times 10^{-3}$ days$^{-1}$ to $0.431 \times 10^{-3}$ days$^{-1}$ did result. Thus, the increase in water movement through the solid waste does have a beneficial effect in stimulating the hydrolysis of particulate material. Unfortunately this did not stimulate methane production, because the natural buffer capacity of the MSW could not control the reactor pH. The subsequent acid formation in this reactor depressed the pH sufficiently so that methanogenesis could not proceed.

The difference between Reactor 2 and 3 was that buffer was added to the leachate in Reactor 3. During the lag phase, the decay rate exhibited by Reactor 3 was nearly the same as that for Reactor 2, $0.494 \times 10^{-3}$ day$^{-1}$ versus $0.431 \times 10^{-3}$ day$^{-1}$, respectively. However, by day 169 in Reactor 3, the pH had risen to above 7.0 and methane production had started. Reactor 3 exhibited a significantly higher decay rate of $1.79 \times 10^{-3}$ day$^{-1}$ during phase 1 and $1.08 \times 10^{-3}$ day$^{-1}$ in phase 2 than was experienced in Reactor 2. By providing an improved environment through buffering, a viable population of anaerobic organisms became established in this reactor, increasing the rate of decay. As shown in Table 3, a total of 4.2 years would be required to effect 80% BVS conversion, as opposed to 10.2 years for Reactor 2 or 19.3 years for Reactor 1.

Reactor 4 received buffer and nutrients (nitrogen and phosphorus) added to the leachate recycle, whereas Reactor 3 received only buffer. The beneficial effect of additional nutrients is most clearly shown in the lag phase. Reactor 4 had a lag phase of 90 days, as opposed to 160 days for Reactor 3. The rate of decay in the lag phase increased from $0.494 \times 10^{-3} \, \text{day}^{-1}$ in Reactor 3 to $0.788 \times 10^{-3} \, \text{day}^{-1}$ in Reactor 4. Thus, the addition of nutrients to the leachate decreased the length of time required for methane production to begin and increased the rate of stabilization during this time. During phase 1, the period of maximum methane production, the decay rates exhibited by Reactors 3 and 4 were nearly the same, $1.79 \times 10^{-3} \, \text{day}^{-1}$ and $1.98 \times 10^{-3} \, \text{day}^{-1}$, respectively. Thus, once a viable microbial population was established, the addition of supplemental nutrients did not significantly improve the rate of decay. In phase 2 the rate of solubilization of particulate material governs the overall reaction rate. In Reactor 4 this rate dramatically decreased to $0.457 \times 10^{-3} \, \text{day}^{-1}$, whereas Reactor 3 maintained a higher rate of $1.08 \times 10^{-3} \, \text{day}^{-1}$.

Reactors 5 and 6 each received microbial inoculum in the form of anaerobically digested sewage sludge and septage, respectively. First, from these data, it is obvious that septage is not an adequate source of microbial inoculum. This reactor experienced a depressed pH, high volatile acid concentrations in the leachate, and low methane production for the majority of the time it was operated. Therefore, septage cannot be recommended as a source of microbial inoculum or utilized to enhance methane production from a landfill.

Anaerobically digested sewage sludge, however, exhibited a significant beneficial effect upon degradation. During the lag phase, Reactors 4 and 5 operated almost identically. The lag phase lasted nearly the same length of time, and the rates were almost identical. Due to the nature of the experimental programme, separating the effects of nutrient addition and sludge addition during the lag phase cannot be made. Sludge contains available nutrients. In Reactor 5, the leachate was supplemented with additional nitrogen and phosphorus as well as sludge and buffer. No optimization was done and, therefore, the necessity of adding nitrogen and phosphorus to this reactor is unclear.

The effect of adding sludge to the leachate is most evident in phase 1. The rate of decay for Reactor 5 during phase 1 was determined at $3.24 \times 10^{-3} \, \text{day}^{-1}$. This rate is nearly double what was experienced in Reactors 3 and 4. It is evident that establishing a viable population of microorganisms in the reactor significantly increased the rate of methane formation over other methods of operation which did not incorporate microbial additions. Again, however, the question of optimization was not addressed. It is possible that addition of less sludge to this reactor may have produced similar results. Also possible is that greater sludge addition, or sludge addition for a longer period of time, may have resulted in increased rates and higher methane production. Over the one-year period of operation, Reactor 5 converted 50% of the BVS to methane and acids lost in the leachate, and 80% of this conversion resulted in methane being produced. Using Equation 1, it is predicted that 80% BVS destruction would occur in approximately 3 years.

Reactors 3, 4 and 5 were operated in a manner which effectively enhanced methane production. In doing so a viable population of anaerobic organisms was established, capable of rapidly solubilizing organic material, converting these products to acids and subsequently converting these acids to methane and carbon dioxide gases. Each reactor was characterized by a lag phase of short duration in which COD concentrations in the leachate ranged from 20 to 30 g/l, the TVA was approximately 15 g/l, and the pH was below neutral. The lag phase was followed by phase 1, a period of maximum methane production. During phase 1, a rapid decline in the leachate concentrations of COD and TVA was noted. In addition, the leachate pH rose above neutral in all cases. In each reactor, phase 1 was followed by phase 2 in which the rate of solubilization controlled the overall reaction. In phase 2 the leachate COD had fallen to 1.0 g/l and TVA was not measurable (essentially zero). The leachate pH remained above neutral. Thus, by establishing a viable anaerobic population within these reactors, an order of magnitude decline in concentrations of organic pollutants in the leachate resulted. In addition, a pH above neutral was established. Although not measured in this study, it is theorized that a neutral pH is beneficial in controlling the leachate of metals from the MSW. Metals will become solubilized in an acidic environment, whereas a neutral or basic pH will prevent this.

Reactor 1 was operated as a control. The leachate produced by this reactor exhibited the same trends as Reactors 3, 4 and 5; over the period of study similar declines in leachate COD and TVA were noted, although it required a longer period to result. Although this trend was expected, it occurred much more rapidly than what was experienced in other studies. As a basis for comparison, data developed by Pohland (1975) can be related to this study. Pohland operated a reactor with leachate recycle and buffer addition and a reactor with leachate recycle, buffer, and sludge addiction. These data compare favourably with the data developed for similar reactors operated in this study (i.e. Reactors 3 and 5). Pohland (1975) also operated a control reactor in the same manner as this study. Pohland's data significantly differ from this study with respect to the period of decline in COD and TVA which was much longer and never reached the levels which were achieved in this study. One possible explanation for this discrepancy may be the quantity of liquid moving through each reactor. Pohland did not present a water balance with his data. If the quantity of leachate generated in this study was much greater than that generated in Pohland's study, then a more rapid decline in COD and TVA concentrations would be expected.

Examining Fig. 4, it is noted that in Reactors 3 through 5, the lag phase is characterized by the presence of all five volatile acids in substantial quantities. At the beginning of phase 1, when methane production started, a dramatic drop in acetic, butyric, valeric and caproic acids, to negligible levels is noted. An equally dramatic rise in propionic acid is also seen. These increased levels of propionic acid continue throughout phase 1. In Reactor 4, the propionic acid levels rose from 2500 mg/l during the lag phase to over 10 000 mg/l during phase 1. It was during this time period that depressed methane production rates occurred in Reactor 4 (Fig. 2). This would indicate that the propionic acid levels caused an inhibition to the methanogens in Reactor 4 during this time. This was not noted in the other reactors due to the fact that propionic acid levels never reached similar levels.

It is known that specific acetogenic organisms convert propionic acid to acetic acid, hydrogen and carbon dioxide (Smith, 1980). Methanogens then utilize these products to form methane. These data suggest that either the acetogenic organisms responsible for propionic acid dissimilation were not present in sufficient numbers to convert the propionic acid, or that they were inhibited by some mechanism which cannot be determined from the data available. Propionic acid levels then sharply declined at the end of phase 1 in each of these reactors again indicating that either a sufficient population of acetogens had become established or that they had recovered from the earlier inhibition. Reactor 4, however, had a very low decay rate in phase 2. It is possible that the effects of the earlier high concentration of propionic acid in this reactor had a continuing effect upon the conversion process occurring, even though the propionic acid was no longer present in the leachate.

## CONCLUSIONS

A summary of the conclusions generated by this study can be summarized as follows.

Enhancement of methane production from MSW was found to occur when leachate was recycled with additions of buffer, nutrients and microbial inoculum.

In general, enhancement was found to have a positive effect on the rate of methane production and the quality of the leachate produced.

Under enhanced conditions, three distinct phases were noted: a lag phase before methane production started; phase 1 where maximum methane production rates were established; and phase 2 where the rate of methane production was controlled by solubilization of organic material.

Buffering of the leachate being recycled established a proper pH within the reactor. Once a neutral pH was reached, rapid methane formation began.

The addition of supplemental nutrients (nitrogen and phosphorus) and buffer significantly shortened the lag phase, indicating a nutrient deficiency during the initial phase of digestion; however, the continued addition of nutrients after methane production started did not improve the rate of methane production above what was experienced through buffer addition alone.

Anaerobically digested sewage sludge was found to be an excellent source of microbial inoculum. The reactor which received sludge, as well as buffer and nutrient addition to the leachate, exhibited a lag phase performance nearly identical to the reactor receiving only buffer and nutrients; however, in both phases 1 and 2, this reactor produced methane at a substantially higher rate than the other reactors, indicating the positive benefit of artificially increasing the microbial population.

Septage was found to be a poor source of microbial inoculum and exhibited a detrimental effect upon methane production and leachate quality.

The use of leachate recycle, without addition, was found to be ineffective in enhancing methane production or improving leachate quality. This method of operation produced a 'sour' digester, which did not recover.

The control reactor was characterized by slow rates of degradation (one order of magnitude less than under enhanced conditions). It did produce methane in very small quantities after nearly one year of operation.

Enhancement was found to have a beneficial effect on leachate quality. Once methane production became established, rapid decreases in leachate, COD and TVA were noted. By phase 2, in each of these reactors, an order of magnitude decrease in each of these parameters was noted.

## REFERENCES

Chandler, J.A. and Jewell, W.J. (1980). 'Predicting Methane Fermentation Biodegradability'. Cornell University, Ithaca, New York.

Leckie, J.D., Pacey, J.G. and Halvadakis, C.P. (1979). 'Landfill management with moisture control', *Journal of the Environmental Engineering Division ASCE*, **EE2**, 337–355.

Pohland, F.G. (1975). 'Sanitary Landfill Stabilization with Leachate Recycle and Residual Treatment'. EPA-600/2-75-043, MERL, Cincinnati, OH.

Smith, P.H. (1980). 'Studies of Methanogenic Bacteria in Sludge'. EPA-600/2-80-093, MERL, Cincinnati, OH.

## 2.5    Enhancement of Degradation: Large-scale Experiments

JOHN PACEY

*EMCON Associates, 1921 Ringwood Avenue, San Jose, California 95131, USA*

### INTRODUCTION

The Controlled Landfill Project at the Mountain View Landfill, California, was a field-scale demonstration project designed to study the process of methane gas generation in municipal solid waste landfills and to evaluate the effectiveness of methods used to enhance methane generation. The fifth and final annual report of the Mountain View Controlled Landfill Project, presenting data from Monitoring Day 0 (1 June 1981) to Monitoring Day 1597 (31 December 1985) was issued in January 1987. A comprehensive discussion of design concepts, construction techniques, and monitoring instrumentation is contained in the First Annual Report (Emcon, 1982).

Laboratory work conducted by Dynatech R/D Company (Buivid, 1980; Buivid *et al.*, 1981) before the start of this project identified the most important factors affecting the landfill gas generation process and provided the basis for the design of the Controlled Landfill Project, which focused on four potential enhancement techniques: the addition of moisture, buffer, anaerobic bacterial seed and leachate recirculation.

Firstly, moisture addition was chosen as an enhancement technique because low moisture content can limit the production of landfill methane by inhibiting bacterial growth. Most refuse will contain some moist refuse components, such as food waste, which have sufficient moisture for decomposition. Relatively dry materials, such as paper, may require additional moisture for complete decomposition. Additionally, a high moisture content may promote the transport and distribution of nutrients and bacteria by diffusion.

Secondly, buffer addition was identified as a potential enhancement

103

technique because the methanogenic bacteria are sensitive to pH. The pH range in which methanogens flourish is approximately 6.5–7.5. However, the production of organic acids by other bacteria could reduce the pH of a landfill to inhibiting levels, thereby inhibiting the methanogenic bacteria. Buffer addition could maintain the pH within the range promoting methanogenic activity, avoiding disruption of methane production.

Thirdly, bacterial seeding was identified as a potential enhancement technique because the standard refuse placement practices tend to bring much of the refuse into contact with oxygen in the air. This contact inhibits methanogenic bacteria. In contrast, acetogenic bacteria flourish in the presence of oxygen and will continue to produce acids. As previously mentioned, excessive production of acids may inhibit methanogenesis. The addition of methanogenic bacteria is thought to promote a balanced bacterial ecosystem, thus preventing concentration of organic acids in the landfill which inhibits methanogenesis.

Fourthly, leachate recirculation may enhance methane production by providing a positive mechanism of moisture nutrient and bacteria transport.

## FIELD LYSIMETERS

Figure 1 shows the test cell construction. The cells differed from one another due to the various enhancement techniques used. Calcium carbonate buffer, water and/or sludge from an anaerobic digester were added to the cells at the beginning of the project (Table 1). The calcium carbonate buffer was applied in powder form to the top of the refuse lifts (3 ft thick). Water was added through the water distribution system, which was installed under the Hypalon cover of each cell. The anaerobic digester sludge was mixed with refuse before placement. Cell A was the most manipulated cell. Cell F was not manipulated and served as the control cell. Cell A had a significant amount of water added, in addition to a calcium carbonate buffer and sewage sludge, which provided nutrients and bacterial seed. Some leachate produced in Cell A was recirculated to provide greater nutrient/waste product transport and to increase the rate of hydrolysis. Cell B contained buffer and sludge with no additional water. Cell C contained the same additives as Cell A, but with no leachate recirculation. Cell D, to which only buffer was added, and Cell E, to which only sludge was added, also received small amounts of added water.

The gas produced within each cell was collected beneath the Hypalon cover and forced through the gas collection pipes under the natural pressure of gas production. A centrifugal blower was operated, as necessary, to reduce the accumulation of gas under the Hypalon cover when the weight of the Hypalon

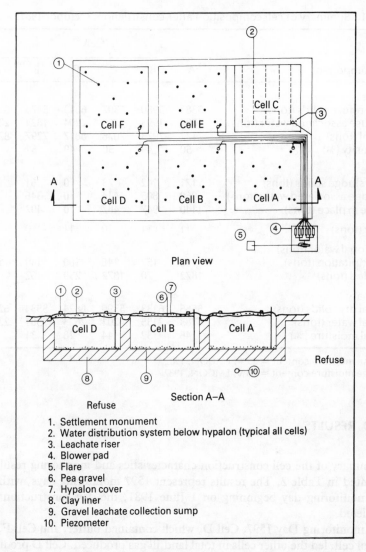

**Figure 1.** Large-scale lysimeters constructed in the Mountain View Landfill, California.

1. Settlement monument
2. Water distribution system below hypalon (typical all cells)
3. Leachate riser
4. Blower pad
5. Flare
6. Pea gravel
7. Hypalon cover
8. Clay liner
9. Gravel leachate collection sump
10. Piezometer

was not sufficient to overcome the head loss associated with high production rates. A slight positive pressure was maintained in the cells at all times during blower operation to prevent the intrusion of air into the refuse cell. Leachate analysis was periodically performed during the project.

106 *Pacey*

**Table 1.** Summary of cell composition after construction, October 1981.

| | Cell | | | | | |
|---|---|---|---|---|---|---|
| Cell component | A | B | C | D | E | F |
| Refuse[a] | | | | | | |
| Dry refuse solids (tons) | 5383 | 5980 | 5307 | 6613 | 5473 | 6223 |
| Refuse associated water (tons) | 1794 | 1993 | 1769 | 2204 | 1824 | 2074 |
| Total (tons) | 7177 | 7973 | 7076 | 8817 | 7297 | 8297 |
| Porosity (%) | 50 | 49 | 50 | 49 | 51 | 48 |
| Sludge[b] | | | | | | |
| Dry sludge solids (tons) | 171 | 142 | 73 | 0 | 61 | 0 |
| Sludge associated water (tons) | 969 | 805 | 414 | 0 | 346 | 0 |
| Sludge in place (tons) | 1140 | 947 | 487 | 0 | 407 | 0 |
| Buffer (tons) | 11 | 11 | 10 | 11 | 0 | 0 |
| Additional water (tons) | | | | | | |
| Precipitation (tons) | 148 | 157 | 146 | 160 | 149 | 154 |
| Added (tons) | 1872 | 0 | 1872 | 259 | 262 | 0 |
| Summary | | | | | | |
| Total dry solids (tons) | 5565 | 6133 | 5390 | 6624 | 5534 | 6223 |
| Total water (tons) | 4783 | 2155 | 4201 | 2623 | 2581 | 2228 |
| Final moisture (%) | 46 | 32 | 44 | 28 | 31 | 26 |

[a] Refuse moisture content = 25%; EMCON, 1982.
[b] Sludge moisture content = 85%; EMCON, 1982.

## FIELD RESULTS

A summary of the cell construction characteristics and monitoring results is presented in Table 2. The results represent 1597 monitoring days, with the first monitoring day beginning on 1 June 1981, the day construction was completed.

By monitoring Day 1597, Cell D, which contained buffer, and Cell F, the control cell, led the other cells in total landfill gas produced. Cell D produced 2.6 standard cubic feet (s.c.f.) of landfill gas per dry pound (1 atm, 32°F) followed by Cell F at 2.2 s.c.f. of landfill gas per dry pound. Landfill gas production in the other cells ranged from a low of 1.1 s.c.f. in Cells B and E to 1.4 s.c.f. in Cell C.

The gas production rate has been variable in all dry cells. The rate in Cell A reached a maximum of approximately 0.76 s.c.f. per dry pound per year between Monitoring Days 464 and 478. Production rates for Cells A, C, D and F were apparently greatest from Monitoring Days 700 to 800, after which the

**Table 2.** Summary of cell construction characteristics and monitoring results.[a]

| | Cell | | | | | |
|---|---|---|---|---|---|---|
| | A | B | C | D | E | F |
| Refuse deposited[b] (refuse + sludge: dry tons) | 5550 | 6120 | 5380 | 6610 | 5530 | 6220 |
| Additions[c] | sbwr | sb | sbw | b(w) | s(w) | (none) |
| Moisture content at construction (%)[d] | 46 | 32 | 44 | 28 | 32 | 26 |
| Moisture content at conclusion (%) | 69 | 54 | 50 | 33 | 45 | 40 |
| Specific landfill gas yield (s.c.f./dry lb) | 1.28 | 1.09 | 1.43 | 2.61 | 1.09 | 2.23 |
| Specific methane yield (s.c.f./dry lb) | 0.72 | 0.62 | 0.81 | 1.48 | 0.61 | 1.23 |
| Conversion[e] (% of ultimate) | 19 | 17 | 22 | 40 | 16 | 33 |
| Average production rate (s.c.f. landfill gas/dry lb/y) | 0.29 | 0.25 | 0.33 | 0.60 | 0.25 | 0.51 |
| Total landfill gas produced (million s.c.f.) | 11.1 | 9.7 | 11.9 | 26.4 | 8.4 | 22.3 |
| Average cell settlement (ft) | 6.7 | 7.2 | 7.6 | 4.2 | 7.7 | 6.1 |

[a] 1597 monitoring days.
[b] Not including buffer. Buffer weight equaled approximately 10 tons.
[c] Additions: s = anaerobic digester sludge; b = buffer (calcium carbonate), approximately 10 tons; w = water, 60 000 s.c.f.; (w) = water, 8300–8400 ft$^3$ (to reach field capacity); r = recirculation of leachate.
[d] After water addition.
[e] Calculated ultimate yield = 3.74 s.c.f. CH$_4$/dry lb refuse.

flow rates decreased. The landfill gas flow rates from Cells A, B, and C decreased to a level undetectable by the gas meters on approximately Monitoring Day 880. The gas flow rate from Cell E also fell below detectable levels on approximately Monitoring Day 1050. Landfill gas production in Cells B and E unexpectedly increased to detectable levels on approximately Monitoring Day 1070. Landfill gas production in Cells A and C apparently resumed on approximately Monitoring Day 1100.

## CLOSURE ACTIVITIES

The Mountain View Controlled Landfill Project is located at the Mountain View Landfill, which continued operations during the five-year construction and operation period of the project. In August 1985, the City of Mountain

View informed the other project participants that the development plans for the landfill required that the remaining volume of refuse necessary to reach final grade over the test cells be added within a few months. As a result, a series of additional tests and analyses were undertaken to gain further insight into data gathered during the project left in order to correlate, if possible, landfill gas production rates with other parameters. These additional investigations included biological and chemical analyses on refuse samples to assess the methane potential remaining in the refuse.

**Refuse Sampling**

As the Controlled Landfill Project drew to a close, plans were developed to conduct refuse sampling in the test cells and to perform various laboratory analyses on the samples. The objective was to further characterize the refuse material and possibly to correlate the methane produced with the methane potential remaining. Several environmental factors possibly limiting the production of methane were also investigated.

Six borings were drilled on day 1553: one each in Cells A and B and two each in Cells D and F. These cells were chosen to investigate a wide range of conditions and landfill gas production. Cell A was the most managed cell with the addition of water, buffer, sludge and some leachate recycle. Cell B produced the least gas. Cell D produced the largest landfill gas amount. Cell F was the control cell. For additional comparison, refuse from the landfill was sampled from the refuse slope directly south of the test cells. This refuse was placed approximately two years after the test cells were constructed.

All samples were collected by drilling into the selected cell with standard 6-inch, continuous flight, hollow stem auger. Two techniques were used for sampling the refuse. The first technique provided samples for chemical and physical analyses. Twenty-eight samples were collected from the auger as they were brought to the surface. Samples were labelled to indicate cell and depth, weighed, and sent to the laboratories for analysis. No attempt was made to maintain anaerobic conditions around the auger samples.

The second sampling technique provided 12 samples of refuse in an anaerobic condition for the assessment of native biological activity. These samples were collected by inserting a Shelby tube or by Pitcher sampling techniques. After drilling to the selected refuse depth, the sampling barrel or tube was inserted through the hollow stem of the auger and forced into the refuse. The sampler was then retrieved and immediately capped to avoid excessive air contact, and the sample placed in a plastic garbage bag filled with landfill gas collected from a nearby cell. Since air exposure had to be avoided, the transfer from sampler to bag did not allow field observation of the sample. Due to the small sample size used for biochemical methane potential (BMP) analysis (approximately 1 g), these samples tended to have a higher organic content and lower grit content than the auger samples. Results were correlated based on volatile solid content.

## Chemical Analyses of Refuse Samples

The chemical analysis objectives were to assess the quantity of biologically degradable material remaining for gas production, to measure the concentration of certain nutrients essential to bacterial growth, and to evaluate the potential for heavy metal inhibition by nickel, which has been identified as possibly inhibitive in landfill environments. The chemical analyses included:

1. moisture content;
2. volatile solid content;
3. cellulose content;
4. lignin content;
5. Kjeldahl nitrogen;
6. nitrate/nitrite nitrogen;
7. ammonia nitrogen;
8. available (soluble) nickel;
9. available (soluble) phosphorus.

Each of the 28 refuse samples were analysed separately, with five replicate analyses per sample. The results are weight-averaged to compensate for differing sampling intervals in the boring.

## Biological Analyses of Refuse Samples

The biological analysis objectives were directly to measure the refuse sample biodegradability and to assess the conduciveness of the landfill environment to biological activity. Additional tests were also performed to facilitate correlation with the chemical analysis findings. The analyses performed included:

1. biochemical methane potential (BMP) test;
2. moisture content;
3. volatile solids;
4. chemical oxygen demand (COD).

Samples were analysed separately for moisture content, volatile solids, and COD, then combined to provide a composite sample from each boring for the BMP test.

In the standard BMP test, conditions are optimized for degradation of the sample's organic material. Nutrients, buffer and anaerobic seed bacteria are added to the refuse samples in an anaerobic environment. The samples are stirred. Temperature is controlled to 35°F. Sample degradation from the project proceeded for 133 days. The gas production volume and rate were measured and calculated based on volatile solids in the original sample. For comparison, the BMP of pure cellulose and newsprint were also measured.

Variations in the standard experimental protocol were applied to investigate the environmental conditions in the Mountain View landfill; namely potential phosphate deficiency and native bacterial activity.

Phosphate deficiency was tested by withholding phosphorus from the reactor mixture; anaerobic bacteria seed and buffer were added as usual. If significantly decreased methane production is compared with the standard BMP test in which sufficient phosphorus is provided, then the results would indicate that phosphorus content is a limiting factor in anaerobic decomposition.

Native bacterial activity was tested by withholding anaerobic digester sludge seed. Bacteria were provided either from the auger sample itself or from anaerobic samples collected from the test cells. Nutrients and buffer were added as usual. Significantly increased time for gas production to begin, compared with the results of the standard BMP test in which bacterial seed is added, indicates that the refuse contained significantly less anaerobic bacteria than anaerobic digester sludge.

## CHEMICAL ANALYSIS OF REFUSE

### Results

A summary of the findings of the chemical analyses conducted on samples from the cells is presented in Table 3. The highest moisture content measured in the samples was in Cell A, followed by Cells B, F and D respectively. These findings correlate well with field measurements and appear to be representative values for characterizing the moisture content of the cells at the conclusion of the five-year project.

The pH Cells B, D and F and of the adjacent refuse sample all fell within the narrow range of 6.6–6.9. The pH of Cell A was somewhat different, measuring 7.9. All of these pH measurements are in the range suitable for anaerobic

**Table 3.** Summary of chemical analyses on refuse samples.[a]

| | Cell | | | | Adjacent | Fresh |
| Component | A | B | D | F | refuse | refuse[b] |
|---|---|---|---|---|---|---|
| Sampling interval (ft) | 0–25 | 0–28 | 0–32 | 0–39 | 0–20 | NA[c] |
| Moisture content (%) | 68.6 | 54.1 | 33.3 | 40.0 | 22.0 | 22.1 |
| pH | 7.9 | 6.9 | 6.6 | 6.7 | 6.8 | NA |
| Volatile solids content (%) | 31.8 | 43.1 | 50.7 | 43.5 | 27.1 | 65.5 |
| Cellulose (%) | 16.3 | 25.6 | 32.8 | 26.6 | 14.6 | 39.7 |
| Lignin (%) | 13.4 | 14.0 | 13.6 | 14.2 | 8.8 | 10.2 |
| Cellulose: lignin | 1.22 | 1.83 | 2.42 | 1.87 | 1.66 | NA |
| Kjeldahl N (mg/kg) | 0.5 | 0.5 | 0.5 | 0.4 | 0.3 | NA |
| $NH_4$—N (mg/kg) | 388 | 196 | 82 | 96 | 246 | NA |
| $NO_3/NO_4$ (mg/kg) | 102 | 33 | 12 | 13 | 32 | NA |
| Phosphorus[d] (mg/kg) | 337 | 332 | 299 | 279 | 185 | NA |
| Nickel (mg/kg) | 6.5 | 8.8 | 13.4 | 6.3 | 3.7 | NA |
| Carbon to nitrogen ratio | 13 : 1 | 20 : 1 | 26 : 1 | 27 : 1 | 20 : 1 | NA |
| Carbon to phosphorus ratio | 593 : 1 | 945 : 1 | 1345 : 1 | 1169 : 1 | 967 : 1 | NA |

[a] Percentage moisture is presented on a wet weight basis; all other parameters were measured on a dry weight basis.
[b] Fresh refuse from Madison, Wisconsin.
[c] NA indicates no data exists for the specific analytical procedure used.
[d] Available (soluble) phosphorus.

bacteria. No apparent effects from the initial buffer addition on the pH of the cells were observed at the project's conclusion.

The highest volatile solids content measured in the samples was for Cell D, followed by Cells F, A and B, respectively. This order is the inverse of the order for the moisture content measurements. The lower the volatile solids content, the greater the gas production expected for that cell. However, the field measurements of gas production were highest for Cell D, followed by Cells F, A and B. On the basis of remaining volatile solids, future gas production is expected to be greatest for Cell D, which exhibited the greatest gas production measured in the project. It is also interesting to note that the volatile solids content of refuse samples from the area adjacent to the test cells is lower than all four cells, despite the fact that it is two years younger. Thus, there are inconsistencies between the laboratory measurements of volatile solids content and the field measurements of gas production.

**Discussion**

The observed gas production rates indicate that less than optimal conditions existed in cells for the bioconversion of refuse to gas. The total gas production rates are lower than rates obtained in many laboratory simulator studies including the Dynatech refuse lysimeters that achieved total gas production rates of 0.14 to 78 ft$^3$/dry lb·y (Halvadakis *et al.*, 1983). The total gas production rates are higher than conventional sanitary landfills that typically produce 0.03–0.15 ft$^3$/dry lb·y (EMCON, 1980).

Distinct differences exist between the measured and calculated gas production values based upon the loss of volatile solids. Measured and calculated yearly gas production rates, where the calculated rates are based upon the volatile solid conversion determined by the refuse analyses data, were 0.23–0.61 and 0.53–0.91 ft$^3$/lb dry refuse per year, respectively.

The relationships between the measured methane production rates and the specific gas enhancement addition(s) employed are generally opposite to those which were expected based upon refuse lysimeters testing. For example, Cell A (with leachate recycle, a high moisture content and sludge and buffer additions) and Cell B (with a high moisture content and sludge and buffer additions) had a lower measured methane production rate than Cell D (with buffer and a lower moisture content) and Cell F (the control cell with no additions). In addition, Cells A and B had higher internal temperatures and higher rates of settlement than Cells D and F. Further, the measured methane production rates conflict with refuse analysis data. Specifically, the cellulose and volatile solids results generally indicate the cells that have lost the most cellulose and volatile solids and so have apparently undergone the most biodegradation, have lower measured methane production rates. Prior studies (Barlaz *et al.*, 1987) have indicated that a clear relationship exists between refuse degradation or gas production and cellulose reduction which is not found here.

The calculated methane gas production rates based upon the loss of volatile solids for Cells A, B and F are higher than the measured methane production rates. The calculated methane production rate for Cell D is slightly lower than the measured rate. The calculated values indicate that a high moisture content, and possibly leachate recycle and the addition of sludge, enhanced methane gas production. The cell temperatures are higher for cells with higher calculated methane production values.

Refuse analysis and leachate level data indicate that significant amounts of precipitation and/or ground water have entered all cells, including the control cell. Cell A shows the greatest water infiltration, increasing from an original moisture content of 46% to a present content of 72%. Cell D had the least water infiltration, increasing from an original moisture content of 28% to a present moisture content of 33%. The other two cells tested, B and F, were between these extremes. The use and interpretation of moisture addition as an experimental design variable is therefore limited.

It is logical that if moisture can enter the cell, gas generated under pressure within the cell can exit (unmetered) via the same route if the route is not under water. Evidence supporting the theory that all generated gas was not measured (in addition to points previously raised in this section) includes:

1. The cells that have the most water infiltration (Cells A and B) have the lowest measured gas production rates.
2. Cells D and F, with the least amount of water infiltration, exhibited the least drop in gas production during a time period (approximately Days 900–1100) in which an unexplained cessation of gas production occurred in Cells A, B, C and E.
3. The calculated gas production values are greater than the measured values except for Cell D, which had little or no water infiltration.

The carbon to nitrogen ratios based upon the Kjeldahl nitrogen and cellulose analyses suggest that a sufficient supply of nitrogen was available for active methanogenesis. Available phosphorus may be present in refuse samples at concentrations less than optimum for cell growth and biogas generation. However, active methanogenesis has been achieved in other refuse lysimeter tests without the addition of phosphorus.

No obvious relationship exists between the nickel concentration in refuse and the measured or calculated gas production rates.

### Conclusions on Chemical Refuse Analysis

Based on the analyses of refuse as samples from Cells A, B, D and F, and adjacent refuse, and on the monitoring data and reports provided, several conclusions have been made:

1. An unexpected relationship exists between the average yearly measured methane gas production rates for Cells A, B, D and F, and the gas enhancement techniques employed. Cells with higher moisture contents, sludge addition, lower amounts of settlement, and lower internal temperatures, had lower measured methane gas production rates.
2. Calculated average yearly methane gas production rates based upon the loss of volatile solids are higher than the measured methane gas production rates for Cells A, B and F, and slightly lower for Cell D. The calculated methane

gas production rates generally indicate that a high moisture content, the addition of sludge, and leachate recycle enhanced methane gas generation. The amount of cell settlement and the cell temperatures were generally higher for the cells with higher calculated gas production values.

3. Refuse analysis and leachate level data indicate that surface water or ground water has entered all cells, including the control cell. The use and interpretation of moisture addition as a methane enhancement agent or experimental design variable is limited.

4. The water balance analysis indicates that the amount of water infiltrating Cells A, B and F is correlated with the difference between the measured and calculated gas production values. Cell D has the least difference and the least infiltration. This suggests that the landfill gas leaked from the cells via the same pathway surface water or ground water infiltrated the cells.

## BIOCHEMICAL ANALYSIS OF REFUSE

A summary of the findings of the biochemical methane potention (BMP) analyses conducted on samples from the cells is presented in Table 4. Before interpreting the results of the tests conducted at Stanford University, a statement from their report should be considered: 'The results from Cell A material were inconsistent with results from analysis of materials from other cells and with the experimental data obtained from the field study; this is believed to have resulted from an unrepresentative composite sample such that results from the analysis for Cell A should be used with caution. Adequate samples were available from Cells B, D and F to prepare a representative sample, and results were consistent for these cells'.

The analyses of moisture content and volatile solids content paralleled the tests conducted at the University of Wisconsin. For purposes of normalizing the BMP analyses, the results from the University of Wisconsin volatile solids content test were used because they were based on larger, more representative samples.

The highest COD measured in the samples was for Cell D, followed by Cells F, B and A. This order corresponds to the order of the volatile solids content measurements. Normalizing the COD values by dividing those values by the volatile solids content (VS) resulted in values lower than expected and in a slightly different order among the cells. The COD/VS values ranged from 0.79 to 1.18. Typical ratios of COD/VS for refuse are 1.1–1.4.

The biological conversion of the organic material in the samples into methane was nearly complete after approximately four months of incubation. The BMP analyses indicated that the composite samples of 1553-day-old Mountain View landfill material have the following remaining potential for

**Table 4.** Summary of biochemical methane potential analyses of refuse samples.

| | | Cell | | | | |
|---|---|---|---|---|---|---|
| | A | B | D | F | Cellulose | Newsprint |
| Sample size (g VS) | 0.51 | 0.54 | 0.62 | 0.78 | 0.15 | 0.54 |
| Incubation time (days) | 133 | 133 | 133 | 133 | 60 | 133 |
| Moisture content (%) | | | | | | |
| Stanford | 66 | 63 | 37 | 48 | — | — |
| Volatile solids (%) | | | | | | |
| Stanford | 49 | 56 | 59 | 58 | — | — |
| COD (g/kg) | 130 | 220 | 440 | 268 | — | — |
| COD/VS (g/g) | 0.79 | 1.05 | 1.18 | 0.89 | — | — |
| Biochemical methane potential | | | | | | |
| ml $CH_4$[a] | 40 | 139 | 166 | 188 | 54 | 66 |
| ml $CH_4$/g VS | 78 | 257 | 268 | 241 | 360 | 122 |
| s.c.f. $CH_4$/lb VS | 1.11 | 3.65 | 3.81 | 3.42 | 5.11 | 1.73 |
| s.c.f. $CH_4$/lb dry refuse | 0.35 | 1.57 | 1.93 | 1.48 | — | — |
| Percent laboratory conversion to $CH_4$ | | | | | | |
| compared with complete conversion | 17 | 55 | 57 | 51 | 77 | 26 |
| compared with cellulose | 22 | 71 | 74 | 67 | 100 | 34 |
| Initial methane potential | | | | | | |
| s.c.f./lb dry refuse | 0.99 | 2.01 | 2.96 | 2.38 | — | — |
| s.c.f./lb VS | 2.58 | 4.18 | 4.81 | 4.45 | — | — |
| Percent field conversion | | | | | | |
| percent VS conversion | 27 | 18 | 35 | 32 | — | — |
| percent of total conversion | 69 | 28 | 49 | 48 | — | — |

[a] Produced in sample.
VS = volatile solids.

methane production: 1.11, 3.65, 3.81 and 3.42 s.c.f. methane per pound of volatile solids in Cells A, B, D and F, respectively. The methane production data support the COD and volatile solids data, with Cell D having the greatest remaining methane potential and Cell A having the least. Based upon the BMP analyses, the percentage of remaining organic material that is readily convertible to methane under suitable environmental conditions is 17, 55, 57 and 51% for Cells A, B, D and F, respectively.

From the BMP analyses and the field measurements of methane production from the landfill, the estimated conversion of landfill organic material into methane gas after 1553 days of operation for Cells A, B, D and F was 27, 18, 35 and 32%, respectively; the fraction this represents of the total original organic material that was readily convertible to methane is 69, 28, 29 and 48% respectively.

Phosphorus was shown to be a limiting nutrient in a landfill material from Cells B, D and F, but not from Cell A; phosphorus limitation significantly reduced the rate of methane production after 133 days incubation in the BMP analyses. Although no specifically determined in the BMP analyses of the landfill materials, together with BMP, results indicate that nitrogen is also likely to be a limiting nutrient for methane production from the landfill material.

## COMPARISON OF LABORATORY ANALYSES AND FIELD MEASUREMENTS

The relative order of various laboratory analyses and field measurements are shown in Table 5. Moisture content measurements were consistent among the three sources, as were the volatile solids content, cellulose content, cellulose-to-lignin ratio and COD all exhibited an inverse relationship to moisture content. Normalization of the COD by the volatile solids content (i.e. COD/VS) resulted in a minor reordering of Cells B and F. The BMP analyses followed the same pattern as COD/VS.

The field measurements of gas production most closely correlated with volatile solids content, cellulose content, cellulose-to-lignin ratio and COD with a variation in the order of Cells A and B. Thus, the highest values for gas production parameters were found in the cells in which gas production measurements were the greatest at the conclusion of the project. This finding would be consistent if indeed the cells with the greatest gas production had the greatest gas production potential at the beginning of the project. Assuming similar values for these gas production parameters at the beginning of the project, one would expect the cells with the lowest gas production measurements to have the greatest gas production potential remaining.

Based upon an assumed initial volatile solids content of 75% and then calculating the volatile solids destroyed, the University of Wisconsin work suggests a different order for gas production than that of the field measurements. This approach indicates that Cell A was the greatest gas producer and Cell D was the lowest. Coincidentally, the order of the cells for the University of Wisconsin gas production calculations is the same order as that of the cell moisture contents.

**Table 5.** Relative order of laboratory analysis and field sampling parameters.[a]

| | Moisture content | | | Volatile solids content | | Cellulose content | Cellulose: lignin ratio | COD | COD/VS | BMP | Gas production | |
| --- | --- | --- | --- | --- | --- | --- | --- | --- | --- | --- | --- | --- |
| | UW | SU | EMCON[b] | UW | SU | UW | UW | SU | SU | SU | UW[c] | EMCON[d] |
| Highest | A | A | A | D | D | D | D | D | D | D | A | D |
| | B | B | B | F | F | F | F | F | B | B | B | F |
| | F | F | F | B | B | B | B | B | F | F | F | A |
| Lowest | D | D | D | A | A | A | A | A | A | A | D | B |

[a] UW represents University of Wisconsin laboratory tests; SU represents Stanford University laboratory tests; and EMCON represents EMCON field measurements.
[b] Based on leachate level.
[c] Based on estimates of volatile solids destroyed.
[d] Based on field measurements.

118

Pacey

**Table 6.** Comparison of field and laboratory methane quantities (s.c.f. CH$_4$/ pound dry refuse).

| | Cell A | | Cell B | | Cell D | | Cell F | |
|---|---|---|---|---|---|---|---|---|
| | UW[a] | SU[b] | UW | SU | UW | SW | UW | SU |
| Methane production measured in field[c] | 0.72 | 0.72 | 0.62 | 0.62 | 1.49 | 1.49 | 1.23 | 1.23 |
| Methane potential remaining[d] | 1.08 | 0.35 | 1.70 | 1.57 | 2.18 | 1.93 | 1.77 | 1.48 |
| Total methane | 1.80 | 1.07 | 2.32 | 2.19 | 3.67 | 3.41 | 3.00 | 2.71 |
| Percent of calculated ultimate yield[e] | 48 | 29 | 62 | 59 | 98 | 91 | 80 | 72 |

[a] UW represents University of Wisconsin laboratory tests.
[b] SU represents Stanford University laboratory tests.
[c] EMCON field measurements.
[d] University of Wisconsin values based upon cellulose analyses; Stanford University values based upon biochemical methane potential (BMP) analyses.
[e] Total methane potential calculated from refuse composition at beginning of project (3.74 s.c.f. CH$_4$/pound dry refuse).

To perform an approximate methane balance on the sampled cells, a comparison was made of the sum of methane production measured in the field and the methane potential remaining (as indicated by each of the laboratory studies) with the calculated ultimate yield based upon the refuse composition at the beginning of the project (3.74 s.c.f. CH$_4$/lb of dry refuse). This comparison is presented in Table 6. The sum of the measured gas production and the remaining methane potential should be approximately equal to the calculated ultimate yield. The results are very close for Cell D, but are off by varying degrees for the other cells. The poorest correlation is for Cell A.

## ACKNOWLEDGEMENT

A number of organizations participated in this project since its inception. The closing project participants include Pacific Gas and Electric Company (PG&E), the Gas Research Institute, and EMCON Associates. Organizations contributing to past phases of the project, including project design and construction, were the US Department of Energy, Southern California Gas Company, Argonne National Laboratory, the City of Mountain View, and Easley & Brassy Corporation (now Laidlaw Solid Waste System).

## REFERENCES

Barlaz, M.A., Milke, M.W. and Ham, R.K. (1987). Gas production parameters in sanitary landfill simulators', *Waste Management and Research* 5, 27–39.

Buivid, M.G. (1980). 'Laboratory Simulation of Fuel Gas Production Enhancement from Municipal Solid Waste Landfills'. Dynatech R/D Company, final report to PG&E and So Cal Gas.

Buivid, M.G., Wise, D.L., Blancher, M.D., Remedios, E.C., Jenkins, B.M., Boyd, W.G. and Pacey, J.G. (1981). 'Fuel gas hnhancement by controlled landfilling of municipal wastes', *Resources and Conservation* 6, 3–20.

EMCON Associates (1980) 'Enhancing Production of Landfill Gas'. Project 343-01.01.

EMCON Associates (1982). 'Controlled Landfill Project. First Annual Report. Construction and Operation' (NTIS).

Halvadakis, C.P. *et al.* (1983). 'Landfill Methanogenesis: Literature Review and Critique'. Technical Report No. 271, Department of Civil Engineering, Stanford University.

## 2.6  Effects of Codisposal on Degradation Processes

R. COSSU and R. SERRA

*Institute of Hydraulics, University of Cagliari, Piazza d'Armi, 09100 Cagliari, Italy*

### PHILOSOPHY OF CODISPOSAL

Codisposal in a landfill is the controlled introduction of waste, even of a hazardous nature, into a municipal solid waste (MSW) landfill. This type of operation aims at reducing the danger of hazardous waste not by simply diluting it with municipal waste but also by taking advantage of attenuation mechanisms present inside the landfill. These mechanisms induce immobilization or degradation of the toxic substances, thereby providing the MSW landfill with an 'attenuation capacity' which may be considered a suitable means of separating some hazardous substances from the environment.

Therefore, in order to take maximum advantage of this possibility the landfill must no longer be considered as a 'black box'. On the contrary, it is necessary to study phenomena occurring inside the landfill in great detail in order to favourably exploit them in the protection of the environment.

Since the early 1970's scientific research has examined aspects linked to the environmental impact of codisposal and has produced useful results for the practical realization of this method for numerous types of hazardous wastes. Based on these scientific results several countries are considering the advantages of codisposal which is recommended also by the World Health Organization (WHO), which considers codisposal a way of limiting environmental impact of monolandfills for chemical waste, particularly when small and poorly managed (WHO, 1983).

121

## ANALYSIS OF FEASIBILITY OF CODISPOSAL

### General Principles

Codisposal cannot be accepted indiscriminately for all types of toxic residues and the environmental situation of the area where the codisposal landfill is to be situated should always be carefully studied (characteristics of the site, rainfall, etc.).

The advantages and disadvantages of this practice are to be found inevitably in the rules and regulations which must be respected in order to limit the danger of disposal. To this regard numerous examples of codisposal carried out during the 1970's in various countries (USA, West Germany, Italy, etc.) should be mentioned as these were carried out without any supervision thus damaging the surrounding area. This has certainly added to the suspicion and worries present in these nations concerning this type of disposal. However, it must be considered that the supervised disposal of MSW and industrial waste in controlled sanitary landfill is foreseeable.

The main problems linked to codisposal may be summarized as follows:

1. The inhibiting effect on biological degradation of MSW due to the presence of toxic compounds in the codisposed hazardous waste;
2. The emission of toxic substances into leachate to a higher degree than that observed during disposal of MSW alone.

The first aspect concerns the so-called 'sterilization of the landfill': the maximum amount of toxic waste which a landfill is capable of assimilating without exceeding the inhibition level of development of microbe populations responsible for the biostabilization of the waste should be verified. Also of great importance is the means of depositing hazardous waste as a concentrated disposal, both with regard to time and space, this could lower the level of the inhibition threshold thus disturbing those phenomena of microbic adaptation which allow the degradation of bio resistant compounds.

With regard to the second aspect the emissions would be influenced not only by the solubility of substances but also by a series of chemico-physical and biological phenomena capable of immobilizing, degradating and generally attenuating the mobility of each single substance. The exploitation of these mechanisms was the basis of the choice of a landfill with 'controlled dispersion of leachate' as the same phenomena takes place in the underlying unsaturated ground.

These attenuating phenomena may be summarized as follows:

1. *Physical and chemico-physical phenomena*: absorption, filtration and dispersion.

2. *Chemical phenomena*: acid–base interaction, oxidization, precipitation, co-precipitation, ionic exchange, formation of complexes.
3. *Biological phenomena*: microbic, aerobic and anaerobic degradation.

## Compatibility

In order to overcome the above-mentioned problems of environmental impact the best relations of codisposal for the various types of toxic wastes must be identified. This is evidently not sufficient to guarantee the successful outcome of the technique if the operational phase is not preceded by *strict control* of the quantity and quality of hazardous waste introduced.

This control is characterized by a more general problem, namely that of compatibility analysis on acceptance of difficult wastes into a landfill. In fact, the knowledge of waste content which is to be codisposed with municipal waste helps to avoid some undesired reactions which may occur when incompatible wastes are mixed: generation of heat due to esothermic reactions (danger of flames or even explosions), generation of toxic gases such as arsine, cyanide, sulphuric acid, etc.; inflammable gas such as $H_2$, acetylene. Figure 1 summarizes the undesired reactions occurring with more common hazardous wastes, when mixed together.

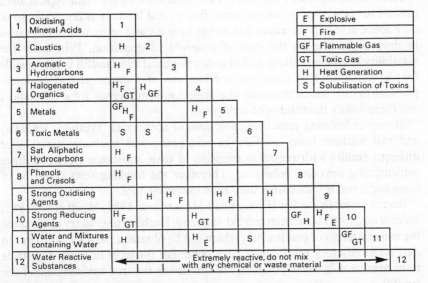

**Figure 1.** Schematic summary of undesired reactions caused when mixing the more common types of hazardous wastes (DoE, 1986).

**Experimental Tests**

When applied in practice, the above-mentioned general considerations may be insufficient and render necessary more specific experimental tests in order to obtain information concerning both the maximum amount with respect to codisposal and the best technique for depositing of waste (in layers, mixed, etc.) and therefore avoid provoking the inhibiting effects or sudden emissions of toxic substances.

However the variability of industrial waste and MSW could suggest the use of experimental tests in specific cases. Scientific research proposes generally three different types of experimental tests for the evaluation of feasibility of codisposal; leaching tests (in batch), lysimeter tests and field cell tests. Leaching tests entail a making of contact between the solid phase from which leachate would derive with an appropriate liquid phase (tap water, distilled water, solution of acetic acid, synthetic or real leachate). In practice, these are tests which examine only the capacity of the waste to emit toxic substances but are limited by the fact that the actual environment present in a landfill is not simulated. The EPA procedures foresee the use of a diluting tool consisting of water to which acetic acid has been added, this being the main constituent representing the organic substances present in leachate obtained from a MSW landfill, thereby simulating the lixiviating action of a 'synthetic leachate'.

These tests represent the worst case situations rather than typical ones present in codisposal for two reasons: firstly, real leachate is not made up of only acetic acid and moreover it does not have a static composition but varies in time depending on the state of anaerobic degradation. Principally the variability of pH conditions and of redox potential in a landfill is responsible for the occurrence of numerous mechanisms of a chemical and chemico-physical nature which attenuate the presence of numerous toxic substances and these justify the validity of codisposal.

Moreover leaching tests using solutions of acetic acid (synthetic leachate) and real leachate from landfills (in acid phase and methane phase) offer different results with regard to emission of toxic substances, even at times evidentiating opposite behaviour. Therefore the leaching tests could be of more use if real leachate were used as an extractive tool.

Recent research studies (Francis and Maskarinec, 1986; Misiti *et al.*, 1988; Baldi *et al.*, 1988) have been carried out using leaching tests to try to simulate the concentration of real leachate which can be obtained from a landfill but to date no satisfactory results have been obtained: the most important obstacle to be overcome is the impossibility of simulating the true environment of the landfill.

A different approach is possible by means of lysimeter tests (Fig. 2): these contemplate the introduction of solid residues into sealed containers (lysimeters), having capacity generally ranging between 100 l and 1 m³. Lysimeters are equipped with a uniform watering system from above and with systems for collection of leachate and biogas. These are made up of reactors in

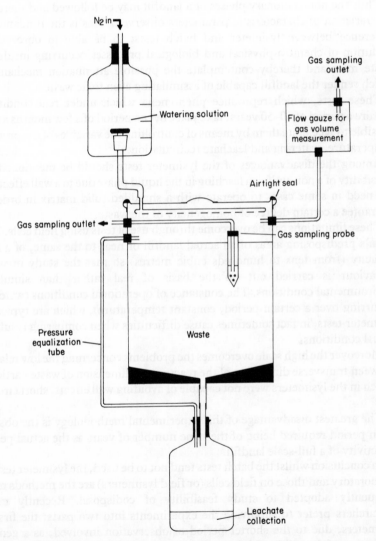

**Figure 2.** Scheme of a lysimeter (reported in Craft and Blakey, 1988).

which the real conditions of a landfill are simulated: anaerobic conditions by means of sealing of the reactor; rainfall through uniform water distribution from above; landfill temperature through thermostating of the reactor and recirculation of leachate by means of pumps introduced into the drain pipes. In order to ensure perfect anaerobic conditions inside the reactor, the water is introduced under a current of inert gas.

Thus the non-stationary phases of a landfill may be followed and therefore the variation of characteristic parameters observed. This is the fundamental difference between lysimeter and batch tests: to be able to observe the evolution of chemico-physical and biological processes occurring inside the waste mass and thereby contemplate the possible attenuation mechanisms which render the landfill capable of assimilating even toxic waste.

These tests, which reproduce phenomena which under real conditions require a period of 10–50 years, take place over a period of a few months as it is possible to accelerate them by means of controlling the variables of the process: temperature, watering and leachate recirculation.

Among the disadvantages of the lysimeter tests should be mentioned the possibility of a non-uniform leaching in the liquid phase due to a wall effect and the need in some cases to operate with a shredded solid matrix in order to guarantee a certain degree of homogeneity of the waste.

These limitations can be overcome through use of tests using field cells. This entails predisposing areas of an actual landfill or near to the same, of a high capacity (from tens to hundreds cubic metres) so that the study of waste behaviour is carried out on the basis of real rather than simulated environmental conditions. The constance of operational conditions (watering occurring over a certain period, constant temperature), which are typical of lysimeter tests, in fact sometimes cause difficulties when comparing results to actual conditions.

Moreover the high scale overcomes the problems concerning the low relation between transverse dimension of the reactor and dimension of waste particles, which in the lysimeters were not capable of avoiding wall effects, short circuits etc.

The greatest disadvantage of this experimental methodology is the observation period required being of the same number of years as the actual period of activity of a full-scale landfill.

In conclusion whilst the batch tests tend not to be used, the lysimeter tests in a laboratory and those on field cells (or field lysimeters) are the methods more frequently adopted to study feasibility of codisposal. Recently many researchers prefer to subdivide the experiments into two parts: the first in lysimeters, due to the shorter period of observation involved, as a general indication of the validity of codisposal followed by a cell experiment for those wastes which gave favourable results in lysimeter tests.

## RESEARCH ON CODISPOSAL OF HAZARDOUS WASTE

During the early 1970's scientific research became interested in the problem of codisposal of municipal waste and industrial residues in sanitary landfill. The majority of tests were started in Great Britain during the research programme financed by the Department of the Environment, the results of which were published in 1978 (Department of the Environment UK, 1978). The reassuring results obtained in many studies and full-scale investigations therefore induced the British politicians and technicians to encourage this type of disposal and to establish practical reference points. These indications were published in many HMSO booklets relating to determined types of waste (mercury and arsenic based, tarry, halogen organic waste, pesticides, etc.) which were distributed from 1976–1980 (Department of the Environment, papers n.1/21).

Moreover during the 1970's several American researchers (Pohland and Harper, 1987) tackled the problem of codisposal by trying to understand the chemico-physical and biological phenomena which accompanied this method. During the 1980's further data have been added to research although full-scale applications have only been carried out in Great Britain. Only recently (Campbell and Pugh, 1988; Barres *et al.*, 1988) other countries, accepting the philosophy of codisposal, have contemplated full-scale experiments. Therefore on the basis of tests carried out over the last 20 years it is now possible to provide some indications regarding the type of hazardous waste for which codisposal is feasible and in some cases it is possible to establish the method for effecting such a practice.

### Arsenic Waste

Several tests carried out in Great Britain and in West Germany have been mentioned (Bentley, 1981) but among the more detailed studies we should mention that of Blakey (Blakey, 1984) which amplified other trials started during the 1970's and succeeded in evidentiating the importance of several reactions, mediated by pH and Eh on the phenomena of immobilization of the arsenic. Firstly, both fresh and aged municipal waste were seen to possess a certain absorption capacity of arsenic (As) in both states of oxidization (III, V) where it can generally be observed (Fig. 3).

It was concluded that, in the range of pH values (5–8) which are typical for landfill leachate, an attenuation factor of at least 2 could be expected: the arsenic-loaded leachate on passing through a layer of MSW (preferably mature) halves its original concentration of As. A considerably important role

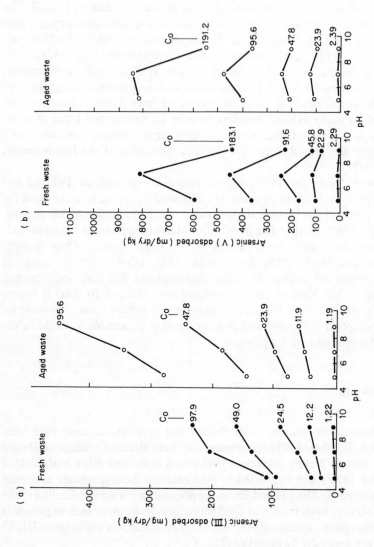

**Figure 3.** Results obtained from experiments regarding the absorption capacity of fresh and aged municipal waste with reference to arsenic (Blakey, 1984). (a) As (III); (b) As (V).

in the controlling of As is played by Eh: under reducing conditions a precipitation of As is obtained in the form of sulphide with a coprecipitation of iron sulphide. These mechanisms are capable of producing attenuation of As concentrations by a factor of 10.

On the basis of these results it may be affirmed that codisposal of As with MSW may generally be allowed at a weight ratio of 10 g As/t MSW which may be increased to 10 kg As/t MSW for waste already containing As in the form of sulphide. It is however necessary to control the amount of As loss in the form of arsine gas which is present when an intense bacterial activity occurs causing production of biogas.

## Mercury Bearing Waste

In tests carried out in Great Britain (Department of the Environment, 1978) numerous codisposal sites for MSW and mercury-based residues did not evidentiate any cases of pollution of ground and surface waters even for those landfills devoid of impermeabilizing systems and although the quantity of this element ranged from 10–2000 mg/Kg.

The more widespread opinion, on the basis of field results, is that the mercury (Hg) is reduced to sulphide in the layers of aged waste and that it remains insoluble in this form in the solid matrix. Absorption and anaerobic biological activity are fundamental components of attenuation mechanisms and the presence of soil with a high capacity of cationic exchange, such as clay, may favour absorption.

Several researchers have observed Hg behaviour in lysimeter tests which would tend to demonstrate the presence of reactions which reduce the Hg to the state of mercury metal with the possibility that the latter may be mobilized in the gas flow especially in the methane phase. The entity of this loss, similar to that already quoted for As, has not yet been quantified.

The currently available indications consider 2 mg/Kg the ratio of mercury to be respected in codisposal with MSW. However a difference is reported between codisposal of waste containing inorganic mercury and that with organic mercury for which a lower ratio is applied due to the higher degree of toxicity involved. To this regard it should be mentioned that the British regulations do not advise the landfilling of waste containing organic mercury.

## Heavy Metal Waste

The behaviour of heavy metals in the solid matrix of MSW in the stabilization phase has been examined in detail in numerous research studies. The earlier

studies (Department of the Environment, 1976/79) investigated the behaviour
of metals in leachate perculating through an unsaturated soil matrix (silica sand
cementified in a limestone matrix) with manifestation of considerable
immobilizing ability for Cu, Pb, Cr, Zn, Cd, Ni and Hg. Among these nickel
was indicated as the more mobile.

During the same years an interesting study (Newton, 1977) concerning the
codisposal of sludges with metallic hydroxides (high levels of Ni and Cr) found
that after 4 years of observation only 0.2% of the amount of metal added had
been lixiviated in leachate thus evidentiating the presence of immobilization
mechanisms in the solid waste. These findings were confirmed in a more recent
study (Young *et al.*, 1984) carried out in lysimeters which observed how the
amount of metals released by the reactors to which metals had been added was
on a par with those containing municipal waste alone. In this case also nickel
was found to be the more mobile of all heavy metals.

Studies performed on field cells (Barber *et al.*, 1985) demonstrated that an
eventual increase in release could be linked to acid pH conditions as neutral pH
conditions induce a reduction of mobility. Table 1 summarizes the data
concerning leachate composition in a first period (A), with MSW alone, and
after addition of waste containing heavy metals (B). B1 indicates the period
immediately after depositing of the waste (of an acid nature) whilst B2 shows
the situation after reinstatement of a neutral pH. Mobilization during the acid
phase is evidentiated but decreases as pH returns to neutral values.

The results of this experiment and of others carried out during the first half
of the 1980's have been summarized in graphs (Blakey, 1987) for each single
element. The most representative is that concerning lead (Fig. 4). It may be

**Table 1.** Concentration of heavy metals observed
in various periods in leachate from field cells used
in experiments performed at Edmonton (Barber *et
al.*, 1985)

| Elements (ppm) | A | B1 | B2 |
|---|---|---|---|
| Cd | 0.002–0.075 | 0.01–0.32 | 0.01–0.02 |
| Cr | 0.02–4.4 | 0.05–60.2 | 0.02–0.11 |
| Cu | 0.02–2.8 | 0.10–18.7 | 0.19–0.49 |
| Ni | 0.03–6.17 | 1.0–669 | 0.82–77.1 |
| Pb | 0.05–1.7 | <0.05 | <0.05 |
| Zn | 0.03–69 | 2.2–299 | 1.39–42.6 |
| pH | 5.8–8.9 | 4.5–7.0 | 6.5–7.0 |

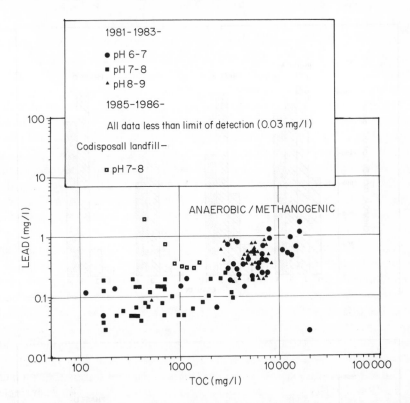

**Figure 4.** Concentrations of Pb observed in leachate from lysimeter tests carried out between 1981–83 and from actual British landfills with regard to TOC (Blakey, 1987). Experimental data indicate correlation to pH.

observed from this figure that at high concentrations of total organic carbon and a low pH, higher concentration values are obtained.

All studies mentioned examined the behaviour of metals in the short and long term. The technique of codisposal is criticized because of the unreliability of a possible immobilization of metals (and other inorganic compounds) over a long period.

An American study (Pohland and Gould, 1986) on the codisposal of galvanic waste with MSW, verified both the possible inhibition caused by an excessive load of toxic metals and the long-term aspect of emission. This study, carried out on a large scale in lysimeters (reactors of 3 m$^3$), based the results on a comparison between the behaviour of a lysimeter containing only MSW (control unit) and a further three lysimeters to which increasing quantities of

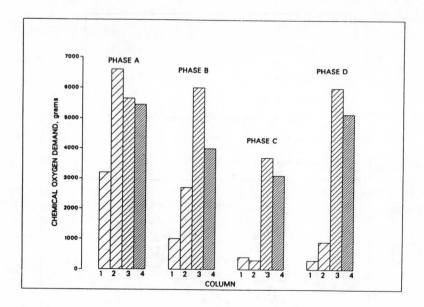

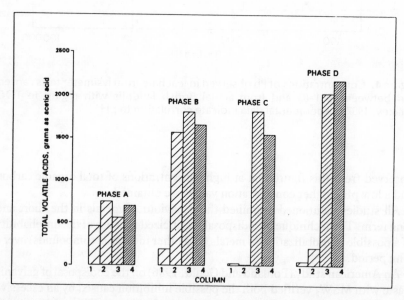

**Figure 5.** Concentrations of COD and TVA in leachate from lysimeters during various phases of codisposal experiments carried out in the USA (Pohland and Gould, 1986).

sludges from a galvanizing industry were added (7.5%; 14%; 25%). This waste was characterized by the presence of zinc (317 g/Kg), chromium (21 g/Kg), cadmium (13.1 g/Kg), copper (0.185 g/Kg) and nickel (0.4 g/Kg). The results, for the sake of simplicity divided into mean values of the four different phases of the test, show (Fig. 5) that the control column and the column with a lower concentration (numbered 1 and 2 respectively in Fig. 5) follow the same trend, albeit with a certain delay, whilst the other two columns show an inhibition of the biostabilization process as COD and total volatile acids do not decrease even in the long-term with a consequent loss of conversion of carbon substances into methane. Therefore this study demonstrates the existence of a load threshold above which the metals exert their toxic activity on biostabilizing microorganisms. With regard to the behaviour of heavy metals (for example the case of cadmium, see Fig. 6) an initial leaching in the three codisposal columns followed by a phase of attenuation for column 2 was found whilst the other columns demonstrated an increase in concentrations during the latter period.

This tendency demonstrates that the landfill possesses a certain capacity of assimilation for metals up until a given load threshold whilst above this, in absence of the above-mentioned inhibitory phenomena and biodegradation processes, the mechanisms for immobilization of the metals are hindered. The increase of concentrations in the last phase has been related to the amount of humic substances (Fig. 6b) which reached exceedingly high levels in the case of those columns with a high degree of codisposal in the final phase: these substances mobilize the metals due to a complexing effect. The results of the various mentioned experiments have permitted a deeper knowledge of the behaviour of heavy metals in MSW landfill. The factors which regulate the possible solubilization of these elements may be summarized as follows:

1. pH
2. Redox potential
3. Complexing
4. Ionic strength

The dependency of pH on solubility of metals is inverted; moreover the pH determines the relative distribution of precipitating anions (sulphide, carbonate, hydroxides). The presence of negative redox potential exerts however considerable influence on the specifying of such anions: conditions of pre-existing reducing situations ($Eh < -200\,mV$), such as those present in a landfill during degradation, induces the presence of sulphide which are most important compounds for the precipitation of metals. At very low sulphide concentrations, carbonates and hydroxy-carbonates play a fundamental role in the precipitation of metals. It has been observed that for chromium, precipitation as $Cr(OH)_3$ is important whilst for Cd, Cu and Pb, the formation

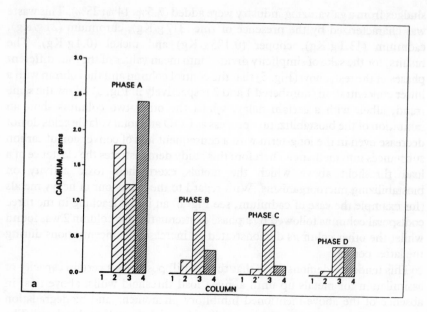

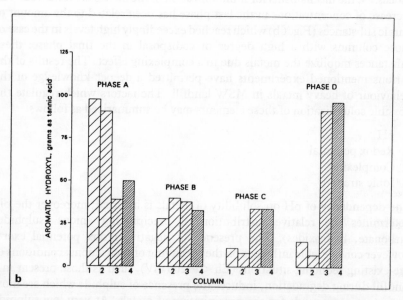

**Figure 6.** Concentrations of cadmium and of aromatic hydroxyl compounds (expressed as tannic acid) observed in leachate from codisposal experiments (Pohland and Gould, 1986).

of $CdCO_3$, $Cu_3(CO_3)_2(OH)_2$ and $PbCO_3$ is important and that in these forms they would be precipitated even in the absence of sulphide (Pohland and Gould, 1986). Zn and Ni tend to precipitate less and therefore can be considered among the more mobile metals. Metals can be immobilized not only by precipitation but also through absorption on the precipitate: these phenomena allow the separation of the metals from the liquid phase and are then constituted into the solid phase which may remain in the waste matrix due to filtration and absorption induced by the same solid matrix; these phenomena are even more efficient for the recirculation of leachate through the landfill.

It has also been observed that the solubilization of metals increases in the case of a high level of organic substances (Fig. 4) and in particular in the presence of humic substances (Fig. 6). This behaviour may be explained by the fact that in leachate the ionic strength should not be considered as null as the activity of ionic species is disturbed by the presence of other ions and is therefore reduced.

These considerations can be translated as the presence of an activity coefficient inferior to one in the formulation of the solubility product:

$$S = [M^+][L^-]$$

where $S$ = solubility product

$\quad [M^+]$ = ion metal concentration

$\quad [L^-]$ = anion linking concentration

must be substituted by:

$$S = \gamma_M [M^+] \times \gamma_L [L^-]$$

Therefore the metal concentration in the liquid phase is:

$$[M^+] = \frac{S}{\gamma_M \times \gamma_L \times [L^-]}$$

If the $\gamma_i$ are reduced the $[M^+]$ increases resolubilization with respect to the ideal solution with absence of ionic strength. The specific conductivity being a parameter which can be directly correlated to the ionic strength of a solution, is therefore correlated to the increase of lixiviation of metals due to the decrease in activity coefficient.

The concentrations of organic substances (TOC) and specific conductivity is linked to the solubilization of metals due to phenomena of complexing which permit the link of metallic ions with non-metallic ions or compounds by means of a coordinate–covalent bond. The linking compounds may be of an inorganic $(Cl^-, NH_4^+, PO_4^{3-}, SO_4^{2-}, OH^-, CO_3^{2-})$ or organic type. With reference to leachate the action of inorganic substances is of little significance both due to the slight action of some of these elements and to the relatively low presence of others ($PO_4$) (Pohland *et al.*, 1981). The monocarboxylic acids are also weak chelating agents whilst the phenolic compounds, humic and fulvic substances are much more effective. This is in accordance with data shown in experimental research studies: if the biostabilization conditions allow a continual presence of these substances, the possibility of solubilization of metals in complex form increases.

Therefore the aforementioned points lead us to conclude that certainly under strict reducing conditions and a pH which is neutral and with a relatively low organic load, the requisites for immobilization of metals are created in the landfill. On an operational level these conditions signify the disposability of wastes with metals with municipal waste which has been left to mature for at least 6 months, so as to guarantee the contemporary presence of all attenuation mechanisms mentioned.

The British guidelines may be considered as a general indication as they contemplate the possibility of disposal of up to 100 g of Cr, Cu, Pb, Zn per ton in mature municipal waste. The amount of cadmium is restricted due to the particular toxicity of this element.

More detailed indications may be deduced from studies of each specific case. Furthermore these considerations are based mainly on codisposal relations aimed at minimizing emissions but sometimes the nature of the inorganic industrial waste may be capable of accelerating the biostabilization of the organic fraction of MSW. To date, several studies carried out in 1980–81 (Stegman, 1981) in Germany indicated that the association of high percentages (10% and 25%) of sludges from activities concerning electrodepositing of metals (containing Cu, Ni, Cr) and MSW produced a more rapid presence of methane. Recent studies carried out by the authors (Cossu and Serra, 1989) confirm these data and correlate this capacity of acceleration to the alkaline nature of the industrial waste co-disposed. Moreover, the fact that the methane phase is reached more rapidly guarantees the presence of those mechanisms which control mobility of metals. In these studies no significant mobilization of heavy metals was observed when compared to that deriving from MSW alone.

### Acid Wastes

Wastes with an acid characteristic have proved to be unsuitable for codisposal with fresh municipal waste due to the capacity to reduce pH values to lower than 5, causing destruction of the buffer capacity and consequent delay in onset of the methane phase.

Moreover this type of waste is capable of increasing the concentration of heavy metals in leachate. Several research studies using stabilized municipal waste have however observed the inferior capacity of the latter in neutralizing acidity of some industrial wastes (residues from electrolytic pickling plants, acid tar wastes, etc.). It has been documented (Knox *et al.*, 1977) that the neutralizing capacity of municipal waste towards acid tar wastes can be quantified to 0.9 meq/g of waste. However it is difficult to attain a practical ratio for codisposal due to the difficulty in guaranteeing the theoretic load for acid tar wastes (a mixture of thixotropic organic substances which would tend to migrate upwards with an uncontrollable emission into the environment (DoE, 1986)).

The current British tendency deems acceptable a codisposal of acids up to 20 kg (as sulphuric acid) per ton of aged waste so long as a sufficient quantity of aged waste lies under this. If hydrogen chloride is to be disposed of, a quantity of 5 kg per ton of mature waste is deemed acceptable due to its greater efficacy as mobilizing agent for metals (DoE, 1986).

## Oil-based Wastes

The more important research studies carried out on oil-based wastes (Newton, 1977; Barber *et al.*, 1983) have shown that waste matter possesses a considerable capacity for retaining oils: in these studies the quantity of oils lixiviated represented no more than 2% of the quantity added. The hydrocarbon fraction of lixiviated oils is the lighter fraction whilst it is the heavier and persistent fractions which, in the long term, create problems in management of sanitary landfill. Mainly on the basis of results obtained in studies already mentioned (Newton, 1977), the British guidelines establish a maximum quantity of oil which may be codisposed in 2.5 kg/t of mature and well compacted waste. Recent studies carried out in France (Barres *et al.*, 1988) examined the possibility of codisposing sludge obtained from grease removal of domestic wastewater plants and sludge from grease extraction of wool-combing works. This study showed that the quantity of polar and slightly polar hydrocarbons in leachate was lower in the reactors with addition of sludge in comparison with MSW alone.

This behaviour is related to the fact that oils and fats are absorbed in the form of a film at the surface of refuse isolating them from the flow of leachate.

## Cyanides

Generally the cyanides can be found in several chemical forms and more precisely as simple cyanides, complex cyanates, thiocyanate and nitrile, but

**Table 2.**  Balance of cyanide-mass carried out on the basis of results obtained from lysimeter experiments performed in Great Britain (Newton, 1977). Percentages in respect of initial quantities observed are reported in brackets.

| | Quantity of cyanide (g CN⁻/m³) | | | |
| | Tank | | Pipe | |
| Experiment | Aerobic | Anaerobic | Aerobic | Anaerobic |
|---|---|---|---|---|
| Initially present | 1100 | 1100 | 4400 | 4400 |
| Removed in | 30 | 1 | 20 | 99 |
| leachate | (2.7) | (0.1) | (0.5) | (2.2) |
| Remaining in | 26 | 21 | — | — |
| landfills | (2.4) | (1.9) | — | — |
| | 1044 | 1078 | | |
| Unaccounted for | (94.9) | (98.0) | — | — |

the fundamental point is the capacity of each residue to form free cyanide. Various studies carried out on the behaviour of cyanides in landfills individuated several mechanisms which are responsible for their control:

1. Volatilization as hydrogen cyanide and dispersion into the atmosphere in trace concentrations;
2. Hydrolysis in the form of ammonia which is favoured by the humidity of the waste;
3. Transformation into thiocyanate in the presence of sulphur compounds deriving from biological, anaerobic and transformation processes;
4. Precipitation in the form of insoluble ferrous compounds (ferrocyanate);
5. Microbic degradation of the cyanides themselves.

The results of lysimeter tests mentioned in Newton (1977) are used as an example. Table 2 shows the most significant data. On observation, it can be calculated that after 3 years of experiments (carried out in anaerobic and aerobic conditions) the loss was between 95–98% which was probably dispersed as hydrogen cyanide or in some way transformed into other nitrogen compounds.

**Phenolic Wastes**

The final report (DoE, 1978) of the British Department of the Environment which concluded the research work carried out in the 1970's which has been already mentioned, referred to numerous cases of phenol-induced pollution due to improper disposal of liquid waste with a high presence of phenol substances. However other situations were referred to in which the phenol concentrations decreased drastically and which could not be explained by only leaching. During the same period several studies were set up to further investigate phenomena such as absorption and biodegradation which were those deemed to be capable of explaining the observed behaviour. These studies (Blakey and Knox, 1978) concluded that in the case of phenols mixed with MSW, a rate of degradation of approximately 100 mg/1 per week and that this value could be maintained also for initial concentrations of 3000–4000 mg/l. Subsequent lysimeter tests (Blakey and Barber, 1980) which were performed to verify the preceeding results obtained from tests on an inferior scale, determined a rate of degradation which was almost double that already observed and which was valid for phenol loads up to 0.6 kg/m$^2$ and for liquid phenol concentrations of up to 25 g/l. Newer and more recent tests (Craik and Senior, 1986) have confirmed the possibility of phenol degradation in a liquid flow which makes contact with the solid matrix of MSW if the period of contact is increased by means of recirculation. These studies also observed that absorption phenomena were present especially in fresh wastes thereby confirming results obtained in previous British studies (Knox and Newton, 1976; Knox et al., 1977). In addition however, these more recent tests showed that the absorption phenomena were reversible.

Thus, it may be concluded that attenuation in the phenol concentrations in leachate from a codisposal site may be explained by the absorption phenomena capable of delaying and therefore increasing the period of phenol presence inside the landfill. This permits a considerable proliferation of acclimatized anaerobic microorganisms which are capable of degradating the phenolic compounds into methane and $CO_2$.

These conclusions have been verified (Barber et al., 1985) during investigations on a wider scale on field cells at Edmonton in the United Kingdom. The phenolic wastes were deposited in 25 cm layers between two layers of municipal waste of 2 and 1 m respectively. Distribution took place on the basis of a superficial load of phenols of 2 kg/m$^2$. Figure 7 shows the levels of phenol present in leachate both with and without the addition of phenolic wastes. The highest concentrations were observed in the periods of a higher infiltration. However from a detailed analysis of the samples it was revealed (Blakey, 1987) that the quantity of lixiviated phenols was equal to 1% of added phenols whilst the quantity still present in both phenolic and municipal wastes was approximately 2% of the total introduced. Therefore it is feasible to assume that the remaining 97% was lost, either totally or in part, due to degradation phenomena.

The latest studies in this field have also shown (Cossu et al., 1987) that the

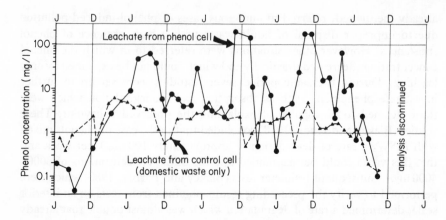

**Figure 7.** Phenol concentration in leachate from control cells containing MSW and from codisposal cells containing also phenolic sludges in field-cell experiments (Blakey, 1987).

association of phenolic sludges may induce acceleration of degradation processes in wastes. The test performed in lysimeters demonstrated that a codisposal ratio of 0.8 kg phenol/ton dry MSW was capable of accelerating biostabilization of the organic fraction of phenol compounds. Figure 8 shows that lysimeters with phenolic sludges (L4) produced a higher quantity of gas with respect to that containing digested domestic sludges (L3) and inorganic fertilizer (L2) as well as those containing MSW alone (L0, L1). The same figure shows the pH trend in leachate which confirmed the acceleration of processes induced by the presence both of phenolic sludges and domestic sludges. This behaviour was related to the alkalinity of waste which increases buffering capacity of the waste mass and therefore favours the establishment of methanogenic microbic populations.

### Wastes with Particular Organic Compounds

The disposal, in a landfill, of wastes with a low degradability such as pesticides, PCB solvents and other hydrocarbon compounds, is not generally advised as other methods of thermodestruction and chemical treatment are more suitable. However several tests regarding the feasibility of codisposing pesticides have been carried out (Bromley *et al.*, 1981) and these were able to provide a ratio for codisposal with MSW. Some as yet incomplete tests (Cossu

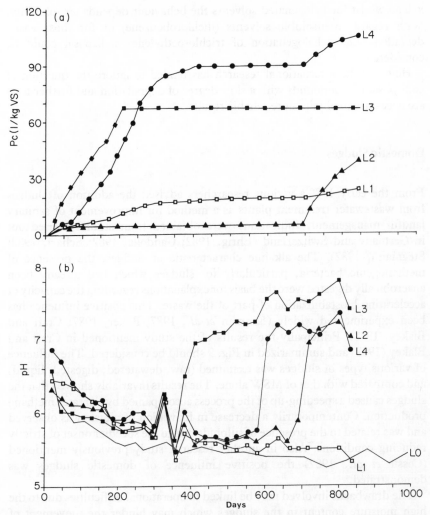

**Figure 8.** Results of lysimeter codisposal tests (Cossu, 1987). (a) The specific cumulative production of biogas for each reactor used (Li). (b) Leachate pH for each lysimeter with regard to time. The meaning of abbreviations is reported in the text.

and Serra, 1989) indicate for organic solvents the possibility of codisposal with a capacity to accelerate biostabilization of the organic fraction (for the same reasons already examined when dealing with codisposal of phenolic wastes). In particular, this is considered valid for solvents such as benzene, toluene and xylene whilst for halogenated solvents the behaviour depends on solubility. With regard to insoluble solvents (dichlorobenzene) or for those easily degraded (e.g. dehalogenation of trichloroethylene) codisposal could be considered.

However the international research has tended to ignore the question of codisposal of compounds with a slow degree of degradation and further tests are necessary to obtain more reliable results.

## Domestic Sludges

From the early 1980's various researchers advised the addition of sludges from wastewater treatment plants as a method for improvement of sanitary landfill management. Indications were obtained from studies carried out in Germany and Switzerland (Ehrig, 1982; Gandolla, 1982; Scharf, 1982; Stegmann, 1982). The alkaline characteristic of sludges, the presence of methanogenic bacteria, particularly for sludges which had already been anaerobically digested were the basis for explanations regarding the capacity of accelerating biostabilization of part of the waste. This positive influence has been experimented widely (Kinman *et al.*, 1987; Beker, 1987; Craft and Blakey, 1988). Principally the results of the study mentioned in Craft and Blakey (1988) and summarized in Fig. 9 should be considered. The influence of various types of sludges was examined (raw, dewatered, digested, liquid) and compared with that of MSW alone. The results invariably showed that the sludges caused a speeding-up of the process accompanied by a greater methane production. Contemporarily a decrease in lixiviation of metals was observed and was related to the presence of sulphide due to the speedier onset of strictly reducing conditions. Also in another research study previously mentioned (Cossu *et al.*, 1987) the positive influence of domestic sludges was demonstrated.

The drawbacks involved may be linked to operational difficulties due to the high moisture content in the sludges which may hinder the movement of vehicles above the waste. A possible solution could be that of mixing the sludges with ashes from power-plants due to the low degree of humidity of the latter and whose environment compatibility has been confirmed in recent studies (Cossu *et al.*, 1988; Eighmy *et al.*, 1987).

## Combustion Residues

The problem of disposal of residues from coal combustion and MSW incineration is of utmost importance. The large quantity of waste involved

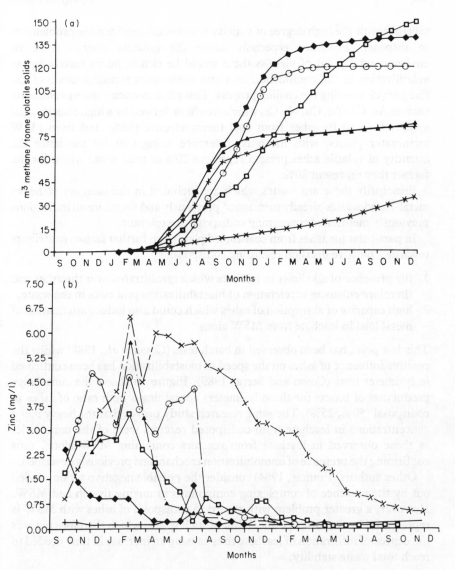

**Figure 9.** Results of lysimeter codisposal tests for domestic sludges. (a) The specific cumulative production of methane obtained for each reactor. (b) The time–trend of zinc concentration present in leachate from each reactor (Craft and Blakey, 1988).

    Domestic waste + raw, dewatered sludge 4.1 : 1 wt/wt.

    Domestic waste + raw, dewatered sludge (homogeneous) 4.1 : 1.

    Domestic waste + raw, dewatered sludge (high infiltration).

    Domestic waste + primary/mixed dewatered sludge 4.8 : 1 wt/wt.

    Domestic waste + liquid digested sludge 9.7 : 1 wt/wt.

    Domestic waste only, control.

together with the high degree of toxicity are causing considerable problems as to disposal. Fly-ashes especially cause the greatest worries from an environmental point of view as these would be rich in heavy metals due to volatilization in the combustion area and subsequent recondensation of the fine particles during the cooling process. This phenomenon concerns elements such as As, Cd, Zn, Cu, Ni, Cr, which are characterized by a high toxicity, and applies both to fly-ashes from coal thermoelectric plants and from MSW incinerator plants, with the only difference being that for the latter the quantity of volatile ashes present represent 20% of total waste whilst for the former they represent 80%.

Principally these are wastes which are included in the category of heavy metal-based wastes already mentioned previously and therefore all indications previously mentioned concerning codisposal are relevant.

In particular for ashes from coal-fueled plants two further factors contribute towards a positive decision regarding codisposal:

1. the presence of alkalinity in leachate which is infiltrated over the ashes and therefore enhances acceleration of biostabilization processes in the waste;
2. high capacity of absorption of ashes which could also induce attenuation of metal load in leachate from MSW alone.

This last point has been observed in batch tests (Cossu et al., 1988) whilst the positive influence of ashes on the speed of biostabilization has been confirmed in lysimeter tests (Cossu and Serra, 1989). Figure 10 shows the cumulative production of biogas for those lysimeters containing a high ratio of ashes in codisposal (50%, 25%). The same research study proves that the heavy metal concentrations in leachate from codisposed reactors were of the same order as those observed in leachate from reactors containing MSW alone, thus confirming the presence of immobilization mechanisms previously described.

Other authors (Francis, 1984) consider the possible negative action carried out by the presence of complexing compounds in mixing fly-ash and MSW. However, a greater problem with regard to codisposal of ashes with MSW is represented by the lower permeability of such wastes which therefore reduces the degree of infiltration of leachate thus causing an increase in time needed to reach total waste stability.

## FULL SCALE EXPERIMENTS

The majority of full scale experiences have been carried out once again in Great Britain.

A considerable number of codisposal sites have been brought to light mainly

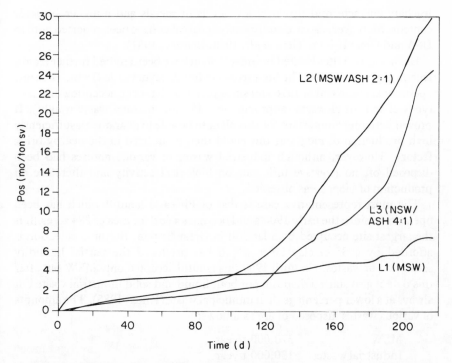

**Figure 10.** Specific cumulative production of biogas in lysimeter tests for codisposal of MSW and coal ashes (Cossu and Serra, 1989).

L1 = control units (only MSW).
L2 = addition of ashes (50%).
L3 = addition of ashes (25%).

through studies carried out by the Department of the Environment, although there are still many such sites which remain as yet unmentioned in literature.

Among the more significant cases, the Bromsgrove landfill should be mentioned. This disposes of industrial waste with a percentage of Zn up to 1.7% but the groundwater of this landfill has invariably maintained a Zn level of 0.1–0.5 mg/l and does not seem to be affected by metal intrusion (Bentley, 1981). Another important site is that of Hammerwich where cyanide-based wastes have been disposed of successfully.

With regard to the disposal of ashes, a Danish landfill has been reported (Hjelmar, 1981) where residues and fly-ashes from a MSW incinerator have been mixed with raw MSW thus creating some 'mixed' sectors which produced leachate characterized by fairly low levels of metals and a higher sulphide content. Moreover, cases of codisposal of coal ashes have been reported both in USA and Great Britain (Gera *et al.*, 1988; Bentley, 1981).

However, two sites located in Great Britain have been studied further from a technical point of view. The Stewartby landfill (Cheyney, 1984) which became operational at the end of 1976 and since July 1978 has been accepting different types of industrial waste (approximately 15% of the total waste weight). It proved extremely important for this plant to be able to guarantee a sufficiently high production of biogas as this could then be utilized in the nearby brick factory. However, although industrial wastes of various natures had been disposed of, no negative influence on biological activity and therefore on production of biogas was observed.

The most representative case is that of Pitsea, a landfill which has been operational since the early 1900's and occupies a surface area of 284 acres. It is the largest site occupied by a landfill in Great Britain. Bottom protection is afforded by a 35 m clay layer which has produced the partial humidity saturation of wastes. Up until 1960 this landfill accepted only MSW but after this date it also started disposing of both liquid and solid industrial waste but always at a lower percentage than municipal waste (Knox, 1985). The amounts of various wastes disposed of are as follows:

| | |
|---|---|
| MSW | 370,000 t/year |
| Industrial wastes | 130,000 t/year |
| Liquid wastes | 65,000 cm³/year |

The technique of leachate recirculation is adopted. To date, no phenomena of inhibition of degradation of household wastes have been observed and the quality of leachate is typical of an outgoing stream from a landfill in the stabilized methane phase. Certainly the fact that the bottom layers are covered by layers of MSW which have been in the methanogenic phase for some time favours the onset of attenuation mechanisms on which the technique of codisposal is based and which justify this method.

It should be mentioned that not all types of industrial waste can be disposed of at Pitsea but only those established by the 'Site Licence'.

## CODISPOSAL SYSTEM

Hazardous wastes can be present in various forms (liquid, sludges, solid and dusts) and for each of these wastes a correct approach to codisposal in landfill can be ascertained (DoE, 1986).

With regard to liquid wastes there should be some less-compacted areas in order to facilitate the absorption of the waste. It has been estimated (Blakey and Craft, 1986) that for wastes with a density of $0.7-0.8\,t/m^3$ the absorption capacity is of $0.16-0.28\,m^3/t$ of dry waste. However, for a density of $1\,t/m^3$ the capacity of absorption decreases to $0.02-0.03\,m^3/t$ of dry waste (Campbell, 1982). Liquid wastes are usually stored in a tank near the site and can be introduced into a landfill according to one of the following four methods:

1. *Trenches or lagoons*: the more commonly used technique for the trench method is that shown in Fig. 11; lagooning is performed in areas excavated inside the waste. Trenches and lagoons should always be clearly marked and fenced. The latter method has the disadvantage of not being suitable for smelly wastes.
2. *Injection*: the liquids can be introduced under the surface of the landfill by means of facilities installed during site operation. These are made up of perforated pipes or by building up columns of tyres. This technique can be used for smelly wastes.
3. *Irrigation*: this method exploits the phenomenon of evaporation to reduce volume and lower the organic load. The spray irrigation method is not acceptable due mainly to the potential danger to the health of operators whilst surface irrigation is intended as a below-surface practice and should be carried out as for 2.

If the industrial waste is in the form of sludge the trench method already mentioned for liquid wastes is used. Also in this case the hydrological balance should be considered prior to introduction of the waste as if the absorption limit is exceeded, rapid leaching of the sludge liquid phase will occur. The passage of vehicles directly above the sludges is to be avoided. Sludges should ideally be collected at the base of working face and covered immediately with solid waste.

This indication is valid also for solid hazardous wastes. Moreover, if this type of waste is characterized by a high degree of impermeability, disposal over large areas is not advised.

On some occasions industrial wastes are lightweight and involve the problem caused by dust during introduction into the landfill. Thus it is necessary to keep the disposal areas wet and to take adequate precautions for the site workers. Dusty wastes characterized by a high degree of toxicity should not be directly landfilled.

## CONCLUSIONS

There are varying opinions regarding codisposal, however numerous experiences and research studies confirm the validity of this technique.

148                                                           *Cossu and Serra*

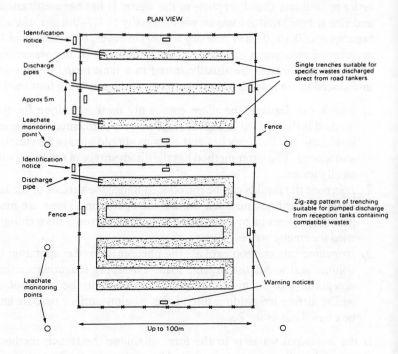

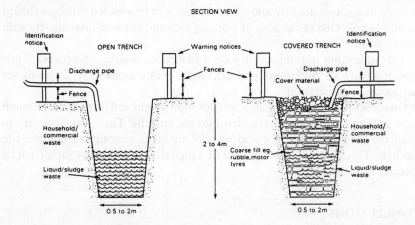

**Figure 11.** Realization scheme for a trench in which to carry out co-disposal of liquid waste and sludges in a MSW landfill as proposed by the British guidelines (reported in DoE WMP No. 26).

For codisposal to be successful a strict control of the quantity and quality of industrial wastes introduced into the landfill is necessary in order to be completely certain that the degradation processes are not inhibited and that uncontrolled emission of toxic substances does not occur. The application of this method could be extended in the future if research studies are able to establish more valid data regarding the ratio of hazardous wastes which can be disposed of with municipal wastes. Thus new methods of application may be determined to be used as guidelines for the landfill operators on the basis of what has already occurred in Great Britain (DoE, WMP No. 1/21), where codisposal is widely applied.

## REFERENCES

Barber, C., Maris, P.J. and Johnson, R.G. (1985). Behaviour of waste in landfills: study of the leaching of selected industrial wastes in large-scale test cells at Edmonton, WLR Technical Note 69, Department of the Environment, UK.

Barber, C. *et al.* (1983). The codisposal of oil waste with domestic solid wastes in landfills: leaching and persistence of oil, ASTM–International Symposium on Industrial and Hazardous Solid Wastes, Philadelphia USA, 7–10 March.

Barres, M. *et al.* (1988). Experimental Studies on Household Refuse and Industrial Sludges Codisposal, ISWA'88 Proceedings, Copenhagen, vol. 1, pp. 169–176, Academic Press.

Bentley, J. (1981). Hazardous Wastes: Landfill Research, Hazardous Waste Disposal (Ed. Lehman J.P.), Plenum Press, New York and London, pp. 231–253.

Beker, D. (1987). Control of acid phase degradation, International Symposium ISWA on Process Technology and environmental impact of Sanitary Landfill, Cagliari 19–23 October, Proc., Vol. 1.

Baldi, M. *et al.* (1988). Valutazioni sul rilascio di metalli pesanti nel percolato di fanghi industriali mediante esperienze in batch, *RS-Rifiuti Solidi*, Vol. 2, No. 3, 233–243.

Blakey, N.C. and Knox, K. (1978). The biodegradation of phenol under anaerobic condition by organisms present in leachate from domestic refuse, Report of an investigation for the Department of the Environment, WLR Technical Note No. 63, Contract report No. 675, November, Water Research Centre, Stevenage, UK.

Blakey, N.C. and Barber, C. (1980). Codisposal of Hazardous Wastes: attenuation mechanisms within a landfill site, Proc. of 5th European Sewage and Refuse Symposium EAS, WEPCA–ISWA, Munich 22–26 June, pp. 715–726.

Blakey, N.C. (1984). Behaviour of Arsenical Wastes Codisposed with Domestic Solid Wastes, *Journal of Water Pollution Control Federation*, 56, 1, 69–75.

Blakey, N.C. and Craft, D.G. (1986). Infiltration and adsorption of water by domestic wastes in landfills, leachate volume changes with time. Proceedings, Harwell Landfill Water Management Symposium, Harwell, UK, pp. 5–18.

Blakey, N.C. (1987). Hazardous Waste codisposal in Municipal Solid Waste Landfills, WPCF 6th Annual Conference, Philadelphia, USA, 5–8 October.

Bromley, J., Stevens, C., Parker, A. and Rees, J.F. (1981). Codisposal of Hazardous Waste: Attenuation Mechanisms Within a Landfill Site, Symposium of The ISWA, Munich, 22–26 June 1981.

Campbell, D.J.V. (1982). Absorptive capacity of refuse, Harwell Research, Proceedings Harwell Landfill Leachate Symposium, Harwell, UK.

Campbell, D.J. and Pugh, M.P. (1988). The Codisposal of Chemical Wastes in Test Cells in Hong Kong: design details and interim experimental results, ISWA '88 Proceedings, Copenhagen, vol. 1, pp. 261–266, Academic Press.

Cheyney, A.C. (1984). Experience with the codisposal of hazardous waste with domestic waste, Chemistry and Industry, 3 september, pp. 609–615.

Cossu, R., Blakey, N.C. and Trapani, P. (1987). Degradation of mixed solid wastes in conditions of moisture saturation, International Symposium on Process Technology and Environmental Impact of Sanitary Landfill, Cagliari, October 19–23, Proc. Vol. 1.

Cossu, R., Serra, R. and Marrocu, M.G. (1988). Indagine sperimentale sullo smaltimento combinato in discarica di rifiuti solidi urbani e ceneri di carbone, Proc. Convegno Biennale ANDIS, Roma, 16–17 December, Vol. 1, pp. 177–186.

Cossu, R. and Serra, R. (1989). Codisposal of coal ash and other waste in landfill, Sardinia 1989, 2nd International Symposium, Porto Conte, 9–13 October.

Craft, D.G. and Blakey, N.C. (1988). Codisposal of sewage sludge and domestic waste in landfills, ISWA '88 Proceedings, Copenhagen, vol. 1, pp. 161–168, Academic Press.

Craik, I.A. and Senior, E. (1986). Landfill codisposal: an option for phenolic wastes, Microbe '86, Manchester, September 1986.

Department of the Environment UK (1976/1979). Uffington Lysimeters Operation and Results (Parts 1 to 4), WLR Technical Notes 36, 40, 42 and 60, London 1976, 1977, 1978, 1979 Part 5 UKAE Harwell, May 1981, HMSO.

Department of the Environment UK (1978). Cooperative Programme of Research on the Behaviour of Hazardous Wastes in Landfill Sites, Final Report of the policy review committee, 169 pages, HMSO, London.

Department of the Environment UK, Waste Management Papers No. 1/.../21, London, HMSO.

Department of the Environment UK (1986). Landfilling Waste. Waste Management Paper No. 826, HMSO Books, London.

Ehrig, H.J. (1982). Results from investigation on degradation processes in lysimeters, Gas and Wasserhaushalt—Veröffentlichungen des Institut für Stadtbauwesen. T.U., Braunschweig Heft 33, Eigenverlag.

Eighmy, T. et al. (1987). Codisposal of municipal solid waste bottom ash and wastewater sludges, International workshop on Municipal Waste Incineration, October 1–2, Montreal, Quebec.

Francis, C.W. (1984). Leaching Characteristics of Resource Recovery ash in Municipal Waste Landfills, DoE project No. ERD-83-289, Draft Final Report.

Francis, C.W. and Maskarinec, G. (1986). A laboratory extraction method to simulate codisposal of Solid Wastes Texting, Fourth Symposium ASTM STP 886, Philadelphia, USA.

Gandolla, M. (1982). Results from lysimeter studies at the Sanitary Landfills of Croglio, Switzerland, Gas and Wasserhaushalt von Mülldeponien, Veröffentlichungen des Institut für Stadtbauwesen, T.U. Braunschweig Heft 33, Eigenverlag.

Gera, F. et al. (1988). Le ceneri di carbone: un'opportunità per operazioni di recupero territoriale, International Congress on Solid Wastes disposal in urban and industrial zones, Aprile, Porto Torres, Italy.

Hjelmar, O. (1981). Codisposal of incineration waste and household waste at a controlled landfill site, ISWA Symposium, Munich, 23–25 June, Proc., pp. 85–102.

Kinman, R.N. *et al.* (1987). Gas Enhancement Techniques in Landfill Simulators, *Waste Management and Research*, 5, 1987, 27–39.

Knox, K. and Newton, J.R. (1976). Study of landfill disposal of acid tars and phenol-bearing lime sludges; I: characterization of water-soluble constituents of wastes and preliminary study of interaction of phenolic with domestic refuse, WLR Technical note No. 19, WRCs report No. 554 R, January, Water Research Centre, Stevenage, UK.

Knox, K., Newton, J.R. and Stiff, M.J. (1977). Study of landfill disposal of acid-tars and phenol bearing lime sludges; II: mathematical model and further laboratory studies, Report of an investigation for the Department of the Environment, WLR Technical Note No. 52, Contract Report No. 635, October, Water Research Centre, Stevenage, UK.

Knox, K. (1985). Monitoring at Codisposal Sites: two Case Studies, Proceeding Landfill Monitoring Symposium, Cockroft Hall, Harwell Laboratory, 23 May, pp. 135–150.

Misiti, A. *et al.* (1988). Comparison between leaching tests performances and toxic waste behaviour in landfill disposal, Proc. ISWA 5th International Solid Wastes Conferences, September 11–16, Copenhagen, vol. 1, pp. 247–256, Academic Press.

Newton, J.R. (1977). Pilot scale studies of the leaching of industrial wastes in simulated landfills, I. *Inst. Wat. Poll. Control.*, 4, 468–480.

Pohland, F.G. *et al.* (1981). Containment of heavy metals in landfills with leachate recycle, Proc. 7th Ann. Research Symposium on Land Disposal: Municipal Solid Wastes, Philadelphia, Pa. USEPA Report EPA 600/9-81-002a, pp. 179–194.

Pohland, F.G. and Gould, J.P. (1986). Testing Methodologies for Landfill Codisposal of Municipal and Industrial Wastes, Hazardous and Industrial Solid Waste Testing and Disposal: sixth volume, ASTM STP 933, Philadelphia, 1986, pp. 45–62.

Pohland, F.G. and Gould, J.P. (1986). Codisposal of municipal refuse and industrial waste sludge in landfills, *Wat. Sci. Tech.*, 18, 177–192.

Pohland, F.G. and Harper, S.R. (1987). Retrospective Evaluation of the effects of selected Industrial Wastes on Municipal Solid Waste stabilization in simulated Landfills, EPA/600/S2-87/044, August 1987.

Scharf, W. (1982). Codisposal of Municipal Solid Waste and sewage sludge; investigation in laboratory scale, Gas and Wasserhaushalt von Mülldeponien, Veröffentlichungen des Institut für Stadtbauwesen, T.U. Braunschweig Heft 33, Eigenverlag.

Stegman, R. (1981). Criteria for the codisposal of municipal and industrial solid wastes, Proc. of 5th European Sewage and Refuse Symposium EAS, EWPRA– ISWA, Munich, 22–26 June, pp. 181–201.

Stegman, R. (1982). Description of biological degradation processes of municipal solid waste in laboratory scale lysimeter, Gas and Wasserhaushalt von Mülldeponien, Veröffentlichungen des Institut für Stadtbauwesen, T.U. Braunschweig Heft 33, Eigenverlag.

WHO (1983). Management of Hazardous Waste, Regional Publication European Series, No. 14, Copenhagen.

Young, P.J., Baldwin, G. and Wilson, D.C. (1984). Attenuation of Heavy metals within Municipal Waste Landfill Sites, Hazardous and Industrial Waste Management and Testing: Third Symposium, ASTM STP 851, Philadelphia, pp. 193–212.

# 3. BIOGAS

# 3.1 Measurement and Prediction of Landfill Gas Quality and Quantity

ROBERT K. HAM* and MORTON A. BARLAZ

*Department of Civil and Environmental Engineering, University of Wisconsin-Madison, Madison, WI 53706, USA

## INTRODUCTION

Knowledge of landfill gas quantity and quality is obviously important in assessing the feasibility of a gas utilization system at a landfill, but it is also important in evaluating the potential for gas migration and as an indication of landfill stabilization.

The quality of the gas depends on the microbiological system, the substrate being decomposed at the time of the measurement and site-specific variables such as oxygen access to the waste and moisture content.

There is little data on the total amount of gas to be generated from solid wastes in a landfill. Available data are largely from lysimeter studies in which refuse is placed in enclosed spaces, ranging from a few litres in volume to contain thousands of tons of refuse, and the gas involved is measured. There is little data from full-scale landfills, in part because of the long period of monitoring required to cover the period of active gas generation and also because of the difficulty of measurement. Theoretical model results, plus data from lysimeter and full-scale landfills, and a mass balance approach can be used to get some idea of total gas generation.

The best data from a practical perspective, however, is likely to come from a full-scale gas utilization project. Such data are beginning to be available and, considering the large number of such projects in operation or being constructed, should be increasingly available over the next several years.

The biochemical processes causing the gas generation are presented by Christensen and Kjeldsen in Chapter 2.1 of this volume and are further discussed here.

155

## QUALITY OF GAS GENERATED DURING METHANOGENIC DECOMPOSITION

The amount and composition of the generated gas can be predicted for different substrates using equation 1 presented by Tchobanoglous *et al.* (1977):

$$C_a H_b O_c N_d + \left( \frac{4a - b - 2c + 3d}{4} \right) H_2O \rightarrow$$

$$\rightarrow \left( \frac{4a + b - 2c - 3d}{8} \right) CH_4 + \left( \frac{4a - b + 2c + 3d}{8} \right) CO_2 + d\,NH_3 \tag{1}$$

The results for several substrates of importance in refuse are shown in Table 1. For example, Table 1 indicates that the gas composition for cellulose decomposition would be 50% $CO_2$ and 50% methane. For protein decomposition, it would be 51.5% methane and 48.5% $CO_2$, and for decomposition of fats, the gas composition would be 71.4% methane and 28.6% $CO_2$. This illustrates that the gas composition is a direct function of the material being decomposed at a specific time. To the extent that one can characterize materials being decomposed at a specific time in a mixed substrate such as municipal refuse, one can predict the gas composition.

Climatic or environmental conditions will also affect the gas composition. Due to the heterogeneous nature of the landfill environment, there will be some acid-phase anaerobic decomposition occurring in any large-scale landfill, along with the methanogenic decomposition. Since aerobic and acid-phase anaerobic decomposition gives rise to $CO_2$ and not methane, there may be a higher $CO_2$ content in the gas as generated than would be predicted from equation 1 alone. Furthermore, depending upon the amount of moisture present and the movement of moisture through the landfill, there will be some

**Table 1.** Theoretical $CO_2$ and $CH_4$ generation.

| | $CO_2 + CH_4$ produced (l/kg decomposed material × 100) | Gas composition (% $CH_4$) |
|---|---|---|
| Cellulose[a] | 8.29 | 50.0 |
| Protein[b] | 9.88 | 51.5 |
| Fat[c] | 14.30 | 71.4 |

[a] $(C_6H_{10}O_5)x$.
[b] Assumed 53% C, 6.9% H, 22% O, 16.5% N, 1.25% S (S not included in calculation) (Sawyer and McCarty, 1978).
[c] Assumed $C_{55}H_{106}O_6$ as typical fat (Wertheim, 1956).

$CO_2$ dissolution. $CO_2$ is much more soluble in water than is methane. The dissolution of $CO_2$ can appear to artificially increase the methane content of the gas measured in a landfill.

Many operators and researchers have measured gas composition in landfills. In general, figures of approximately 55% methane and 45% $CO_2$ are common; however, the authors are aware of landfills gas methane contents ranging from 35 to 75% for sites which can be characterized as undergoing methane formation. Lower concentrations of methane are often a result of landfills that are either early in the decomposition process, so a mature methanogenic population has not been developed, or are late in the decomposition process, so that other organisms dominate. Other factors giving rise to unusual gas composition would include unusual refuse composition or environmental factors adverse to methane formation, such as extremely dry conditions. Dry conditions reduce the activity of almost all organisms, but because some organisms are more susceptible to moisture content than others, the microbial balance necessary for methane formation may be disrupted. Dry conditions can also lead to increased air access to the interior of the landfill, reducing methane formation.

It should be noted that in some landfills, gas concentrations with up to 20% hydrogen have been reported. Hydrogen is converted to methane in a well-balanced biological system actively producing methane, and so should be present at very low concentrations (well under 1%). Measurable hydrogen levels over a period lasting several months or more indicate a biological imbalance in which organisms are converting refuse components to organic acids and hydrogen, but the methanogens are not converting these materials to methane and $CO_2$. In the authors' experience, dry conditions or unusual refuse compositions appear to have been the original cause for high hydrogen contents, but other factors inhibiting the methane formers could also be important. Note that elevated hydrogen contents over a short period are common and simply indicate the transition from acid-phase to methanogenic decomposition.

Measuring gas composition is readily accomplished with portable or laboratory gas chromatographs, explosimeters or portable gas analysers, etc. The procedures for gas sampling and analysis are routine and so will not be discussed here.

## QUANTITY OF GAS GENERATED IN A LANDFILL

Equation 1 can also be used to estimate the total amount of gas to be produced in a landfill. If the amount of refuse ultimately forming methane and $CO_2$ and the elemental composition of the refuse is known, this equation can be used to obtain an estimate of the maximum amount of gas to be generated. This was done in Table 1, which indicates that the amount of gas produced per kilogram of substance decomposed depends on the substance decomposed.

Table 2 provides typical values that have been observed for the quantity of gas generated from refuse. This table is derived from the literature and experience and includes observations based on predominantly methanogenic conditions. The first entry is the theoretical amount of methane and $CO_2$ which will be formed based on the use of equation 1 for 'typical' US municipally collected refuse. This amount of $CO_2$ and $CH_4$ would be the maximum that could be formed. It assumes a perfect biological system and perfectly degradable materials which are converted to $CO_2$ and $CH_4$. More realistically, the second entry indicates typical values that would be obtained if it is assumed that not all of the refuse will be decomposed to methane and $CO_2$. Typically, it may be assumed that only the food waste content of refuse and perhaps only two-thirds of the paper content of refuse would actually be decomposed to form methane and $CO_2$. These assumptions allow for the fact that the refuse will not be completely decomposed, that some of the refuse will be decomposed to end products other than methane and $CO_2$ (such as cell mass) and that there will be some materials which will not be decomposed under methanogenic conditions at all. Such resistant materials would include plastics, lignin under anaerobic conditions, and certain other materials that may be protected physically or not decomposable biologically under conditions found in a sanitary landfill. These assumptions lead to values

**Table 2.** Total gas ($CO_2$ + $CH_4$) production from municipal refuse.

| Conditions | Gas production refuse[a] (l/kg) |
|---|---|
| 'Typical' US municipal refuse, theoretical estimate (eq. 1)[b] | 520 (53% $CH_4$) |
| Weight organic components by degradability, theoretical estimate (eq. 1) | 100–300 |
| Anaerobic digestion of refuse with sewage sludge, laboratory measurement | 210–260 |
| Lysimeters or closed container, varying success in obtaining $CH_4$, periods approximately 1–3 years | 0.5–40 |
| Full-size landfills projected from existing short-term data | 50–400 |

[a] Rounded to two digits and adjusted to 21% $H_2O$ (wet weight) and 53% $CH_4$.
[b] 28% C, 3.5% H, 22.4% O, 0.33% N, 24.9% noncombustibles, 20.7% $H_2O$ (Corey, 1969).

ranging from one-third to two-thirds of the total amount of gas which could be generated, as shown in the first entry.

Several investigators have utilized what may be considered to be the ideal anaerobic system in order to determine the amount of methane and $CO_2$ which can be formed from refuse. In this case, anaerobic digesters, usually stabilized using sewage sludge as a seed and substrate source, have been fed known amounts of municipal solid wastes or specific refuse components, and the additional amount of gas attributable to the solid wastes calculated. Gas production figures based on such an approach may not be observed in a landfill, because a landfill would not have the closely controlled conditions of the digester. A well-mixed, properly controlled digester would provide virtually optimal methane formation conditions. In practice, therefore, the values shown for the digesters with sludge addition in Table 2 may be considered the best that could be achieved in a landfill.

A significant number of research experiments have involved placing known quantities of refuse in an enclosure, allowing decomposition to proceed, and monitoring the amount and composition of the gas generated. These experiments are termed lysimeter experiments in Table 2. These experiments are the best sources of information giving exact values for gas generated in a landfill. However, it must be noted that the act of placing the refuse in an enclosure virtually assures that the refuse does not decompose in the same fashion that it would in a full-scale landfill. In fact, the history of such lysimeter experiments has been that substantially less methane has been formed than observed in full-scale landfills. Such factors as oxygen access and moisture routeing are probably significant in changing the full-scale landfills. Because of the difficulties in setting up and operating such lysimeters, a large range in gas generation rates has been observed. The values shown in Table 2 indicate the range of typical observed values, but note that the lower values usually reflect the difficulty in achieving good $CH_4$ (not just $CO_2$) generation in lysimeters. Higher values have been obtained, often by using modifications to the refuse or special experimental designs.

Measurement of gas at a full-scale landfill is very difficult. The usual procedure is to withdraw gas from a well or trench placed in the landfill, measure the amount and composition of the gas, and then attempt to identify the volume of refuse from which the gas is being withdrawn using a series of pressure monitoring probes. If gas is withdrawn under vacuum, the pressure at locations in the landfill close to the withdrawal point will decrease. This decrease will be less at increasing distances from the withdrawal point, and conceivably, at some distance and beyond, there will be no change in pressure regardless of whether gas is being withdrawn from the landfill. This distance is termed the 'radius of influence', and the volume of the landfill within this distance is the 'zone of influence' for that withdrawal well or trench at that withdrawal rate or vacuum. It is required to know the zone of influence in order to obtain the amount of refuse producing the measured amount of gas withdrawn from the system. Such measurements have been obtained from numerous landfills. Typical values which have been obtained are shown in Table 2. It is observed that measurement of the amount of gas withdrawn in such a test is not difficult; the inaccuracy is in determining the zone of influence and, therefore, the weight of refuse producing the gas.

A major problem in determining the total amount of gas produced in all of the lysimeter and landfill experiments is that the gas has to be monitored over the total decompositional life of the landfill or lysimeter. In some lysimeter experiments, the rate of generation of gas has decreased at the end of monitoring to virtually insignificant levels, and a more or less 'complete' history of the amount of gas generated has been obtained. In no case of full-scale landfill measurement, however, has a complete decompositional history been obtained, so the results may be considered low just on the basis of not having measured the total decompositional life of the refuse.

Another approach to determining the amount of gas generated from a landfill is to measure the loss of refuse from the landfill. To the extent that the loss of refuse represents gas generation and components in leachate, it is possible to estimate the amount of gas generated. This is done by subtracting the weight of materials measured in the leachate from the weight loss of refuse to obtain the weight of the gas generated by difference. To use this concept, it is first necessary to determine that decomposable materials can be measured in refuse and that these materials, in fact, do decrease in concentration as refuse decomposition proceeds. A summary of a study in which cellulose concentration was observed for different refuse samples taken from actual landfills throughout the US is presented by Bookter and Ham (1982). These refuse cellulose concentrations were compared to typical fresh refuse values and to values obtained from an old well-decomposed landfill. The results indicated a clear decrease in cellulose with time or refuse age. Cellulose is thought to be the largest chemically identifiable component in refuse involved in methanogenic decomposition.

Further verification of this concept requires use of mass balance in strictly controlled volumes of refuse that have been decomposed under known conditions. Such a mass balance has been completed recently at the University of Wisconsin. The data are summarized in Table 3. A series of 55 gallon drums were filled with selected shredded municipal solid wastes, totalling approximately 45 kg dry weight in each drum. Samples of the refuse were analysed for cellulose, hemicellulose, protein and other components. The drums were provided with various treatments in order to promote decomposition, including different moisture contents, leachate recycling, leachate neutralization and recycling, and seeding with old refuse. Leachate and gas leaving each drum was collected and analysed, and the refuse

**Table 3.** Mass balance on two test containers.

| | CH$_4$ or CH$_4$ equivalent (l) | | | |
| | Drum S | | Drum 0 | |
| | Initial | Final | Initial | Final |
|---|---|---|---|---|
| Cellulose | 4465 | 2499 | 4427 | 2331 |
| Hemicellulose | 1003 | 698 | 1052 | 667 |
| Protein | 1149 | 965 | 950 | 612 |
| Volatile acids in leachate | — | 122 | — | — |
| Measured methane | — | 2593 | — | 2098 |
| Total | 6617 | 6877 | 6429 | 5708 |
| Unaccounted for | 260 | | | 721 |
| Recovery (%) | $\frac{6877}{6617} \times 100 = 103.9$ | | $\frac{5708}{6429} \times 100 = 88.8$ | |

remaining at the conclusion of the experiment was sampled and analysed. The experiment is described more completely by Barlaz *et al.* (1987). The results from two of the five most active drums producing methane are summarized in Table 3.

At the conclusion of monitoring, when gas generation rates were at a relatively low level, it is observed that the cellulose, hemicellulose, and protein contents decreased and that for the most part this loss of decomposable material resulted in measurable methane and CO$_2$. The recovery was 104% in one case and 89% in the other. It appears that, to the extent that the different components in refuse can be measured accurately, the loss of these components does, in fact, translate to gas being generated. This mass balance technique is probably the most accurate method for measurement of the total amount of gas generated over the decompositional life of a landfill, assuming that anaerobic decomposition to form methane and CO$_2$ is the predominant mechanism by which refuse is converted to gaseous end products.

## THE RATE AND DURATION OF GAS GENERATION

Theoretical approaches to the rate of gas generation in a landfill generally involve developing models based on first-order kinetics, expressed as:

$$-\frac{dc}{dt} = kc \tag{2}$$

This equation states that the rate of loss of decomposable matter is proportional to the amount of decomposable matter remaining. If the decomposable matter can be translated directly into gas generation, such as equation 1 indicates, Equation 2 can be rewritten in terms of the rate of methane formation and the total amount of methane to be formed. This first-order expression is the most widely used model of gas generation rates at a landfill. It assumes that the factor limiting the rate of methane generation at a landfill is the amount of material remaining in the landfill that will ultimately form methane. It assumes that other variables and factors affecting the decomposition process are not limiting the rate of methane generation. Such factors include the availability of moisture or the biological availability of measured substrate, the absence of inhibitory concentrations of toxic substances, etc. This equation models accurately laboratory experiments in which the growth conditions are such that the availability of substrate does, in fact, control the rate of decomposition. The BOD determination is a good example of a microbiological system that is modelled by a first-order expression. In a landfill, however, many complicating factors may reduce the rate of gas generation to a lower value than would be expected based on the availability of substrate alone.

A zero-order kinetic expression indicates that the rate of methane generation is independent of the amount of substrate remaining. In the authors' experience, many landfills appear to produce gas according to zero-order kinetics, at least during periods of the most active gas generation. It appears that factors such as the availability of moisture, nutrients, organisms, and substrate are limiting the amount of methane to be formed (or the loss of decomposable material) to a relatively constant value independent of time. It is interesting that many full-scale landfills where gas is recovered produce approximately the same amount of gas on a yearly basis. This suggests that the rate of change of gas generation rates over at least some parts of the landfill and over at least some portions of the total decomposition process is not so great as to suggest that zero-order kinetics is not working. Measurements from full-scale landfills are complicated by the effect of climatic variations plus the fact that refuse has been placed at different times, so there are many different refuse ages within the volume of refuse producing the gas. This makes developing a kinetic expression difficult and, in fact, the kinetic expression which best models the decomposition of a small volume of refuse within a landfill may be very different from the kinetic expression describing the rate of gas generation for the total landfill with time.

Table 4 presents the range of typical gas generation rates cited for different landfill situations in the literature, along with information gathered from experience. Gas rates from lysimeters vary widely, even when only lysimeters producing reasonable $CH_4$ concentrations are included, as was done in

**Table 4.** Total gas ($CO_2$ and $CH_4$) production rate from municipal refuse.

| Conditions | Gas production rate refuse per year[a] (l/kg) |
|---|---|
| Lysimeters, average production rate observed during periods of $CH_4$ generation | 0.001–30 |
| Pilot-scale or test landfills | 15–60 |
| Full-size landfills, field tests | 0.8–40 (typically 10–20) |

[a] Rounded to two digits and adjusted to 21% $H_2O$ (wet weight) and 53% $CH_4$.

compiling Table 4. Lysimeters producing methane at rates over a wide range are shown in Table 4. Further, lysimeter experiments tend to be of short duration compared to full-scale landfill decomposition periods, and the lag period prior to methane generation may be a significant portion of the total period of monitoring. The result is different gas rates depending on experimental conditions and the vagaries of methane generation in lysimeters, compounded by the changing rates throughout the experiment and the method of reporting the data. The rates in Table 4 are generally averages over the period of maximum methane generation, which are probably the most comparable to what might happen in an actual landfill.

Test landfills are small sections of larger landfills or separate areas specially designed for gas measurements. Gas rates are probably more realistic in that these test areas were subjected to landfill conditions, but gas measurements would be less accurate than with enclosed lysimeters. The values shown in Table 4 are averages over periods of active methane generation.

The most reliable data on gas generation rates for application to full-scale landfills are probably from measurements on full-scale landfills. Many landfills have been so tested by withdrawing gas and measuring the volume of the landfill influenced by the withdrawal system, as was discussed previously. Typical gas generation rates range from 10 to 20 l/y kg refuse. Since the gas composition during such measurements indicates active methanogenesis (typically 50–60% methane), and these are full-scale landfills subject to climatic variations, etc., the results should be comparable to other landfills in similar situations. Gas rates less than 10 or over 20 l/kg/y tend to be from landfills subjected to dry conditions which reduced the rate, or from landfills containing unusual wastes or subjected to large amounts of moisture.

Knowing the total amount of gas to be produced (Table 2) and the rate of gas production over some period of time (Table 4), it is possible to take a first-order model to project the gas rates for a landfill. It is common practice to make such a projection for design and feasibility studies for landfill gas utilization systems. A typical approach is to assume a lag phase or start-up period over which little methane is produced, followed by a period of active methane generation modelled with a first-order equation (equation 2) using assumed half-lives of refuse components of interest, such as paper, to compute the constant $k$. Each year's production of refuse is modelled separately, and the total gas production of each of the years of refuse placement in the landfill is computed on a yearly basis to give total landfill gas projections. The useful life of methane gas generation by such modelling appears to be 5–20 years, depending on the model used and the landfill under study. Note that if a landfill continues to take refuse, portions of the site older than 5–20 years would be of marginal interest for gas generation, but this would be off-set by gas production from new refuse so the total landfill production would remain constant (assuming constant yearly tonnage of refuse).

It is the authors' opinion that, during the period of most active methane generation, overall generation rates within each year's placement of refuse do not change substantially for most landfills. This is because of moisture limitations (largely uneven distribution of moisture) and the likelihood that most of the methane generation is concentrated within the landfill in small active sites, which change continually. If desired, a first-order build-up may be used to model the initial lag period, and a first-order drop-off may be used to model the decline in gas generation as portions of the landfill die off after the period of most active generation. The period of active gas generation has constant gas rates (zero order). Its duration appears to range from 5 years in warm, moist climates to 20 years in dry climates for sites which can be characterized as producing methane. This simple approach produces results consistent with the data in Tables 2 and 4 and the periods of active methane generation observed. For example, if a gas rate of 15 l/kg/y (Table 4) is measured and a total amount of gas of 150 l/kg is assumed (i.e. 30% of the total potential $CH_4 + CO_2$, or 60% of the gas produced in anaerobic digestion experiments, and in the conservative end of the range projected from full-sized landfills), a zero-order rate equation (no change in rate over the most active period) gives a life of 10 years. Given the inaccuracies in estimating the total gas to be produced, neglecting gas generated in the initial lag period and in the final die-off period provides a conservative projection consistent with available data.

## CONCLUSIONS

It is observed that factors affecting gas composition are known, numerous values exist from full-scale and test landfills, and these values are consistent with theoretical projections. There are complications which occur as a result of multiple stages of biological decomposition occurring simultaneously in any given landfill and from the effect of each given volume of refuse going through different stages of decomposition. These complications make it difficult to predict gas composition. These complications make it difficult to predict gas composition, especially over the first few years. However, once methanogenic decomposition has been obtained generally, throughout the portion of the landfill of interest, the data obtained correspond very well with what would be expected from theoretical considerations. The total quantity of gas to be obtained in a full-scale landfill is difficult to project from theoretical considerations alone. This is because competing reactions and resistant materials which will not necessarily form methane. Taking the various measurements into account from different types of experiments and comparing these values to total mass balances which are just now beginning to be published, it appears that the total amount of landfill gas to be formed will be in the range of 100–250 l/kg. Values within this range reflect the degree to which the refuse is decomposable, the lag phase prior to the onset and activity of methane formation during the decompositional life of a landfill. The predominant variables appear to be associated with the moisture conditions of the refuse and incorporate both moisture content and moisture flow through the refuse mass. Little data exists, however, regarding the total amount of methane to be formed in full-scale landfills. This is because of the difficulty in monitoring the methane generated by a landfill over its complete decompositional cycle.

The rate of gas generation from a landfill over at least parts of the decompositional life of a landfill is easier to obtain than is the total amount of gas to be generated. This is because it involves discrete measurements and not continuous monitoring over many years. Gas generation rate values have been obtained from lysimeters and full-scale landfills. These values are relatively consistent with each other. It may be expected that as more full-scale landfills are developed for gas withdrawal and utilization, data will become available which will further our ability to predict the rate and amount of gas generation at landfills. Unfortunately, much of this data is proprietary and is not given in the open literature. However, such numbers are beginning to be available and should be increasingly available in the future.

## REFERENCES

Barlaz, M.A., Milke, M.W. and Ham, R.K. (1987). 'Gas Production Parameters in Sanitary Landfill Simulators', *Waste Management and Research* 5.
Bookter, T.J. and Ham, R.K. (1982). 'Stabilization of Solid Waste in Landfills', *Journal of Environmental Engineering Div.*, ASCE, 108, EE6, 1089–1100.

Corey, R.C. (1969). 'Principles and Practices of Incineration'. Wiley-Interscience, New York.

Sawyer, C.N. and McCarty, P.L. (1978). 'Chemistry for Environmental Engineering', 3rd edn. McGraw-Hill, New York.

Tchobanoglous, G., Theisen, H. and Eliassen, R. (1977). 'Solid Wastes'. McGraw-Hill, New York.

Wertheim, E. (1956). 'Introducing Organic Chemistry', 3rd edn. McGraw-Hill, New York.

## 3.2 Landfill Gas Extraction

R. STEGMANN

*Technical University of Hamburg-Harburg, Eißendorfer Str. 40, 2100 Hamburg 90, W. Germany*

## INTRODUCTION

Landfill gas extraction with energy utilization has been carried out in West Germany for more than seven years. However, it has not been realised until recently that gas must be extracted from all landfills. In the opinion of the author each landfill should be equipped with a gas extraction system and the gas should, if possible, be used for the production of energy. Landfill gas should also be extracted from landfills that are still in operation. In this case horizontal gas extraction systems are available that can be installed during operation and have proven to be successful (Stegmann, 1985).

The necessity of early gas extraction is obvious from the point of view of emission minimization. Besides methane and carbon dioxide, organic trace components that are toxic, and in some cases carcinogenic, are emitted with the gas from the landfill. In addition explosive mixtures can build up.

So landfill gas extraction systems are as much a part of a landfill as for example the bottom liner and the leachate collection system.

## GAS PRODUCTION AND QUALITY RELATIVE TO TIME

The author estimates that the total gas production from municipal solid waste produced in West Germany is approximately 120–150 m$^3$ per ton of dry solids (Stegmann, 1982). These results are gained from municipal solid waste lysimeter tests. From this investigation it was also found that paper and cardboard result in higher gas production rates than sawdust.

167

**Table 1.** Concentration range of halogenated hydrocarbons in landfill gas (Müller and Rettenberger, 1987).

| Gas | Formula | Concentration range |
|---|---|---|
| Trichlorofluoromethane | $CCl_3F$ | 1–84 |
| Dichlorodifluoromethane | $CCl_2F_2$ | 4–119 |
| Chlorotrifluoromethane | $CClF_3$ | 0–10 |
| Dichloromethane | $CH_2Cl_2$ | 0–6 |
| Trichloromethane | $CHCl_3$ | 0–2 |
| Tetrachloromethane | $CCl_4$ | 0–0.6 |
| 1,1,1-Trichloroethane | $C_2H_3Cl_3$ | 0.5–4 |
| Chloroethane | $C_2H_3Cl$ | 0–264 |
| Dichloroethene | $C_2H_2Cl_2$ | 0–294 |
| Trichloroethene | $C_2HCl_3$ | 0–182 |
| Tetrachloroethene | $C_2Cl_4$ | 0.1–142 |
| Chlorobenzene | $C_6H_5Cl$ | 0–0.2 |

The methane content of landfill gas is somewhere between 55 and 65% with the rest mainly $CO_2$. The proportion of air in landfill gas is dependent upon the type of gas extraction operation and is caused by air being sucked into the landfill. In addition due to leaks in the gas extraction pipes, valves and tubings, oxygen and nitrogen can be measured. Trace organics in landfill gas have also to be taken into account. Besides $H_2S$ and the fluorinated chlorinated hydrocarbons, the chlorinated hydrocarbons and aromatics may be relevant (Table 1). In addition metals at low concentrations have been detected in landfill gas (Young and Parker, 1983; Rettenberger, 1985). Trace organics are relevant with regard to emission from landfills into the atmosphere, as well as corrosion problems during landfill gas utilization (Dernbach, 1984). In addition, the possibly harmful effects of some trace organics as well as their behaviour during incineration have to be considered.

It is still very difficult to predict the total period of landfill gas production. In most cases the time period during which landfill gas could be utilized is calculated as up to ten years after the landfill is closed; keeping this number in mind it has to be noted that the gas production is initially higher and decreases with time, obeying a nonlinear equation against zero (Tabasaran and Rettenberger, 1987; Stegmann, 1978/79).

It is common to estimate gas production through time by using the following equation:

$$G_t = G_e(1 - e^{-kt})$$

$G_t$ = cumulative gas production with time $t \, (m^3/t \, MSW)$
$G_e$ = total gas production $(m^3/t \, MSW, \text{dry solids})$     (1)
$t$ = time in years
$k$ = 2/H
$H$ = half life, in years

In lab-scale test facilities by far the main biological degradation has taken place within a period of 6–9 months; in this case the optimum milieu conditions for biological processes were present (Stegmann, 1982).

## BASIC DESIGN OF LANDFILL GAS EXTRACTION SYSTEMS

Landfill gas is extracted out of the landfill by means of gas wells (Fig. 1). These gas wells are normally drilled by an auger (diameter > 50 cm) or rammed into the landfill at spaces of between 40–70 m (Stegmann *et al.*, 1987). In addition, or as an alternative, horizontal systems (Fig. 2) can be installed during the operation of the landfill (Stegmann, 1988). The gas wells consist mainly of perforated plastic pipes (HDPE) surrounded by coarse gravel and are connected with the horizontal gas transportation pipes by means of flexible tubings (Fig. 3). A vacuum is necessary for gas extraction and transportation and is created by means of a blower that is, in most cases, installed in a building at the border of the landfill.

The most important influencing factors concerning planning and construction of landfill gas extraction systems are:

– settling
– water tables in landfills
– condensate
– gas quality

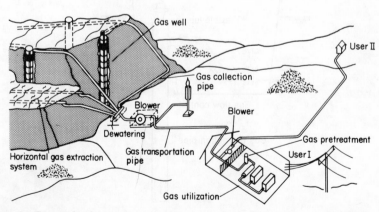

**Figure 1.** General scheme for a gas extraction and utilization plant (Müller and Rettenberger, 1987).

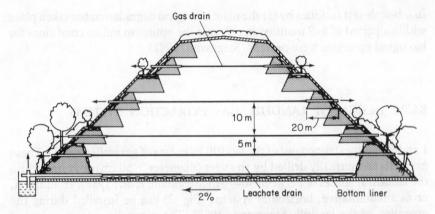

**Figure 2.** Possible design of a horizontal gas extraction system (Boll *et al.*, 1988).

The amount of the settling to be expected is mainly dependent upon the degree of compaction during the implacement of the waste. Wiemer (1982) has reported about the settling behaviour of landfills by means of data both from the literature and from his own measurements. Based on this work it can be expected that for example at a 20 m high refuse mound the initial average density (based on wet weight) of about $\gamma = 0.8 \, t/m^3$ can be increased to about $1.0 \, t/m^3$, if the waste load increases by 10 m of refuse. This means

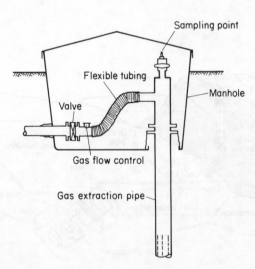

**Figure 3.** Gaswell head.

that, in this case, where a relatively high implacement density could be achieved, a total settling of about 2–3 m can be expected. In this regard it has to be remembered that severe settling differences in the landfill often take place, mainly caused by different local implacement densities and different waste loads on top.

Concerning the planning of gas extraction systems it ought to be noted that, for the above-mentioned reasons, the horizontal gas tranportation pipes and the connecting tubings to the vertical gas wells are especially endangered. In order to avoid pipes breaking, the flexible tubings built out of non-corrosive plastic material should be chosen (Fig. 3). Severe settling in the area of the transporation pipes can, in specific cases, result in the breaking of the pipes; in most cases, where the pipes have settled locally, condensate will accumulate, and that will result in pressure loss and fluctuations and in severe cases may cause the clogging of the whole pipe. For this reason the following design criteria are recommended for the emplacement of landfill gas pipes:

1. Emplacement of the pipes with relatively steep slopes in the same direction of the gas flow (if condensate flow and gas flow have different directions, the pipes must have larger diameters, in order to achieve low gas velocities).
2. Gas transportation pipes should be installed mainly in virgin soil; this means that the amount of pipes placed in the landfill should be minimized.
3. If necessary gas transportation pipes with low slopes in the landfill site may be supported by steel or wood constructions underneath the pipes in order to avoid local settling. Overall settling cannot be avoided by those means (Stegmann and Dernbach, 1982).
4. Settling of pipes can be supervised by connecting steel rods at certain distances directly to the pipe, so that they reach the landfill surface. By means of periodically surveying the pipe length above the landfill settlings can be supervised.
5. Dewatering installations to remove condensate from the pipes have to be installed in the landfill area and outside.
6. The slope of the gas transportation pipes in the landfill area should reach at least 2%; steeper slopes should be chosen if possible.

From many landfills—including landfill mounds—high watertables inside the landfills have been reported. These watertables, that may vary within short distances, may result in filling up the gas wells with leachate, so that gas extraction becomes impossible. This means that, in most cases, deep vertical gas wells down to the bottom of the landfill are not necessary since they may fill up with water. It should always be kept in mind that installations to collect gas from landfills (e.g. gas wells) also have the effect of dewatering the landfill. For this reason installations to encourage water percolation out of the gas wells should be constructed. The wells should be built so that pumping-off of water by means of submersible pumps is made possible. These measures have not always been successful since the wells can be emptied in a very short time, whilst water migrates into the wells of low velocities. For this reason only low efficiencies can be achieved.

If horizontal landfill gas extraction systems are used, it must be remembered that these systems can be filled partially or completely with water. For this reason horizontal ditches that are filled with coarse gravel should be put in place with relatively high slopes. In addition every 20–30 m drainage areas should be installed. Those areas are about the size of $1\,m^3$, also filled with coarse material and are connected to the drainage ditches. Since there are no long-term experiences with this system these ways of construction might be modified in the future (Boll *et al.*, 1988).

If transportation pipes are put into place on the landfill site the condensate should be able to drain out of the pipes back into the landfill by means of syphons. It is important to take care that the condensate has enough room to migrate back into the waste. In many cases the condensate is led back into the vertical gas wells. The condensate also has to be removed from the transportation pipes outside the landfill. In this case the system becomes more complicated and in general the condensate has to be removed and treated separately.

If landfill gas has to be transported over long distances the possibility of dewatering the gas by cooling should be considered. The dewatering of the gas has a positive effect on the lifetime of the machines in contact with it. In addition dried gas has lower flow resistance in the pipes than wet gas, which often results in small drops developing on the inside of the pipes which causes higher flow resistance.

Experiences with landfill gases show that they are very corrosive; this has to be considered when materials for pipes and engines are selected. HDPE–material and carbon steel have proven adequate; but other materials are often used. The aggressiveness of the gas has also to be taken into account when sealings get into contact with the gas; rubber and silicone should not be used. In selecting the appropriate materials the gas analyses from the literature should be considered.

## FLARING OF LANDFILL GAS

The extracted gas should not be discharged into the air (Stegmann, 1985). If landfill gas is not utilized, it has to be burnt. In West Germany there is discussion as to whether a complete incineration of the gas components in flares is feasible; if no complete incineration takes place, based on theoretical reflections, the production of dioxins and furans cannot be excluded. Investigations on the emissions of flares that have been operated at different temperatures (900°C and 1200°C) showed no significant dioxin emission (Engel and Rettenberger, 1988). However, utilization of landfill gas for energy production should be investigated. The costs for sophisticated flares may be better invested in gas utilization equipment.

## CONCLUSIONS AND FURTHER DEVELOPMENTS

In the opinion of the author, landfill gas extraction should be practiced at each landfill site in order to minimize the emission of gaseous compounds. Landfill gas should be used for the production of energy; this should be practiced even if no profits are foreseen. The demands, developments and problematics of particular relevance if landfill gas extraction is practiced, are mentioned briefly in the following paragraphs:

1. The procedures for landfill gas extraction should be simplified (for example avoidance of condensate problems by means of increasing gas velocity in the pipes; simpler gas well constructions).
2. The question of emissions will be of more significance in the future; as a result new technical requests may come up.
3. Landfill capping by means of clay or plastic material after the landfill is closed will result in higher gas extraction rates.
4. In the future also longer gas transportation pipes will be constructed outside the landfill, in order to increase the possibilities of landfill gas utilization.
5. New landfill techniques in order to enhance biochemical degradation processes in the landfill will be practiced. As a result landfill extraction has to be practiced during operation. This should also be practiced at conventional landfills.
6. The quality of incineration of landfill gas in flares and burners will be improved.

These points do not cover all the new developments that can be expected in the future. The interest that can be found today in the field of landfill gas extraction and utilization indicates that new developments and modified procedures will come up in the future. Those modifications may also be necessary due to more stringent requests for emission and safety standards.

## REFERENCES

Boll, F.W., Doedens, H., Hebbelmann, H., Schlüter, U. and Weber, B. (1988). 'Auswirkungen der aktiven Entgasung auf die Möglichkeit der Rekultivierung von Hochdeponien'. *Müll und Abfall*, Heft 3.

Dernbach, H. (1984). 'Deponiegasnutzung in Braunschweig—Gasqualität und Korrosionsfragen—Deponiegasnutzung'. Dokumentation einer Fachtagung, Ed. BMFT, Projektträger Umweltbundesamt.

Engel, H. and Rettenberger, G. (1988). 'Untersuchungsergebnisse von Verbrennungsversuchen mit unterschiedlichen Fackeltypen', *Hamburger Berichte* 1, Abfallwirtschaft, Economica–Verlag, Bonn.

Müller, K. and Rettenberger, G. (1987). 'Sicherheitstechnische Aspekte von Entgasungs- und Gasverwertungsanlagen', *Beinhefte zu Müll und Abfall*, Heft 26, Erich Schmidt Verlag.

Rettenberger, G. (1985). 'Trace Compounds in Landfill Gas'. Proceedings from the GRCDA 8th International Landfill Gas Symposium, April 9–11, 1985, San Antonio, Texas, Ed. GRCDA, GLFG-12.

Stegmann, R. (1978/79). 'Gase aus geordneten Deponien', *ISWA-Journal* 26/27.

Stegmann, R. and Dernbach, H. (1982). 'Deponieentgasung mittels Gasbrunnen zum Zwecke optimaler Gasnutzung am Beispiel der Deponie Braunschweig', *Beihefte zu Müll und Abfall*, Heft 19.

Stegmann, R. (1982). 'Erbegnisse von Abbauversuchen im Laborversuch, Gas und Wasserhaushalt von Mülldeponien', *Veröffentlichung des Instituts für Stadtbauwesen*, TU Braunschweig, Heft 33, Eigenverlag.

Stegmann, R. (1985). 'Deponieentgasung und Gasnutzung', *Kommunalwirtschaft*, Heft 4.

Stegmann, R., Heidenreich, E., Grosstück, W. and v. Borck, K. (1987). 'Entgasung der Deponie Georgswerder', *Beihefte zu Müll und Abfall*, Heft 26, Erich Schmidt Verlag.

Stegmann, R. (1988). 'Kriterien zur Auswahl und Dimensionierung des geeigneten Entgasungssystems unter Berücksichtigung des Deponiealters, der -geometrie, der -oberflächenverdichtung und des -wasserstandes'. *Hamburger Berichte Abfallwirtschaft*, Heft 1, Economica–Verlag, Bonn.

Tabasaran, O. and Rettenberger, G. (1987). 'Grundlagen zur Planung von Entgasungsanlagen. *In* Kumpf, Maas, Straub; Müll—und Abfallbeseitigung'. Loseblattsammlung, Erich Schmidt Verlag.

Wiemer, K. (1982). 'Qualitative und quantitative Kriterien zur Bestimmung der Dichte von Abfällen in geordneten Deponien'. Dissertation an der TU Berlin.

Young, P.J. and Parker, A. (1983). 'The identification and possible environmental impact of trace gases and vapours in landfill gas'. *Waste Management and Research*, 1, No. 3, Academic Press.

## 3.3  Landfill Gas Utilization

R. STEGMANN

*Technical University Hamburg-Harburg, Eißendorfer Str. 40,*
*2100 Hamburg 90, W. Germany*

### INTRODUCTION

In order to achieve emission control, to avoid gas migration into the near surroundings of the landfill and uncontrolled migration through the landfill surface, landfill gas (LFG) has to be collected and either be flared or utilized. Hazards from LFG may result from:

1. Methane, which is supposed to participate in destroying the ozone layer of the atmosphere.
2. Trace components in LFG that are toxic.
3. Explosions if LFG should accumulate, for example in manholes in or around a landfill. LFG can also migrate into houses near the landfill (severe accidents have already occurred).
4. The inhibition of plant growth on or around the landfill.

On the other hand about 120–150 m$^3$ LFG/t dry municipal solid waste (MSW) are produced with a high heating value of about 5.9 kWh/m$^3$ so that LFG utilization should be practised in order to save fossil fuels. To optimize LFG collection the biological processes in a landfill should be enhanced, so that gas production takes place over a shorter period of time (Stegmann and Spendlin, 1987). In addition LFG should be extracted as soon as possible (during the operation phase of the landfill), so that high gas yields can be achieved (the highest gas production takes place in the first two years after disposal).

### POSSIBILITIES AND TRENDS OF LFG–UTILIZATION

In West Germany two main areas have been established where LFG is utilized.

175

(i)  Production of electricity and heat in internal combustion engines with heat recovery.

(ii) Using LFG for direct heat production in industry or for heating.

In industry LFG is, for example, used to produce steam in a knacker's yard (Schneider, R., 1988) or to heat the oven in brick kilns (Sperl, 1988). LFG is upgraded to pipeline quality by the removal of $CO_2$ and trace organics (Snyder, 1984). There is also one example in Austria where LFG is used for the heating of private houses (Tscherner, 1985); district heating is practised in Sweden.

## PRODUCTION OF ELECTRICITY AND HEAT IN INTERNAL COMBUSTION ENGINES

Production of electricity in internal combustion engines is practised in West Germany at about 35 landfill sites; 8 more are in the planning stage. The oldest plants have now been running over a period of 8 years. The utilization of the heat produced by the cooling water of the engines is not possible at all the plants. Experiences show that this kind of LFG–utilization is a proven option and can be recommended; of course a number of specific points have to be respected. Electricity can be either used by the producer or sold to the public utility company. Small systems that are installed in containers can be bought from the shelf.

Landfill gas is very corrosive and water saturated. This has to be borne in mind when the materials for gas pipes, valves and seals are chosen. In addition, trace components in the LFG may be converted in combustion engines into strong acids that may cause corrosion (Reinicke, 1988; Dernbach, 1985). Monitoring programs show that in municipal solid waste (MSW) fluorinated hydrocarbons and chlorinated hydrocarbons (CHC) may also be present, and their concentrations may even rise to about $200 \, mg/Nm^3$ (Laugwitz *et al.*, 1988). These concentrations decrease progressively, so that at landfills older than 3 years the concentrations are in general $< 50 \, ppm$. The halogenated hydrocarbons may be converted into a strong acid in the combustion cylinders. The first problem occurred at the sanitary landfill at Braunschweig (co-disposal of industrial and municipal waste) where LFG is used in internal combustion engines and one engine broke completely (Dernbach, 1985). At that time there was no information about the concentrations of halogenated hydrocarbons available and only the $H_2S$ concentrations were considered with regard to corrosion ($H_2S$ can be converted in the engine to $H_2SO_4$). After this event the problem of the trace organics became obvious, since in the landfill gas at Braunschweig approximately $700–1000 \, mg/Nm^3$ of halogenated hydrocarbons have been detected.

This problem can be solved when specific oils developed for LFG-engines are used and the oil is measured for metal content and pH as well as acidity on a routine basis; oil exchange rates should be higher at the beginning of the operation and may be extended due to experience (Reinicke, 1988). In addition the total oil volume of the engines should be enlarged.

These measures are common practice for the operators of internal combustion engines run with LFG from MSW landfills. When the above-mentioned measures are taken the utilization of LFG in internal combustion engines is a routine proven operation process which can be recommended.

As emission control becomes more significant in West Germany, in order to meet the standards for $NO_x$ and CO (500 mg/Nm$^3$ resp. 650 mg/Nm$^3$ if engines produce more than 1 MW electricity and heat), the tendency is towards the utilization of gas internal combustion engines where the engine is run with excess air. These engines are in most cases turbocharged. The lean gas engines meet the above-mentioned emission standards without using a catalyst. This type of engine is commonly used today (Reinicke, 1988).

## PRETREATMENT OF LANDFILL GAS

For LFG with higher concentrations of halogenated hydrocarbons (due to a specific waste composition) there are also tests underway (in a very few cases these are already practised) to use activated carbon columns or organic solvents to remove the trace organics before LFG is fed into the internal combustion engines. Since the activated carbon is loaded after 12–24 hours, the desorption of the adsorbed halogenated hydrocarbons has to occur at this frequency. In general the desorbed halogenated hydrocarbons are concentrated by means of cooling. Due to the loss in effectiveness with each desorption cycle and the high price of activated carbon, costs are high. In addition the pretreatment becomes even more ineffective if low concentrations of CHC < 50 ppm have to be treated, since the maximum loading rate of the activated carbon is in this case also lower. Two full scale facilities are in operation in West Germany, where LFG is pretreated (Kewitz, 1988; Schneider, J., 1988). Practice will show the economics of this system.

Pretreatment may also be necessary if high $H_2S$ concentrations are present. Treatment has been practised successfully in test facilities by using activated carbon and iron compounds (Mollweide and Rettenberger, 1988).

Pretreatment should only be considered in specific situations as it is well known that the halogenated hydrocarbon concentrations decrease with the time of active gas extraction and in some cases the concentrations have decreased dramatically after the pretreatment plant was built. An interesting way of dealing with those high concentrations in the early phase of gas extraction is the installation of two parallel transportation gas pipes, one connected to those wells with high, and the other to those with low, trace component concentrations. As long as high concentrations ($>100 \, mg/m^3$) are extracted, this gas portion should be flared, whereas the other part—the 'clean gas'—can be used.

The economics of LFG–utilization in internal combustion engines depend very much on the specific situation. The overall efficiency of the conversion of LFG into electricity is about 33%; this can be increased to about 80–85% if the heat from the cooling water is also used. On the other hand the production of electricity can, in general, occur at most of the landfills; on the contrary, direct heat utilization is not possible that often. If the produced electricity can be used by the producer himself, the economics are usually satisfactory; if the electricity is sold to the public utility company lower incomes are generally obtained.

## OTHER WAYS OF LANDFILL GAS UTILIZATION

The efficiency of using LFG for heat production (for either buildings or industry) would increase to 85–90%, if it were possible to use the LFG over the total period of a year. This is not possible if LFG is used for heating; in addition utilization over the entire year isn't often possible even for industry. Of course the specific gas emission standards have to be respected.

In the USA, at huge landfills, LFG is treated to pipeline quality using $CO_2$–absorption processes. There are quite a few plants in operation where 'Selexol' is mainly used for the absorption of $CO_2$. These processes are expensive and only economical if landfill gas from comparatively young landfills of $>10$ Million tons is used.

In the USA and Switzerland membranes are also used for the separation of $CO_2$ and $CH_4$ (Gandolla and Simonet, 1985). Due to these experiences, this process is already operative, although it has to be stated that until today this process has no multiplication factor in practice. It is not clear until now, why this is the case. It might be the price of the membranes and the presence of clorinated hydrocarbons.

In West Germany the separation of $CO_2$ and $CH_4$ is tested in activated carbon columns using the differential pressure method. This process is still in the test phase. Because of the relatively high investment and operation costs for $CH_4$ and $CO_2$ separation, this process may only be valid at very large landfills, but this is not very often the case in Europe.

In addition gas turbines are in operation for bigger landfills in the USA. The advantage may be that this method creates less problems with chlorinated hydrocarbons and is characterized by a simpler technique. On the other hand, the electrical efficiency is lower when compared with internal combustion engines. Experience and economics will show in the future if this is a real alternative for the internal combustion engines under European conditions.

Another possibility is the production of steam through LFG, where the steam runs a turbine which generates electricity. This system would operate at a low electrical efficiency, but it could be economical if the excess heat could be utilized. This development has to be seen under the problem of the existence of chlorinated and fluorinated hydrocarbons in the LFG. Using the steam to run a turbine the problem of corrosion in internal combustion engines is not so critical.

A new development is in the test phase in Austria: LFG is upgraded to pipeline quality (ca. 95% $CH_4$) using a molecular sieve. The gas is pressurized to 200 bar and then used to feed the engine of a compactor at the same landfill (Tscherner, 1988).

## SITUATION OF LFG–UTILIZATION IN WEST GERMANY

LFG is used exclusively for electricity, heat and steam production—in some cases together with other fuels. A pretreatment of the LFG is practised in only one case (Kewitz, 1988). In general the LFG–condensate is removed and the LFG is filtered. In order to avoid corrosion the LFG is often preheated before it is injected into a specifically designed burner. The uniform combustion should be practised at high temperatures and detention times. In addition installations for explosion control have to be made. In West Germany about 30 LFG–utilization plants are in operation, where heat is utilized, whereas in about 43 plants electricity is produced and in most cases sold to public utility companies (Rosenbusch, 1988). Many new plants are planned for both electricity and heat production.

Economics are not discussed in this paper, because the basic concepts are different in each country and the price of the purchased heat and electricity is also different. In addition specific regulations must often be respected in each country. If electricity is used for own purposes, the money saved from the otherwise purchased electricity can be considered as an income; of course stand-by electricity has to be purchased when the LFG–engine is out of operation. If the produced electricity is sold to the electric company income is gained directly. There is a law in West Germany stating that the public utility companies have to accept LFG utilization for energy production. More and more companies run LFG plants themselves.

180                                                              *Stegmann*

## SAFETY ASPECTS

If methane concentrations in the atmosphere rise to between 5–12%, explosive mixtures are present. This is also true if the oxygen concentration in LFG is greater than 11% vol. (i.e. if this mixture is ignited an explosion occurs). Precautions therefore have to be installed so that explosions can be excluded. The best way to do this is to avoid the build-up of explosive mixtures which can be done by measuring the gas quality in different sections of the plant; if set concentrations are exceeded an alarm starts or the plant is shut down. Further installations may have to be made depending on the specific situation.

In West Germany a handbook has been developed to help planners with the design and operation of safety installation in LFG plants (Müller and Rettenberger, 1986). In order to show how far these safety installations have to be carried out, Fig. 1 shows an example of a typical plant.

## CONCLUSIONS

Biogas utilization in West Germany is a standard technology and is practised at more than 60 landfills. About half of the utilization rate is for electricity production, while in the other plants LFG is used directly for heat production. These technologies have proven controllable and the technical installations are adapted to LFG. Further development is underway especially in the area of gas engines, where turbocharged lean gas internal combustion engines are used today. This is due to an improvement in exhaust gas emission concentration.

The discussion about the organic trace components and their effect on gas utilization and flare emissions is ongoing in West Germany. Flares are constructed to have more controlled burning and higher temperatures and detention times of the gas in the burning area; the reason for these developments is due to the theoretical possibility of the production of dioxines during incineration. In fact first measurements at flares have not detected PCDD and PCDF–concentrations.

Gas purification before utilization is already practised, but results are not yet available. Research shows that the elimination of organic trace components using activated carbon is possible but expensive.

The separation of $CO_2$ from $CH_4$ to improve pipeline gas quality is possible, but it will only be economical at very large landfills; at full-scale this technique is only practised in the USA.

Landfill gas extraction and utilization should be practised where possible at each landfill, including those where no revenue from energy utilization would be made.

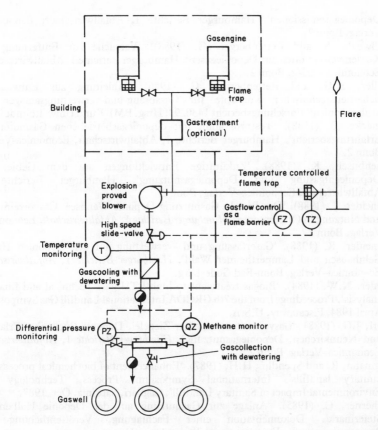

**Figure 1.** Example of a safety concept for a landfill gas utilization plant (Müller and Rettenberger, 1986).

## REFERENCES

Dernbach, H. (1985). 'Korrosionsprobleme beim Betrieb der Blockheizkraftwerke auf der Deponie Braunschweig'. Dokumentation einer Fachtagung, Veröffentlichung des Bundesministers für Forschung und Technologie (Hrsg.).

Gandolla, M. and Simonet, R.A. (1985). 'Einsatz der Membrantechnologie zur Gastrennung. BMFT-Forschungsreihe, Deponiegasnutzung'. Dokumentation einer Fachtagung 1984, Hrsg. Umweltbundesamt.

Kewitz, J.J. (1988). 'Gasvorreinigung und—nutzung in Gasottomotoren auf der Deponie Kapiteltal/Kaiserslautern'. Hamburger Berichte 1, Abfallwirtschaft, Economica-Verlag, Bonn 2.

Laugwitz, R., Poller, T. and Stegmann, R. (1988). 'Entstehen und Verhalten von Spurenstoffen im Deponiegas sowie umweltrelevante Auswirkungen von

Deponiegasemissionen'. Hamburger Berichte 1, Abfallwirtschaft, Economica-Verlag, Bonn 2.

Mollweide, S. and Rettenberger, G. (1988). 'Versuche zur Entfernung von Kohlenwasserstoffen aus Deponiegasen'. Hamburger Berichte 1, Abfallwirtaschaft, Economica-Verlag, Bonn 2.

Müller, K.G. and Rettenberger, G. (1986). 'Anleitung zur Entwicklung sicherheitstechnischer Konzepte für Gasabsaug-und-verwertungsanlagen aus Mülldeponien'. Forschungsbericht 1430293, Hrsg. BMFT und Umweltbundesamt.

Reinecke, B. (1988). 'Testprogramm mit deponiegasbetriebenen Gasmotoren— Erfahrungsbericht'. Hamburger Berichte 1, Abfallwirtschaft, Economica-Verlag, Bonn 2.

Rosenbusch, K. (1988). 'Zukünftige Entwicklungen auf dem Gebiet der Deponieentgasung und Deponiegasnutzung'. Hamburger Berichte 1, Abfallwirtschaft, Economica–Verlag, Bonn 2.

Schneider, J. (1988). 'Das Projekt Gasnutzung Deponie Wannsee. Gasvorreinigung und Nutzung in Ottomotoren'. *Hamburger Berichte 1, Abfallwirtschaft*, Economica–Verlag, Bonn 2.

Schneider, R. (1988). 'Gaserfassung und -verwertung auf den Deponien Hailer-Gelnhausen und Lampertheimer Wald'. *Hamburger Berichte 1, Abfallwirtschaft*, Economica–Verlag, Bonn-Bad Godesberg.

Snyder, N.W. (1984). 'Biogas treatment to high-BTU gas. Technical and financial analysis'. Proceedings from the 7th GRCDA International Landfill Gas Symposium, April 1984, Piscataway, U.S.A.

Sperl, J.G. (1988). 'Gasverwertung in einer Ziegelei, Deponie Buckenhof/Erlangen und Neunkirchen. Deponiegasnutzung'. *Hamburger Berichte 1, Abfallwirtschaft*, Economica–Verlag, Bonn 2.

Stegmann, R. and Spendlin, H.H. (1987). 'Enhancement of biochemical processes in sanitary landfills'. International Symposium 'Process, Technology and Environmental Impact of Sanitary Landfill', Cagliari, Sardinia, Oct. 1987.

Tscherner, C. (1985). 'Anlage zur Gasnutzung auf der Deponie Halbenrain/ Steiermark'. Dokumentation einer Fachtagung, Veröffentlichung des Bundesministers für Forschung und Technologie (Hrsg.).

Tscherner, C. (1988), personal communication.

# 4.  LEACHATE

# 4.1 Landfill Hydrology and Leachate Production

R. CANZIANI* and R. COSSU**

*Institute of Sanitary Engineering, Polytechnic of Milan, Via Fratelli
Gorlini 1, 20151 Milano, Italy
**Institute of Hydraulics, University of Cagliari, Piazza d'Armi,
09100 Cagliari, Italy

## INTRODUCTION

The effect of uncontrolled infiltration of leachate into the environment is the
biggest environmental impact that a sanitary landfill can have. For this reason
legislation, at both the local or national level, tends to define methods to
restrict leachate contamination of either surface or ground waters. More
frequently landfills are sited in areas that are naturally impermeable or areas
which have been rendered impermeable by artificial means. It is evident that,
under these conditions, the management of leachate during its percolation,
collection and disposal assumes a critical part in the design of the landfill both
from the technical and from the economic standpoint. In this chapter, factors
governing the production of leachate in controlled landfill sites are examined
and discussed.

## HYDROLOGICAL BALANCE AND LEACHATE PRODUCTION

Figure 1 gives an overall summary of the components which make up the
hydrological balance of a controlled landfill (Cossu, 1982). It is immediately
obvious that whatever the methodology adopted for the design of the landfill,
only some of the components of the hydrological balance can be influenced.

The principal factors governing the formation of leachate are as follows (Lu *et al.*, 1981):

1. *water availability*-rainfall, presence of surface water, water content of sludges, recirculation of leachate or irrigation of final cover;
2. *characteristics of final cover*-type of soil and vegetation, presence of impermeable cover material, slopes or other topographical characteristics;
3. *characteristics of tipped waste*-density, tipping methods, moisture content of waste when landfilled;
4. *method of impermeabilization*-of the site and/or soil characteristics of site.

Results of studies conducted in West Germany (Ehrig, 1983) on numerous controlled sites in different phases of operation are reported in Table 1.

From this it can be immediately seen that the effect of compaction in reducing the amount of leachate produced is significant. In fact it is 25–50% of precipitation for wastes compacted by crawler tractors and only 15–25% of precipitation for wastes compacted by steel wheeled compactors. It can also be seen that the final cover (depth and soil type are not well specified) has an effect of reducing leachate formation only in sites that are not well compacted.

The maximum seasonal peaks of leachate production are always observed at the end of winter and in spring emphasizing, indirectly, the significance of evapotranspiration in reducing leachate production during summer months.

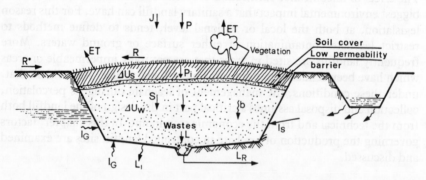

**Figure 1.** Schematic of the general hydrological balance in a completed sanitary landfill with leachate drainage system. P: precipitation; J: irrigation or leachate recirculation; R: surface runoff; R*: runoff from external areas; ET: actual evapotranspiration; $P_i = P + J + R^* - R - ET \pm \Delta U_S$; $U_S$: water content in soil; $U_w$: water content in wastes; S: water added by sludge disposal; b: water production (if > 0) or consumption (if < 0) caused by the biological degradation of organic matter; $I_S$, $I_G$: water from natural aquifers; $P_1 = P_i : S + I_S + I_G$;     $L = P_1 \pm \Delta U_w + b$ : total     leachate     production; $L_I$ = infiltration into aquifers; $L_R$: leachate collected by drains.

**Table 1.** Rainfall and leachate production in some sanitary landfills in the Federal Republic of Germany (from Ehrig, 1983).

| Landfill | Precipitation (mm year$^{-1}$) | (%)* | Leachate (m$^3$ ha$^{-1}$ day$^{-1}$) | Remarks |
|---|---|---|---|---|
| *Compaction with steel wheel compactors* | | | | |
| 1 | 652 | 15.1 | 2.7 | |
| 2 | 651–998 | 12.2–29.8 | 3.2–8.1 | Covered and planted: last 1.5 years flow increases |
| 3 | 651–998 | 16.9–21.6 | 3.0–5.9 | |
| 4 | 632 | 16.3–18.3 | 2.8–3.2 | |
| 5 | 509 | 16.8 | 2.3 | |
| 6 | 556–1057 | 15.6–19.6 | 2.6–5.1 | |
| 7 | 770 | 3.3–7.2 | 0.7–1.1 | Very young landfill |
| 8 | — | 22 | 3.8 | Covered with loam: leachate recycle |
| 9 | — | 38 | 6.7 | Leachate recycle |
| *Compaction with crawler tractor* | | | | |
| 10 | 571 | 31.3 | 4.9 | |
| 10 | 571 | 4.4 | 0.4 | Covered with loam |
| 11 | 501–729 | 25–48.2 | 5.3–8.3 | |
| 12 | 662 | 58.2 | 10.6 | |
| 13 | 632 | 32.3 | 5.9 | Covered and planted |
| 14 | 565–655 | 39.2–42.0 | 6.1–7.5 | |
| 15 | 636 | 19.9–21.4 | 3.5–3.7 | |

* Percentage of precipitation.

Research conducted in England on pilot-scale cells, each 50 × 25 m and at a depth of 3.5 m, located in old clay workings, produced data shown in Table 2 (Campbell, 1982). Contrary to what has been reported above, the production of leachate was shown to be more dependent upon site slope and thickness of cover material than on density of compacted waste.

## CALCULATION OF HYDROLOGICAL BALANCE

In the following section each of the component parts of the hydrological balance (see Fig. 1) are examined together with details of how each component can be calculated.

Canziani and Cossu

**Table 2.** Production of leachate (expressed in percentage rainfall) measured in experimental cells at Stewartby, England. (Dimensions of each cell $50 \times 25 \times 3.5$ m). From Campbell (1982).

| Cell number | | 1 | 2 | 3 | 4 | 5 | 6[b] |
|---|---|---|---|---|---|---|---|
| Initial waste density $(kg/m^3)$ | | 670 | 950 | 650 | na[a] | 1010 | 790 |
| Thickness (t) and slope (s) of the clay cover material | t (cm) = | The same but not specified | | 30 | 30 | 50 | 50 |
| | s (%) = | 0 | 0 | 0 | 0 | 2[c] | 0[d] |
| Production of leachate (yearly average, as % of rainfall) | | Two years after closure | | One year after closure | | | |
| | | 57.9 | 57.3 | 60.4 | 39.6 | 24.3 | 42.0[e] |

[a] Not estimated with adequate precision.
[b] Cell maintained with a water content equal to its field capacity.
[c] Cover shaped to shed rainwater towards surrounding land.
[d] Cover shaped to hold rainwater within perimeter of cell.
[e] During first year 75% was registered. The subsequent reduction is explained by the natural consolidation of the clay cover. Immediately after application this cover developed fissures during dry periods.

## Rainfall

This represents the largest single contribution to the production of leachate. As with all cases of infiltration, the most critical situation occurs during periods of light rainfall over a long lapse of time; short bursts of heavy rainfall during a storm result in a quick saturation of the cover material with the result that the remainder is shed as run-off, so there is little net infiltration.

To complete calculations on the hydrological balance all forms of rainfall have to be considered and the run-off calculated as detailed in the following paragraphs. In this context the effect of thawing snow has not been well defined in the literature (Lu *et al.*, 1982) and, for the present, the traditional criteria of 1 cm fresh snow equals 1 mm effective rainfall should still be used.

Data should be preferably obtained from measuring apparatus on site or alternatively from the nearest meteorological station. For order of magnitude calculations, for example for feasibility studies, national hydrographic maps can be used.

## Surface Run-off

The most important factors which influence the run-off of rainwater, either towards the internal area of the landfill or towards the land surrounding the landfill, are the topography, the morphology of the soil, type of soil cover, vegetation, permeability of the soil and the drainage systems installed. Included in the term topography is the slope of the site, the form and configuration of the surface of the site (roughness etc).

Estimation of run-off can be undertaken by various methods. The two most significant are reported below.

## Rational Method

This is based on the general formula $R = c \times P$, where $R$ and $P$ are the run-off and rainfall respectively (each expressed in the same units, for example mm/day or mm/month) and $c$ is an empirical coefficient which varies in function with the nature of the soil, the slope of the site, the presence and type of vegetation.

There are numerous papers detailing values for $c$ but, for the major part, these refer to the peak flow during short intense rainfall. It is obvious that this leads to an overestimation of the run-off from the site if monthly rainfall figures are used. Similarly, if only peak rainfall data is used, run-off during long periods of light rain would not be calculated, though these waters can be significant.

Values of $c$ adopted for the evaluation of the hydrological balance around a landfill site have been proposed by Salvato *et al.* 1971 (in Lu *et al.*, 1982): see Table 3.

Values obtained from Table 3 can be used for an order of magnitude calculation in which the moisture content of the soil, the duration and intensity of rainfall (the type of table assumes that coefficients refer to global annual values) and the roughness of the terrain are not considered. Similar data are reported in other papers (Cossu, 1982; Ciaponi, 1984).

**Table 3.** Run-off coefficients for the rational method (Salvato *et al.*, 1971).

| | | Soil texture | | |
|---|---|---|---|---|
| Soil cover | Slope (%) | Sandy loam | Loamy clay | Clay |
| Grassed soil | 0–5 | 0.10 | 0.30 | 0.40 |
| | 5–10 | 0.16 | 0.36 | 0.55 |
| | 10–30 | 0.22 | 0.42 | 0.60 |
| Bare soil | 0–5 | 0.30 | 0.50 | 0.60 |
| | 5–10 | 0.40 | 0.60 | 0.70 |
| | 10–30 | 0.52 | 0.72 | 0.82 |

**Table 4.** Monthly variations of the average yearly run-off coefficient, (Ciaponi, 1984).

| | | | |
|---|---|---|---|
| January | 1.60 | July | 0.29 |
| February | 1.80 | August | 0.29 |
| March | 1.43 | September | 0.46 |
| April | 0.97 | October | 1.20 |
| May | 0.89 | November | 1.40 |
| June | 0.37 | December | 1.60 |

To take account of the moisture content of the soil and of the vegetation growing on the soil, at least approximately, monthly coefficients can be adopted obtained from experimental studies by Pasini during work on recovery of land in the Lower Po Valley (Table 4).

## Infiltration Curve Method

This method, formulated in 1947 by the United States Department of Agriculture and Soil Conservation Service (USDA-SCS), calculates the run-off taking into account various factors such as the permeability of the top soil strata, type of vegetation growing on this strata, the slope of the site and the moisture content of the soil at the moment for which the estimation is to be made (Noble, 1976). It is possible to calculate the run-off as a function of the intensity of rainfall in any rainfall event.

For correct use of this method it is necessary to have rainfall data on at least a daily basis but preferably for 12 or 6 hourly periods. Use of the USDA-SCS method is recommended for calculating daily fluctuations in run-off whereas the rational method is more use for calculations of run-off over longer periods of time.

## Evaporation and Evapotranspiration

From hydrological data (Luisley *et al.*, 1982; Shaw, 1983; Maione, 1978) it becomes clear that the vegetation growing on the final cover causes a water loss to atmosphere by evapotranspiration which is greater than that which could be lost by evaporation from the soil alone without any vegetation.

It is evident that distinctions will have to be made between landfills still in operation with intermediate cover and finished landfills with final cover and landscaping. In the majority of cases technical literature only considers evapotranspiration as the references are usually only for completed landfills (Lu *et al.*, 1982; Ciaponi, 1984; Noble, 1976).

The production of leachate during operation of the landfill is not, however, insignificant (Ehrig, 1983; Campbell, 1982) if this phase lasts for several years, or the landfill is situated in an area of high rainfall, or a significant quantity of sludge with a high moisture content (e.g. from wastewater treatment plants) is disposed of. For a correct appraisal of the problem both situations should be considered.

### Evapotranspiration

Empirical and semi-empirical formula to calculate potential and effective evapotranspiration are numerous (Luisley *et al.*, 1982; Shaw, 1983; Maione, 1978).

For European climates the formula developed by Thorntwaite, corrected by coefficients to take into account latitude, seems best to adhere to reality, particularly if the Thorntwaite–Mather method of calculating the rate of effective evapotranspiration is used with the hypothesis developed by Benfratello and others (Benfratello, 1961; Melisenda, 1964, 1966) and verified for the Italian climate (Galbiati and Gruppo, 1979; Cavazza, 1979). In addition it has been proposed to apply this theory to completed landfills which have been covered with soil and reinstated as agricultural land (Ciaponi, 1984). The method of calculation is given below.

### Calculation of Evapotranspiration

1. Calculation of the potential evapotranspiration by using the Thorntwaite formula

$$PE_i = 16\left(\frac{10T_i}{I_t}\right)^a C_i \qquad (1)$$

where:

$PE_i$ = potential evapotranspiration in the $i$-th month (mm/month)

$T_i$ = monthly average temperature (°C)

$a = 6.75 \times (10^{-7} \times I_t^3) + 7.71 \times (10^{-5} \times I_t^2) + 1.79 \times (10^{-2} \times I_t) + 0.49239$

$I_t$ = annual thermal indice = $I_t = \sum_1^{12} i\left(\frac{T_i}{5}\right)^{1.514}$

$C_i$ = correction coefficient relative to the month which takes account of the variation in length of day at various latitudes; it's obtained by dividing the value $N$ reported in Table 5 by 10.

**Table 5.** Mean daily duration of maximum possible sunshine ($N$, hours). From Shaw, 1983.

| Latitude °North | Jan | Feb | Mar | Apr | May | Jun | Jul | Aug | Sep | Oct | Nov | Dec |
|---|---|---|---|---|---|---|---|---|---|---|---|---|
| 60 | 6.7 | 9.0 | 11.7 | 14.5 | 17.1 | 18.6 | 17.9 | 15.5 | 12.9 | 10.1 | 7.5 | 5.9 |
| 58 | 7.2 | 9.3 | 11.7 | 14.3 | 16.6 | 17.9 | 17.3 | 15.3 | 12.8 | 10.3 | 7.9 | 6.5 |
| 56 | 7.6 | 9.5 | 11.7 | 14.1 | 16.2 | 17.4 | 16.9 | 15.0 | 12.7 | 10.4 | 8.3 | 7.0 |
| 54 | 7.9 | 9.7 | 11.7 | 13.9 | 15.9 | 16.9 | 16.5 | 14.8 | 12.7 | 10.5 | 8.5 | 7.4 |
| 52 | 8.3 | 9.9 | 11.8 | 13.8 | 15.6 | 16.5 | 16.1 | 14.6 | 12.7 | 10.6 | 8.8 | 7.8 |
| 50 | 8.5 | 10.0 | 11.8 | 13.7 | 15.3 | 16.3 | 15.9 | 14.4 | 12.6 | 10.7 | 9.0 | 8.1 |
| 48 | 8.8 | 10.2 | 11.8 | 13.6 | 15.2 | 16.0 | 15.6 | 14.3 | 12.6 | 10.9 | 9.3 | 8.3 |
| 46 | 9.1 | 10.4 | 11.9 | 13.5 | 14.9 | 15.7 | 15.4 | 14.2 | 12.6 | 10.9 | 9.5 | 8.7 |
| 44 | 9.3 | 10.5 | 11.9 | 13.4 | 14.7 | 15.4 | 15.2 | 14.0 | 12.6 | 11.0 | 9.7 | 8.9 |
| 42 | 9.4 | 10.6 | 11.9 | 13.4 | 14.6 | 15.2 | 14.9 | 13.9 | 12.6 | 11.1 | 9.8 | 9.1 |
| 40 | 9.6 | 10.7 | 11.9 | 13.3 | 14.4 | 15.0 | 14.7 | 13.7 | 12.5 | 11.2 | 10.0 | 9.3 |
| 35 | 10.1 | 11.0 | 11.9 | 13.1 | 14.0 | 14.5 | 14.3 | 13.5 | 12.4 | 11.3 | 10.3 | 9.8 |
| 30 | 10.4 | 11.1 | 12.0 | 12.9 | 13.6 | 14.0 | 13.9 | 13.2 | 12.4 | 11.5 | 10.6 | 10.2 |
| 25 | 10.7 | 11.3 | 12.0 | 12.7 | 13.3 | 13.7 | 13.5 | 13.0 | 12.3 | 11.6 | 10.9 | 10.6 |
| 20 | 11.0 | 11.5 | 12.0 | 12.6 | 13.1 | 13.3 | 13.2 | 12.8 | 12.3 | 11.7 | 11.2 | 10.9 |
| 15 | 11.3 | 11.6 | 12.0 | 12.5 | 12.8 | 13.0 | 12.9 | 12.6 | 12.2 | 11.8 | 11.4 | 11.2 |
| 10 | 11.6 | 11.8 | 12.0 | 12.3 | 12.6 | 12.7 | 12.6 | 12.4 | 12.1 | 11.8 | 11.6 | 11.5 |
| 5 | 11.8 | 11.9 | 12.0 | 12.2 | 12.3 | 12.4 | 12.3 | 12.3 | 12.1 | 12.0 | 11.9 | 11.8 |
| Equator 0 | 12.0 | 12.0 | 12.0 | 12.0 | 12.0 | 12.0 | 12.0 | 12.0 | 12.0 | 12.0 | 12.0 | 12.0 |

2. Calculation of the monthly deficit/excess this is obtained by subtracting the potential evapotranspiration $(PE_i)$ from the rainfall $(P_i)$ less run-off $(R_i)$.
A deficit occurs if $P_i - R_i - PE_i < 0$ (dry periods)
An excess occurs if $P_i - R_i - PE_i > 0$ (wet periods)
3. Calculation of effective evapotranspiration. During periods of water deficit (dry periods) the evapotranspiration is generally less than the potential evapotranspiration* and is called effective evapotranspiration. This is calculated taking into account the progressive dehydration of agricultural soil using the following formula of Benfratello:

$$\frac{d\alpha}{d\lambda} = \alpha^m \tag{2}$$

where:
$m$ = "resistance-to-desaturation" coefficient

$\alpha = A/U$ = ratio between the effective influx of water $A$ into the soil and the maximum value $U$ that could enter the soil
$U$ represents the maximum volume of water in mm that the strata of soil (of thickness $h$) exploited by roots could contain (by capillary action) after water had infiltrated under gravity; $U$ reflects the field capacity $\theta_{ms}$** of the soil from the formula

$$U = \theta_{ms} \frac{\gamma_i}{\gamma} h$$

where $\gamma_i$ and $\gamma$ are the specific weights of the dry soil and of water.

$\lambda = L/U$ where $L = (P - R) - PE$ represents the deficit as a continuous function defined during the dry season and is therefore always negative.

The integration of Equation 2 gives

$$\alpha = [\alpha_0^{(1-m)} + (1 - m)\lambda]^{1/1-m}, \quad \text{for } m \geq 1 \tag{3}$$

$$\alpha = \alpha_0 e^\lambda, \quad \text{for } m = 1 \tag{4}$$

$\alpha_0$ represents the value of $A/U$ at the start of the dry season.
Calculating $\alpha$ and therefore $A$, the difference '$\Delta A$' between two consecutive months during the dry season summed with the rainfall less run-off $(P - R)$ represents the actual evapotranspiration $ET$:

$$ET_{\text{dry}} = P - R + \Delta A$$

---

* In cases where the ground is kept wet by an aquifer or where it is regularly irrigated, evapotranspiration is maintained close to the maximum, even during dry periods.
** $\theta_{ms}$ = water content on a dry weight basis = mass of water/mass of dry soil.

## Observations on the Calculation

For practical use Table 6 reports some typical values of $U$ for various vegetation and for several types of agricultural soil.

It can be seen that values shown are very different. In general it can be concluded that high values of $U$ indicate a high buffering capacity during the course of a year, attenuating the flux of water that infiltrates the top layer of soil. Low values of $U$ characterize a large deficit of water during the dry season and an excess of water during the wet season (which means that a larger quantity of water infiltrates). The value of $U$ reflects the storage capacity of the soil.

The procedure illustrated in the previous section presupposes that the value of $\alpha_0$ (and consequently $A$ at the start of the dry season) is known. Usually it is assumed that $\alpha_0 = 1$ which means that $A_0 = U$, and that the excess from the previous wet season has filled the volume $U$. This hypothesis is generally true for all European climates for values of $U$ not greater than 300 mm, whereas it is not true for semi-arid or tropical climates.

Experimental studies (Galbiati and Gruppo, 1979) have demonstrated the validity of this approach, taking into account the capillary action of water from the groundwater which may partially or completely supply the ground with water lost by evapotranspiration. The study also evaluated that the value of the exponent $m$ lies between 1.25 and 1.75 for agricultural land of medium sandy soil cultivated with maize ($U = 105$ mm).

Cavazza (1979), furthering investigation into the exponent $m$, has demonstrated the dependence of $m$ on $U$ and $A$. In particular, $m = 0$ immediately after rainfall and increases as $L$ increases. This dependence renders $m$ variable with $\alpha$, invalidates the integration of (2) which was undertaken based on the assumption that $m$ is constant. However, the application of equations (3) and (4) for limited periods of time can justify, even if only approximately, the assumption that $m$ is constant for some periods.

Formulae such as Penman (Luisley *et al.*, 1982; Shaw, 1983; Maione, 1978) based on more detailed meteorological data (such as number of daylight hours, air humidity, wind velocity and average temperature at that latitude) can give good results on the basis of local data. This information can be obtained from local meteorological stations or from instruments installed *in situ*; the latter having the additional benefit of providing a form of control between measured and predicted values.

## Evaporation

There are not many detailed formulae in the literature to calculate the

**Table 6.** Maximum water retention capacity (in mm) for various species of vegetation as a function of soil conditions. The two values indicated for every species refer to different conditions of root system development (very developed or little developed). From Ciaponi, 1984.

|  | Soil use | | | | | | | | | | | | | | |
|---|---|---|---|---|---|---|---|---|---|---|---|---|---|---|---|
| Soil texture | 1 | 2 | 3 | 4 | 5 | 6 | 7 | 8 | 9 | 10 | 11 | 12 | 13 | 14 | 15 |
| Loamy clay | 100 / 70 | 110 / 80 | 55 / 35 | 125 / 90 | 145 / 110 | 145 / 90 | 125 / 110 | 145 / 90 | 160 / 100 | 250 / 160 | 270 / 180 | 250 / 160 | 250 / 160 | 125 / 110 | 110 / 90 |
| Clay loam | 175 / 115 | 190 / 125 | 90 / 70 | 230 / 135 | 230 / 155 | 230 / 145 | 230 / 155 | 230 / 115 | 250 / 125 | 350 / 270 | 365 / 290 | 350 / 270 | 350 / 270 | 230 / 155 | 215 / 135 |
| Sandy clay loam | 140 / 100 | 155 / 110 | 70 / 50 | 155 / 100 | 230 / 140 | 195 / 130 | 155 / 100 | 230 / 130 | 245 / 140 | 285 / 220 | 295 / 230 | 285 / 220 | 285 / 220 | 155 / 100 | 140 / 90 |
| Sandy clay | 150 / 110 | 165 / 120 | 75 / 55 | 165 / 110 | 250 / 150 | 210 / 140 | 165 / 110 | 250 / 140 | 260 / 150 | 305 / 235 | 315 / 250 | 305 / 235 | 305 / 235 | 165 / 110 | 150 / 110 |
| Sand; sandy loam | 65 / 40 | 70 / 40 | 45 / 30 | 95 / 60 | 135 / 70 | 115 / 70 | 95 / 60 | 135 / 70 | 140 / 70 | 210 / 155 | 220 / 165 | 215 / 155 | 215 / 155 | 95 / 60 | 85 / 55 |
| Silt clay loam | 180 / 130 | 200 / 140 | 115 / 75 | 215 / 165 | 280 / 180 | 265 / 165 | 215 / 165 | 280 / 150 | 300 / 155 | 365 / 280 | 380 / 300 | 365 / 280 | 365 / 280 | 215 / 165 | 200 / 150 |
| Silt loam | 180 / 130 | 200 / 140 | 115 / 75 | 215 / 165 | 280 / 180 | 265 / 165 | 215 / 165 | 280 / 150 | 300 / 155 | 365 / 280 | 380 / 300 | 365 / 280 | 365 / 280 | 215 / 165 | 200 / 150 |
| Loam | 205 / 150 | 225 / 160 | 105 / 75 | 225 / 150 | 340 / 190 | 285 / 170 | 225 / 150 | 340 / 190 | 360 / 205 | 415 / 320 | 435 / 340 | 415 / 320 | 415 / 320 | 225 / 150 | 205 / 130 |
| Loam; sandy loam | 100 / 55 | 105 / 65 | 65 / 45 | 140 / 90 | 195 / 105 | 170 / 100 | 140 / 95 | 195 / 105 | 205 / 110 | 310 / 230 | 320 / 240 | 310 / 230 | 310 / 230 | 140 / 90 | 125 / 80 |
| Sandy loam | 100 / 55 | 105 / 65 | 65 / 45 | 140 / 90 | 195 / 105 | 170 / 100 | 140 / 95 | 195 / 105 | 205 / 110 | 310 / 230 | 320 / 240 | 310 / 230 | 310 / 230 | 140 / 90 | 125 / 80 |

1 = Seeded dry ground
2 = Grass seeded
3 = Vegetables
4 = Vineyard
5 = Olive grove
6 = Vineyards/olive
7 = Orange or lemon orchard
8 = Fruit orchard (soft fruit)
9 = Fruit orchard (hard fruit)
10 = Deciduous trees
11 = Tall trees
12 = Mixed woodland
13 = Chestnut trees
14 = Grass and lawned area
15 = Meadow

evaporation from land without vegetation. For saturated soils, such as land after heavy or prolonged rain, it has been proposed to use the evaporation rates calculated for shallow water basins multiplied by the following coefficients (Maione, 1978):

$$1 - \text{fine sand}$$
$$0.9 - \text{loams}$$
$$0.75 - 0.95 - \text{clays}$$

Assuming that the ground is saturated is probably correct for winter months or during rainy seasons. During the dry seasons however, or periods of little rainfall, the top strata of ground will tend to dry in a short period of time, causing a reduction in evapotranspiration. An unsaturated ground can be represented by Turc's formula for ground without vegetation (from Shaw, 1983, modified):

$$E = \frac{P + w}{\sqrt{\{1 + [(P + w)/L]\}}}$$

where:

$E$ = evaporation in 10 days (mm)
$P$ = rainfall occurring in 10 days
$w$ = maximum quantity of water that could evaporate from the soil in 10 days in the absence of rainfall; varying from 10 mm for moist ground to 1 m for dry ground;*

$L$ = factor = $L = \dfrac{(T + 2)/I_g}{16}$

   in which $T$ is the median air temperature in °C
$I_g$ = effective solar radiation (cal/cm$^2$ per day); in the absence of direct determinations the effective solar radiation, $I_g$, can be estimated from the theoretical $I_0$ by relating it to the number of effective daylight hours ($n$) with respect to those theoretical ($N$) reported in Table 5 (Shaw, 1983; Maione, 1978).

$I_g = I_o (0.135 + 0.68\ n/N)$ for foggy areas and for values of $n/N \leqslant 40$
$I_g = I_o (0.155 + 0.69\ n/N)$ for latitudes between 54.5 and 56 deg North
$I_g = I_o (0.16 + 0.62\ n/N)$ for latitudes between 45 and 54.5 deg North
$I_g = I_o (0.18 + 0.662\ n/N)$ for latitudes between 35 and 45 deg North

---

* On the basis of the theory of soil drying (Gardner, 1974) in which the cumulative evaporation from the ground proceeds in a square root relationship with time, it is possible to estimate $w$, at least to a first approximation in the following manner:

$$w = w_0 - 0.01\ \sqrt{t}, \text{ with } 1 \leqslant w \leqslant 10\,\text{mm, and } w_0 = 10\,\text{mm}$$

$t$ = time with no rain after last rainfall, in seconds, before the considered 10-day period.

Values of $I_o$ are reported for various latitudes in Table 7.

The estimates obtained from the above formula can be no more than roughly approximate, especially if considering the intermediate cover of a landfill. For cover principally containing clay, during dry periods, cracking often occurs allowing easy infiltration of water which, once penetrated the strata underneath, is unlikely to be removed from the site by evaporation consequently increasing the likelihood of leachate formation (Campbell, 1982). *Vice versa*, compaction of waste and levelling of the cover (with low clay content) should make infiltration of water more difficult and increase the chance of disposing of the water by evaporation. To date there have been few experiments undertaken to establish formula or models to predict this effect accurately.

## Infiltration and Filtration Through the Cover Material

The quantity of water that infiltrates through the cover material can be simply estimated using the theory developed in the preceding paragraphs. The effective rainfall (net rainfall, i.e. less run-off) will go to fill the volume $A$ and, when this has reached its maximum retentive capacity, water will start to percolate into the strata underneath.

If a low permeability layer (e.g. clay) is present, this will limit the flow of water to the waste underneath forming a saturated layer through which liquid will flow according to Darcy's Law:

$$Q = K_B S_i$$

where:

$K_B$ = the permeability of the soil $(m/s)$
 $S$ = the sectional area of flow
 $i$ = the hydraulic gradient, obtained by dividing the hydraulic head of the groundwater $H_s$ by the thickness of the strata $H_B$
 $Q$ = the flow across the strata

Expressing $Q$ in $l$/month and $S$ in $m^2$ it is possible to write $Q/S = K_B$ 86400 s/d 1000 mm/m 365 d/yr : 12 months/year = $2.628 \times 10^9 K_B H_S/H_B$
which represents the hydraulic flux towards the waste expressed in mm/month (note that $1 \text{ mm} = 1 \text{ l/m}^2$).

The value of $H_S$ can be easily obtained by dividing the average volume of water percolating from upper layers during the time considered (for example) by the effective porosity of the soil.

Ciaponi (1984) has demonstrated that the dispersion of a water table by surface flow above a low permeability layer is negligible, with the exception of

**Table 7.** Maximum possible solar radiation at different latitudes (cal/(cm$^2$ day)), assuming latent heat of water = 590 cal/g. From Luisley, 1982.

| Latitude °North | Jan | Feb | Mar | Apr | May | Jun | Jul | Aug | Sep | Oct | Nov | Dec |
|---|---|---|---|---|---|---|---|---|---|---|---|---|
| 60° | 80 | 210 | 415 | 655 | 860 | 970 | 920 | 745 | 500 | 280 | 120 | 5 |
| 50° | 220 | 354 | 545 | 750 | 915 | 980 | 950 | 810 | 615 | 420 | 260 | 180 |
| 40° | 365 | 495 | 655 | 815 | 940 | 985 | 960 | 865 | 715 | 550 | 400 | 330 |
| 30° | 480 | 620 | 755 | 870 | 950 | 975 | 955 | 900 | 800 | 660 | 535 | 465 |
| 20° | 640 | 730 | 825 | 900 | 925 | 930 | 930 | 910 | 850 | 760 | 665 | 615 |
| 10° | 755 | 820 | 870 | 900 | 885 | 870 | 880 | 885 | 870 | 840 | 770 | 755 |
| 0° | 860 | 885 | 900 | 870 | 820 | 790 | 800 | 845 | 880 | 885 | 860 | 845 |

sites where there is a steep slope or high permeability in the cover soil. In the same work (Ciaponi, 1984) it was emphasized that a layer of easily draining soil placed above the impermeable layer and below the cover material will allow water to drain out of the site effectively eliminating any permeation of water into the waste. Without this drainage the disposal of water from the water table will principally be through capillary action into the soil above and from there by either evaporation or evapotranspiration.

In practice it can be assumed that for average soil conditions the transport of water by capillary action can account for all the water loss $(P - R - PE)$ until the level of the water table is located at least 50–60 cm under the soil exploited by roots. For lower levels the removal of water by this method rapidly falls off with the result that virtually no water from 80–100 cm below the root zone is removed in this manner (Galbiati and Gruppo, 1979).

## Movement of Water Through Wastes

The mass of waste can be considered a soil with particular characteristics (e.g. high concentration of putrescible substances) to which global parameters usually applied to soils (specific weight, density, water content, etc.) can be, applied.

With regard to water content (the definition of which is given in the Appendix) it is possible to make an analogy to loose soil (Cancelli and Cossu, 1985). The measurable difference between the field capacity and the initial water content of the waste after completion of the landfill is defined as the absorptive capacity or hydraulic capacity of the waste. The loss of water from the field capacity until only that present from capillary action, i.e. that held by surface tension, can only occur through evaporation (assuming that no mechanical dewatering is obtained through compaction, etc.).

Water percolating through from the surface of a landfill tends to be absorbed by the waste until the field capacity is reached; it is only when the infiltration of water exceeds this value that movement of water through the waste occurs, initially under unsaturated conditions and, finally, if sufficient water is present, under saturated conditions.

## Permeability of Waste

Whilst the movement of fluids through soils has been thoroughly studied and expressions have been formulated to model the flux of water, what little work that has been done on the movement of water through waste (Fungaroli *et al.*, 1979 from Lu *et al.*, 1982) has not allowed a correlation between permeability and density to be established.

The range of values varies from a maximum of $10^{-2}$ cm/s for non-compacted and non-shredded waste (density up to 300 kg/m$^3$) down to $10^{-4}$ cm/s for fine compacted waste. Similar values have been observed more recently also by Ettala (1987).

One aspect that cannot be ignored in the evaluation of the movement of water in a landfill is the presence of intermediate cover. Intermediate cover can provide a barrier sufficiently impermeable to cause water to accumulate above the cover producing water tables within the landfill or, better still, zones of saturated waste. A situation such as this was encountered at a site in England where leachate was recirculated onto waste (Barber and Maris, 1983). The water table over the intermediate cover increased by about 1 m per year with recirculation at a rate of 500–600 mm. Saturation led to quick settling of anaerobic conditions and leachate from the site after one or two years had characteristics similar to those of the leachate from a site of 5 years or more in age. It was necessary to take action to prevent this water from diffusing out of the site horizontally and polluting the surrounding land.

Formation of saturated zones can be prevented if facilities for draining water or leachate are installed for every strata of waste deposited. When wastes are not saturated with water the liquid flow should be modelled as flow through a porous unsaturated medium (Korfiatis *et al.*, 1984; Demetracopulos *et al.*, 1986). However, this leads to the integration of differential equations making calculation more difficult, particularly for routine design purposes.

As a first approximation a simplified model can simply take into account the water content of the waste and the water retention capacity. It can be assumed that leachate production will occur at a certain water content (generally less than the field capacity, due to short circuiting and channelling effects through the waste layers) and then it will gradually increase, reaching its maximum rate when field capacity is approached.

## Field Capacity and Water Retention Capacity of Waste

Knowledge of the field capacity and of the initial moisture content of the waste together with the determination of the water retention capacity of waste allows an approximate evaluation of the time required before leachate will reach the drainage system.

In Table 8 various values of initial moisture content are reported together with field capacity and retention capacity that have been determined in trials in England (Barber and Maris, 1983; Holmes, 1980). The variation of initial moisture content is between 15–25% in volume, whereas field capacity is between 20 and 50%.

Experiments on full size landfills and pilot scale tests have verified that the

**Table 8.** Water content, field capacity and absorption capacity of solid waste (Holmes, 1980).

| Type of waste | Density (on wet basis) (kg/m³) | Initial water content (1) $\theta_{mti}$ (% weight) | Initial water content (2) $\theta_i$ (% volume) | Field capacity (1) $\theta_{mtc}$ (% weight) | Field capacity (2) $\theta_c$ (% volume) | Absorption capacity (1) $(\theta_{mtc} - \theta_{mti})$ (% weight) | Absorption capacity (2) $(\theta_c - \theta_i)$ (% volume) |
|---|---|---|---|---|---|---|---|
| Raw | 282 | 14 | 4 | 101 | 29 | 87 | 25 |
| Pulverised | 370 | 39 | 14 | 72 | 27 | 33 | 13 |
| Raw | 523 | 37 | 19 | 75 | 39 | 38 | 20 |
| Pulverised | 504 | 26 | 13 | 42–75 | 38 | 16–49 | 15 |
| Raw | 624 | 28 | 18 | 59 | 37 | 31 | 19 |
| Raw mixed with liquid sludge after: | | | | | | | |
| a. 4 yrs | 638 | 31.13 | 19.14 | 60.78 | 38.60 | 30.65 | 19.47 |
| b. 10 yrs | 814 | 30.63 | 24.83 | 49.57 | 40.19 | 18.94 | 15.35 |
| c. 17 yrs | 960 | 33.05 | 31.68 | 44.50 | 42.65 | 11.45 | 10.97 |

Notes (1) = on a wet basis (referring to total weight = wt water (kg)/wt of waste (kg) as %)
(2) = on a wet basis (referring to total volume = m³ water/m³ waste as %)

field capacity of pulverized waste is not significantly different to that of non–pulverized waste (Stegmann, 1982).

## Accumulation of Water in Landfills and Reaction Times

The values for field capacity and for absorption capacity defined above are theoretical; they represent the values reached by waste after a long period of contact with water. Because of the presence of preferential pathways through the landfill along which water can flow easily, in practice, leachate will find its way into the drainage system quicker than in the theoretical time based on absorption capacity.

A study in England (Blakey, 1982) has confirmed that the first appearance of leachate is when the moisture content reaches between 51 and 58% by weight on a dry weight basis. Taking the initial moisture content as approximately 35% by weight of the initial dry weight, the absorptive capacity results as 16–27% by weight on a dry weight basis.

For calculations Campbell (1982) advised to use a value for the initial moisture content of 25% by weight on a moist weight basis (for wastes with average densities of $650 \, \text{kg/m}^3$) and to consider an absorption capacity of $100 \, \text{l/t}$.

The relationship between the quantity of water absorbed at the time of the first appearance of leachate and that absorbed in the long term is approximately $1:4$. The time necessary for leachate to fill the absorptive capacity is easily determined. Given that the monthly infiltration originates from the surface it is sufficient to estimate the number of months needed to supply enough water equal to the absorptive capacity of the waste and the intermediate cover.

The calculation to determine the first appearance of leachate can be used as the basis to evaluate, knowing the volume available for waste disposal, the quantity of waste that must be tipped daily to complete the landfill before the waste reaches its field capacity in the short term and therefore avoiding the production of leachate (Campbell, 1982).

$$P = \frac{Q_R}{t_{pa}} = \frac{V_R \times r_{bt}}{t_{pa}}$$

where:
$P$ = daily capacity (t/day)
$Q_R$ = quantity of waste for disposal (t)
$t_{pa}$ = time before leachate first appears (days)
$V_R$ = volume available for waste disposal
$r_{bt}$ = density of waste (t/m$^3$)

Capacities less than this value require more time to complete the landfill and will allow the formation of leachate during the phase of landfill operation. Values in excess of this value mean that the landfill will be completed before the waste reaches its field capacity.

It should be noted that although the time when leachate will first appear is calculated taking into account the intermediate cover, and eventually part of the final cover, the volume $V_R$ is that available for waste excluding cover volume.

## Other Sources of Leachate

As already shown in Fig. 1 rainfall is not the only source of water in the landfill. The eventual disposal of sewage to the final cover or the recycling of leachate constitute water volumes that have to be added to the volume of rainfall and can be encountered, particularly in arid climates. Recirculation of leachate has already been shown to be advantageous independent of the type of climate. Another significant source of leachate is from the disposal of non-dewatered sewage sludge to the landfill.

## HYDRAULIC MODEL BASED ON ALGORITHMS

The calculation of single element hydrological balances can be easily organized into a model to simulate the overall hydrological balance of a landfill. In the USA, for example, it has been shown for some years that a program called Hydrological Simulation on Solid Waste Disposal Sites (HSSWDS) is capable of estimating the hydrological situation on a daily basis (Perrier *et al.*, 1980 from Lu *et al.*, 1982).

On the basis of meteorological and geo-morphological data for the site under consideration the model can calculate the run-off (using the curve number method), evapotranspiration (using Penman's formula), evaluation of the capillary action of the top cover and the infiltration into the waste.

Assuming that the water content of the waste approaches the field capacity the program hypothesizes that all infiltration water in excess of this value is transformed into leachate. Given that waste has a residual retentive capacity, it is also possible to evaluate the delay in the appearance of leachate ignoring the time taken for the leachate to flow between strata in the landfill.

Based on similar assumptions a simplified model has been developed by the authors with monthly time intervals to evaluate the possibility of utilizing the model as a tool in the design of landfills. In particular, the model can be applied to landfills in operation (i.e. without final cover) and to completed landfills (with final cover and with a low permeability layer under final cover). A hydraulic representation of the model referring to a completed landfill is shown in Fig. 2 where each term of the water balance has been converted to inflows and outflows through reservoirs and valves.

Figure 3 summarizes the flow sheet of the algorithm highlighting the main steps. The topsoil acts as a reservoir: the higher the field capacity the lower the infiltration of water into the lower layers through the clay barriers. The model assumes, as a first approximation, that the leachate will appear a short term after the field capacity has been reached, and that the flow rates will increase until long term field capacity has been reached.

**Sensitivity Analysis**

The parameters which mostly influence the model are, in order of importance, the permeability of the capping layer, the method of calculating evapotranspiration, the capacity of the cover material and, finally, the water retention capacity of the waste.

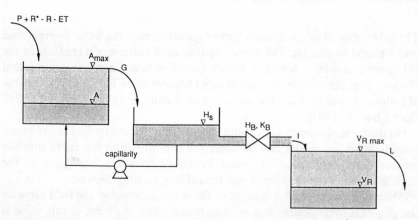

**Figure 2.** Hydraulic model as a simulation of the water balance in a sanitary landfill. P: precipitation; ET: actual evapotranspiration; R*: runoff from external areas; R: runoff; A: water content of soil cover; U: field capacity of soil cover; G: water infiltration under soil cover; KB: Darcy's hydraulic transmissivity $(L \cdot T^{-1})$; L: leachate; $H_B$: thickness of the low-permeability barrier; $H_S$: saturated layer; $V_R$: water content of wastes; $V_{Rmax}$: field capacity of wastes.

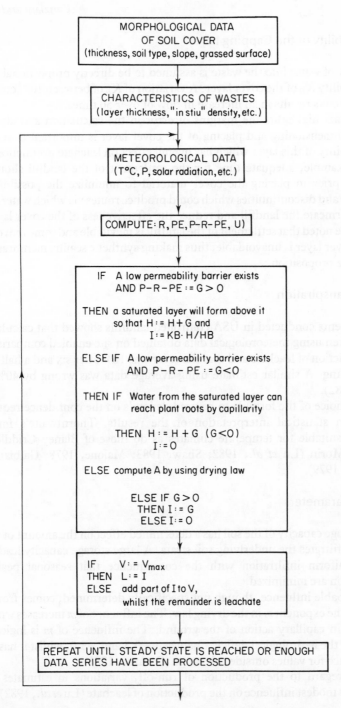

MORPHOLOGICAL DATA
OF SOIL COVER
(thickness, soil type, slope, grassed surface)

CHARACTERISTICS OF WASTES
(layer thickness, "in stiu" density, etc.)

METEOROLOGICAL DATA
(T°C, P, solar radiation, etc.)

COMPUTE: R, PE, P–R–PE, U)

IF   A low permeability barrier exists
     AND P–R–PE := G > 0

THEN a saturated layer will form above it
     so that H := H + G and
     I := KB· H/HB

ELSE IF  A low permeability barrier exists
         AND P–R–PE := G < 0

THEN IF  Water from the saturated layer can
         reach plant roots by capillarity

         THEN H := H + G (G<0)
              I := 0

ELSE compute A by using drying law

         ELSE IF G > 0
         THEN I := G
         ELSE I := 0

IF     V := V_max
THEN   L := I
ELSE   add part of I to V,
       whilst the remainder is leachate

REPEAT UNTIL STEADY STATE IS REACHED OR ENOUGH
DATA SERIES HAVE BEEN PROCESSED

**Figure 3.** Schematic flow sheet of leachate hydrological balance algorithm (symbols as in Fig. 2).

Permeability of the Capping Layer

The flow of water into the waste is assumed to be directly proportional to the permeability $K_B$ of the cover layer. A variation of $K_B$ of between $10^{-7}$ cm/s and $5 \times 10^{-6}$ cm/s results in an increase in flow of over 50 times.

This fact highlights the margins of error in the calculation and also that correct dimensioning and placing of the cover layer is important as it is the permeability of this layer which has overall effect on leachate production.

For example, adequate time for the settlement of the landfill should be allowed prior to placing the cover material to minimize the possibility of cracking and discontinuities which could produce routes via which water could easily permeate the landfill and reduce the effectiveness of the cover layer. It should be noted that settlement of the landfill is inevitable and some movement in the cover layer is unavoidable, thus making synthetic sealing membranes an attractive proposition.

Evapotranspiration

Experiments conducted in USA controlled landfills showed that calculations undertaken using meteorological data obtained on site enabled comparison of the prediction of leachate and that the error was 16% after six and a half years of sampling. A similar estimate using average data was wrong by 40% (Lu et al., 1982).

The choice of the formula to be used depends on the confidence required, based on statistical interpretation of the results. Thorntwaite's formula appears suitable for temperate climates as do those of Blaney–Criddle and Blaney–Morin (Lu et al., 1982; Shaw, 1983; Maione, 1978; Galbiati and Gruppo, 1979).

Other Parameters

The storage capacity of the soil has a determined effect on the amount of water which infiltrates the underlying soil strata. A large storage capacity leads to a more uniform infiltration with the consequence that seasonal peaks of infiltration are minimized.

A probable influence, though not specifically determined, comes from the value of the exponent $m$ in the drying law. The influence of $m$ increases with the increase in capillary action of the ground. The influence of $m$ is logical for values of the hydrological deficit $\lambda$ between $-0.5$ and $-3.0$, whilst it has little importance for values outside this range (Cavazza, 1981).

With regard to the production of run-off, variations in estimates have relatively modest influence on the production of leachate (Lu et al., 1982). The design and morphology of the surface cover is, however, of fundamental importance in the production of run-off. Accentuating the slope of the site, consequently increasing the drainage of surface water, the amount of surface run-off can be increased signficantly and, especially in wet climates, the value of $(P - R - PE)$ can be minimized.

The consequence of this is that there is greater control over the formation of water tables within the landfill although there is a potential detrimental effect on vegetation planted on the cover material from the formation of small watercourses in the depressions of the site.

In dry climates the converse is true where reduction of surface water run-off can help minimize the amount of irrigation that the top cover material needs to keep vegetation healthy.

## Landfills in Operation

It is difficult to predict the formation of leachate in an operating landfill as there are too many factors to take into account; each site should be considered on its own merits.

One of the most important factors to consider is action taken to prevent rainwater falling on areas of the site mixing with leachate produced from other areas. In addition the method of landfill operation—the ways in which completed tipping has intermediate cover placed over it, the portion of the exposed refuse, the compaction method, etc.—all have an effect on the amount of leachate produced during landfill operation.

The prediction of the quantity of leachate produced is possible within acceptable limits using a simple balance of the type:

$$\text{leachate} = \text{rainfall} - \text{run-off} - \text{evaporation}$$

The result by this method is just as accurate as a result obtained by a much more detailed analysis. However, if in specific cases other parameters are known in sufficient detail a more comprehensive analysis can be undertaken.

High compaction of wastes and careful preparation of the daily cover with earth channels and slopes are essential for low leachate production.

## MODEL CALIBRATION

Calibration of the model has been carried out using data from experiments carried out at five experimental points on the Tilburg landfill site (NL).

This research was conducted between 1978 and 1984 by the Dutch Health and Environmental Hygiene Institute (R.I.V.M., Bilthoven) and by other researchers (Rijtema *et al.*, 1986).

Calibration has been performed for the complete landfill, since data for the operating period were insufficient. The model described in the Dutch report was implemented on a VAX 750 computer, and both measured data and simulated data were compared with those obtained through this model which has been implemented on a personal computer (IBM-XT or equivalent).

## Description of the Experimental Sectors

A typical section of an experimental sector is shown in Fig. 4. Note that the drainage of leachate relies on a fine sandy layer between the waste and the LDPE liner. Deposition of waste, which was completed within 18 months, was carried out in different ways in each sector.

The density of wastes was higher than 700 kg/m³ (wet basis) in all sectors.

In sectors 1, 2 and 4, two intermediate sandy layers (0.15 m thick) were put among the waste layers but were not connected to the drainage system. On top of the tipped waste (max weight 5.5 m) cover material in the form of a 1 m thick sandy loam was laid. The upper half was enriched with humus to meet agricultural requirements. The northern side was grassed and regularly mowed, whilst the southern side was grassed, kept uncut, and bushes were planted.

Since sector 1 showed considerable leakage from the sides, only sectors 2 to 5 were considered for the calibration. Physical properties of waste and of the soil are summarized in Table 9.

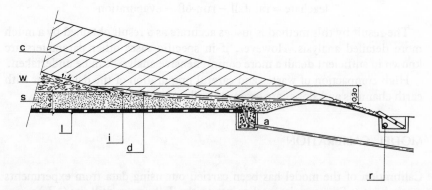

**Figure 4.** Schematic of an experimental sector at Tilburg landfill site. (a) Leachate collection; (r) run-off collection; (c) top soil; (w) wastes; (s) sand layer; (l) LDPE liner, 2.5 mm thick; (i) liner, 0.5 mm thick, 5 m long; (d) collection of subsurface infiltration in the lower section.

**Table 9.** Characteristics of wastes and of the top soil used at Tilburg landfill site (Rijtema *et al.*, 1986).

| Sectors | 2 | 3 | 4 | 5 |
|---|---|---|---|---|
| *Top soil* | | | | |
| thickness, H (m) | 1.0 | 1.0 | 1.0 | 1.0 |
| field capacity, $A_{max}$ (mm) | 145 | 136 | 171 | 162 |
| effective porosity, f | 0.25 | 0.25 | 0.25 | 0.25 |
| *Intermediate cover* | | | | |
| thickness, HI (m) | 0.30 | 0 | 0.30 | 0 |
| water content, $\theta_i$ | 0.10 | — | 0.10 | — |
| field capacity, $\theta_c$ | 0.16 | — | 0.16 | — |
| absorption capacity, $HI \cdot (\theta_c - \theta_i)$ (m³/m²) | 0.018 | — | 0.018 | — |
| *Waste layer* | | | | |
| av. thickness, HR (m) | 3.0 | 3.0 | 3.0 | 3.0 |
| water content, $\theta_i$ | 0.247 | 0.241 | 0.210 | 0.269 |
| field capacity, $\theta_c$ | 0.420 | 0.390 | 0.380 | 0.420 |
| absorption capacity, $HR \cdot (\theta_c - \theta_i)$ (m³/m²) | 0.519 | 0.447 | 0.510 | 0.453 |
| specific mass (kg/m³) | 840 | 839 | 712 | 936 |

## CALCULATED LEACHATE PRODUCTION RATES AND CONCLUSIONS

The application of the model for estimating leachate production has been described in detail elsewhere (Cossu *et al.*, 1988). Figure 5 shows measured and calculated values of monthly leachate production rates, according to the Dutch model and to the model described in this chapter. The Dutch model also estimated the losses that occurred at the experimental landfill sectors which were partly unlined. In contrast our model estimates the overall leachate production.

From the results the following observations can be drawn.

1. It is confirmed that if the '*m*' coefficient is equal to 1, leachate productions are generally underestimated.

2. If it is assumed that $m = 3$ the largest error on the cumulative production over three years never exceeds 25% of the actual leachate production (including losses) and exceeds 20% in 3 cases only (column 5, sector 2 and column 9 and 10, sector 5).

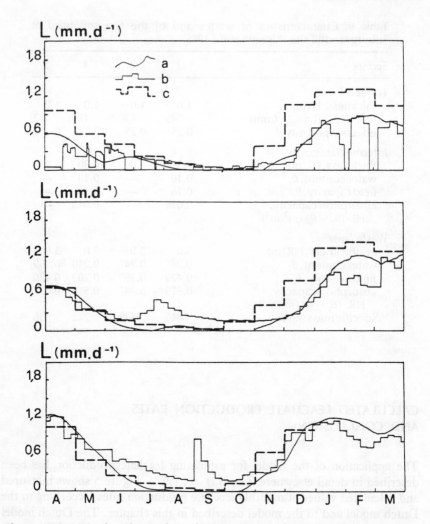

**Figure 5.** Estimated and actual production rates of leachate (L) in sector 2 at Tilburg landfill site, from April 1979 to March 1982.
(a) Estimated, excluding losses;
(b) measured, excluding losses;
(c) estimated, including losses (this work).

3. The best results for single years are generally obtained for the last (third) year, after the wastes have reached their absorption capacity.
4. Different hypotheses about the residual absorption capacity of wastes after tipping affected only the production rate values of the first year. In this

simplified model these assumptions caused only longer or shorter delays in the appearance of leachate.

5. The model can estimate with sufficient approximation monthly variations.
6. Finally it has to be observed that the good fit depends above all on the climatic parameters and on the waste characteristics which should be strictly measured on site.

## REFERENCES

Barber, C. and Maris, P.J. (1983). 'Recirculation of leachate as a landfill management option, benefits and operational problems'. Water Research Centre, Stevenage Laboratory, UK.

Benfratello, G. (1961). 'Contributo allo studio del bilancio idrologico del terreno agrario'. *L'Acqua* 2, 34–53.

Blakey, N.C. (1982). 'Infiltration and absorption of water by domestic wastes in landfills. Research carried out by the Water Research Centre. Landfill Leachate Symposium, Harwell Laboratory, 19 May 1982.

Campbell, D.J.V. (1982). 'Absorptive capacity of refuse – Harwell research'. Landfill Leachate Symposium, Harwell Laboratory, 19 May 1982.

Cancelli, A. and Cossu, R. (1985). 'Problemi di stabilità negli scarichi controllati'. *Ingegneria Ambientale* 13, 11–12, 632–642.

Cavazza, L. (1979). 'Relazioni tra leggi di evaporazione dal terreno agrario'. *In* AIGR, 'L'impegno e il contributo del genio rurale allo sviluppo del Mezzogiorno', Atti del 3° Convegno Nazionale, Catania I Vol., 306–329.

Cavazza, L. (1981). 'Fisica del terreno agrario', UTET.

Ciaponi, C. (1984). 'Bilancio idrologico nelle discariche controllate di rifiuti solidi urbani finalizzato alla determinazione dei quantitativi di percolato prodotto'. Proceedings SEP Pollution 'Discariche: impatto ambientale e criteri di progettazione di discariche controllate per rifiuti solidi, speciali, tossici e nocivi', Fiera di Padova, April; 50–75.

Cossu, R. (1982). 'Alternative di smaltimento del percolato negli impianti di scarico controllato dei rifiuti solidi urbani'. *Ingegneria Ambientale* 11, 7, 564–572.

Cossu, R., Canziani, R. and Gadola, G. (1988). 'Hydrologic Model for Leachate Production in Sanitary Landfill', ISWA Specialized Seminar on Sanitary Landfilling, Amsterdam, 19–21 September.

Demetracopoulos, A.C., Sehayek, L. and Erdogan, H. (1986). 'Modeling leachate production from municipal landfills'. *ASCE, J. Env. Eng.* 112(5).

Ehrig, H-J. (1983). 'Quality and Quantity of Sanitary landfill leachate'. *Waste Management & Research* 1, 53–68.

Ettala, M. (1987). 'Infiltration and hydraulic conductivity at a sanitary landfill'. *Aqua Fennica*, 17, n.2, 231–237.

Galbiati, G.L. and Gruppo, M. (1979). 'Verifica della validità a livello locale di una nota legge di essiccamento del terreno agrario'. *In* AIGR, 'L'impegno e il contributo del genio rurale allo sviluppo del Mezzogiorno', Atti del 3° Convegno Nazionale, Catania I Vol., 7–23.

Holmes, R. (1980). 'The water balance method of estimating leachate production from landfill sites'. *Solid Wastes*, n.1.

Korfiatis, G.P., Demetracopoulos, A.C., Bourodimos, E.L. and Nawy, E.G. (1984). 'Moisture transport in a solid waste column'. *ASCE, J. Env. Eng.* 110(4).

Luisley, K.R. Jr *et al.* (1982). *'Hydrology for Engineers'*, McGraw Hill International Book Company, International Student Edition, Tokyo.

Lu, J.C.S. *et al.* (1981). 'Leachate production and management from municipal landfills: summary and assessment'. Report in fulfillment of US–EPA contract no. 68-03-2861, US–EPA Cincinnati, Ohio.

Lu, J.C.S. *et al.* (1982). 'Leachate production and management from municipal landfills: summary and assessment'. Report in fulfillment of US–EPA contract no. 68-03-2861, US–EPA Cincinnati, Ohio.

Maione, U. (1978). 'Bacino idrografico', *Lessons of Applied Hydrology*, Technical University of Milan.

Melisenda, I. (1964). 'Sui calcoli idrologici per il terreno agrario—influenza del clima'. *L'acqua* 4, 3–20.

Melisenda, I. (1966). 'Possibilità di deduzione delle dotazioni irrigatorie dalla legge di essiccamento'. First National Conference of the Italian Agricultural Engineering Association, Portici, 14–16 April.

Noble, G. (1976). *'Sanitary Landfill Design Handbook'*, Technomic Publishing Co., Westport, USA.

Rijtema, P.E., Roest, C.W.J. and Pankow, J. (1986). 'Onderzoek naar de Waterbalans van Vuiltstortplaatsen', Instituut voor Cultuurtechniek en Waterhuishouding (ICW).

Shaw, M.E. (1983). *'Hydrology in Practice'*, Van Nostrand Reinhold Co. Ltd., UK.

Stegmann, R. (1982). 'Absorptive capacity of refuse'. Landfill Leachate Symposium, Harwell Laboratory, 19 May 1982.

## 4.2   Leachate Quality

HANS-JÜRGEN EHRIG

*ITW Ingenieurberatung GmbH, Erich-Nörrenberg-Str. 5, D-5860 Iserlohn, West Germany*

### INTRODUCTION

Leachate quality is influenced in a complex way by the biological, chemical and physical processes which take place in a landfill. Experience during the past 15 years has resulted in a good knowledge of the main parameters of pollution and of some effects of operation modifications. But also the importance of micro-pollutants, the reason for modification effects and the duration of leachate pollutant emissions should be highlighted. In this report the results from the German experience at laboratory, pilot and full scale are summarized.

### LEACHATE COMPOSITION

Up to now the most important analytical parameters taken into consideration are organics such as $BOD_5$ or COD as traditional sewage pollutant indicators.

An important fact is the variation of organic compounds concentration ($BOD_5$, COD) with the change from the acetic to the methanogenic phase in landfills. Figure 1 presents the results of laboratory scale experiments showing the acetic phase during the first 100 days and the following methanogenic phase. The acetic phase is characterized by high organics with $BOD_5$/COD-ratio > 0.4 and low pH, methane content and gas production. After the transition to the methanogenic phase the methane content and pH are high, but $BOD_5$, COD and the $BOD_5$/COD-ratio are low. The results of leachate analysis from more than 15 landfills and over several years is shown in Fig. 2.

213

214                                                                            Ehrig

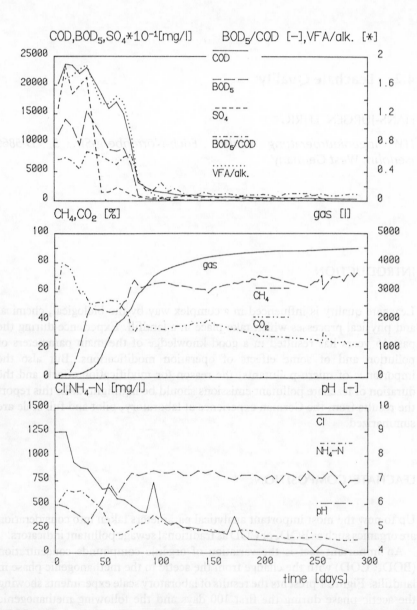

**Figure 1.** Leachate composition and gas production versus time from a laboratory scale test cell with domestic waste (VFA/alk. = volatile fatty acids/ alkalinity).

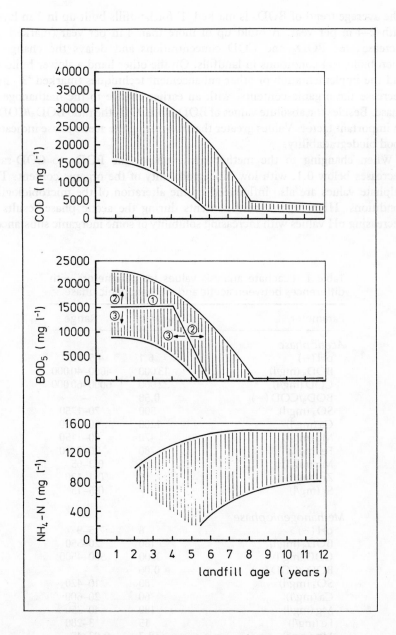

**Figure 2.** General tendencies of COD, BOD₅ and NH₄ versus time.

The average trend of $BOD_5$ is marked '1' for landfills built up in 2 m layers with 2–4 m per year. A build up of more than 4 m per year (marked '2') increases the $BOD_5$ and COD concentrations and delays the change of microbiological conditions in landfills. On the other hand a slower build up and the implementation of other enhancement techniques (marked '3') may decrease the organic contents, with an earlier change to the methanogenic phase. Besides the absolute values of $BOD_5$ and COD, the ratio $BOD_5$/COD is an important factor. Values greater than 0.4 during the acetic phase indicate a good biodegradability.

When changing to the methanogenic phase, the $BOD_5$ to COD-ratio decreases below 0.1, with low biodegradability of the organic content. The sulphate values are also influenced by the alteration of the microbiological conditions. High organic acid contents during the acetic phase results in decreasing pH values with increasing solubility of some inorganic substances,

Table 1. Leachate analysis values for parameters with differences between acetic and methanogenic phase.

| Parameter | Average | Range |
|---|---|---|
| *Acetic phase* | | |
| pH (−) | 6.1 | 4.5–7.5 |
| $BOD_5$ (mg/l) | 13 000 | 4000–40 000 |
| COD (mg/l) | 22 000 | 6000–60 000 |
| $BOD_5$/COD (−) | 0.58 | – |
| $SO_4$ (mg/l) | 500 | 70–1750 |
| Ca (mg/l) | 1200 | 10–2500 |
| Mg (mg/l) | 470 | 50–1150 |
| Fe (mg/l) | 780 | 20–2100 |
| Mn (mg/l) | 25 | 0.3–65 |
| Zn (mg/l) | 5 | 0.1–120 |
| Sr (mg/l) | 7 | 0.5–15 |
| | | |
| *Methanogenic phase* | | |
| pH (−) | 8 | 7.5–9 |
| $BOD_5$ (mg/l) | 180 | 20–550 |
| COD (mg/l) | 3000 | 500–4500 |
| $BOD_5$/COD (−) | 0.06 | – |
| $SO_4$ (mg/l) | 80 | 10–420 |
| Ca (mg/l) | 60 | 20–600 |
| Mg (mg/l) | 180 | 40–350 |
| Fe (mg/l) | 15 | 3–280 |
| Mn (mg/l) | 0.7 | 0.03–45 |
| Zn (mg/l) | 0.6 | 0.03–4 |
| Sr (mg/l) | 1 | 0.3–7 |

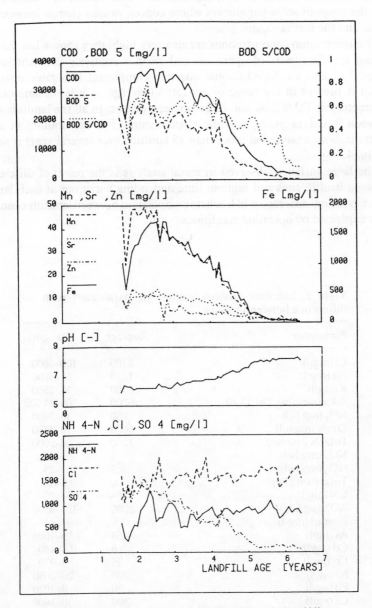

**Figure 3.** Leachate composition versus time of a full scale landfill.

as shown in Fig. 3 from an operating landfill. Table 1 contains average values
and the range of some parameters whose concentrations change between the
acetic and the methanogenic phase.

The ammonium concentrations are also very high; they show a low increase
during the first years of operation and further relatively constant values'
(Figs. 1, 2 and 3). An additional nitrogen component is organic nitrogen,
which is present in the range of 30–150% (average = 70%) of ammonium.
Parameters in Table 2 do not differ between phases. In all the landfills aging
between 0 and 12 years, no time slope occurs. The distribution of metal
concentrations analysed at more than 15 landfills over several years is shown
in Fig. 4.

The large variation observed in metal analysis is the result of differences
between landfill sites and random time depending variations at each landfill
site. Up to now differences in leachate composition between landfills cannot be
fully explained by operation conditions.

**Table 2.** Leachate analysis values for parameters with no
difference between phases.

| Parameter | Average | Range |
|---|---|---|
| Cl (mg/l) | 2100 | 100–5000 |
| Na (mg/l) | 1350 | 50–4000 |
| K (mg/l) | 1100 | 10–2500 |
| Alkalinity (mg CaCO$_3$/l) | 6700 | 300–11 500 |
| NH$_4$ (mg N/l) | 750 | 30–3000 |
| OrgN (mg n/l) | 600 | 10–4250 |
| Total N (mg N/l) | 1250 | 50–5000 |
| NO$_3$ (mg N/l) | 3 | 0.1–50 |
| NO$_2$ (mg N/l) | 0.5 | 0–25 |
| Total P (mg P/l) | 6 | 0.1–30 |
| CN (mg/l) | – | 0.04–90 |
| AOX ($\mu$g Cl/l)* | 2000 | 320–3500 |
| Phenol (mg/l) | – | 0.04–44 |
| As ($\mu$g/l) | 160 | 5–1600 |
| Cd ($\mu$g/l) | 6 | 0.5–140 |
| Co ($\mu$g/l) | 55 | 4–950 |
| Ni ($\mu$g/l) | 200 | 20–2050 |
| Pb ($\mu$g/l) | 90 | 8–1020 |
| Cr ($\mu$g/l) | 300 | 30–1600 |
| Cu ($\mu$g/l) | 80 | 4–1400 |
| Hg ($\mu$g/l) | 10 | 0.2–50 |

* Adsorbable organic halogen.

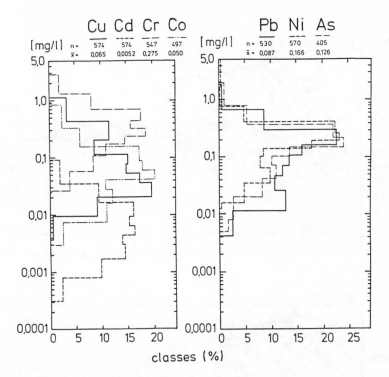

**Figure 4.** Distribution of leachate metal concentrations.

## DURATION OF LEACHATE EMISSIONS

Figure 5a presents COD data of different leachates. Organic carbon was only measured at larger intervals, but it was possible to find a correlation between COD and TOC. After changing to the methanogenic phase most values change to a more homogenous level of 3000 to 4000 mg COD/l. The level of methanogenic phase leachate shows also a slight time slope which could be extrapolated (Fig. 5b). Much more carbon is transferred though from the solid phase into the gas phase (Fig. 5c). The main carbon transfer from the solid phase could be estimated to take place in a period of one century.

In contrast to carbon transfer the transfer of chlorides is based mostly on chemical processes, even if it is influenced by biological processes. When estimating the transfer rates two statements are necessary:

1. Observations at landfills with a maximum age of 20 years show no decrease of chloride concentrations.

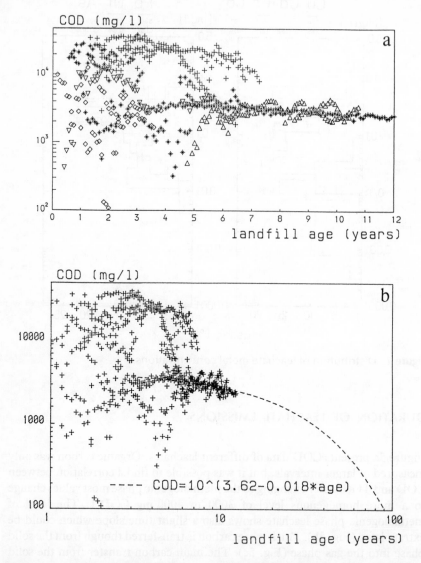

**Figure 5.** (a) Measured COD–values of different landfill leachates, (b) COD-extrapolation (calculated from 5a), (c) estimated carbon potential and transfer by gas and leachate from a refuse column of area = 1 m² and height = 20 m (transfer by leachate calculated after Fig. 5b).

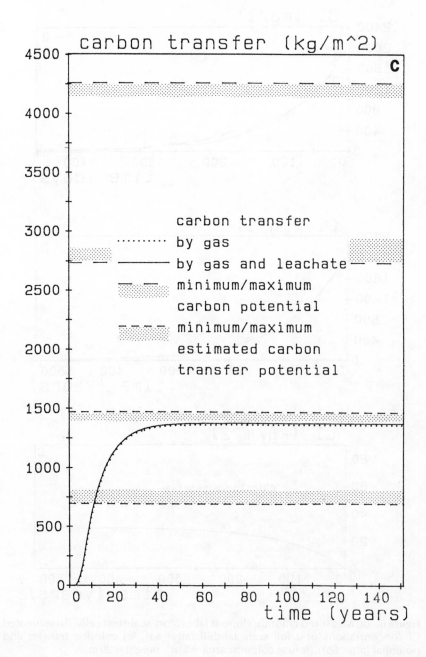

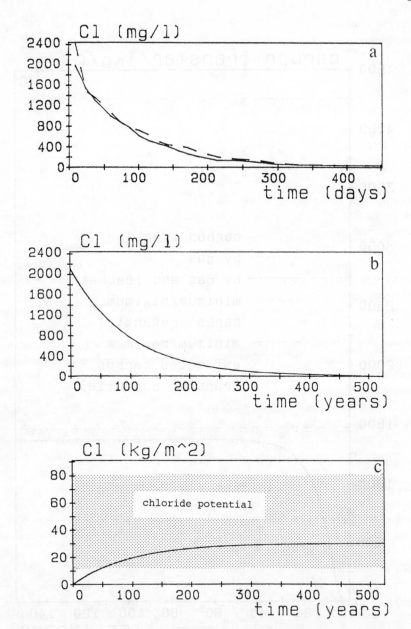

**Figure 6.** (a) Measured chloride slope at laboratory scale test cells, (b) estimated Cl concentrations of a full scale landfill (after 6a), (c) chloride transfer and potential (after 6b); (refuse column: area = 1 m², height = 20 m).

2. The chloride emission potential of landfills as batch reactors is limited and therefore a decrease in concentration must occur.

To close these differences the results of laboratory scale experiments, shown in Fig. 6a have been used, but the operation conditions were completely different from those of the landfills. One connection point was the water regime of both, using two possible parameters:

1. water contact per unit area;
2. water contact per unit volume.

Water contact per unit volume showed a better relationship with 1 day on laboratory scale = 1.16 years on full scale (leachate = 20% of precipitation). The possible general chloride concentration in leachate from full scale landfills is shown in Fig. 6b and the chloride transfer in Fig. 6c. Both lower graphs are derived from a calculation based on Fig. 6a. Referring to this calculation only a part of the measured chloride potential could be transferred.

The nitrogen transfer is a combination of carbon and chloride transport processes, because most of the nitrogen must be converted from organic nitrogen into ammonia by biological processes. As a result the tendency of ammonium concentration in Fig. 7a is less defined when compared with chloride. As for nitrogen, the calculated concentration values for ammonium are shown in Fig. 7b and the transfer function in Fig. 7c. Differences to measured nitrogen potential are much greater than for chloride. This could be a result of the much less uniform composition and also different solubility conditions of nitrogen in refuse. But as with chloride the lack of knowledge is too large to explain such differences in a precise way, because the transferable part of both components is unknown.

The principal transfer functions must be nearly the same as for other compounds. However, the relationship between measured initial concentrations and the potential in landfills is much higher. This means that longer halftimes must be estimated. As an example: the potential of organic chloride as part of the organic halogens from a refuse column with a base of 1 m$^2$ and 20 m height is 19.2 kg. The total organic chloride transfer by gas is 0.053 kg org. Cl/m$^2$ in 100 years. Similar relationships are shown for some metals in Table 3. For halogens and for metals very long emission times with concentrations coming near to the initial concentrations could be expected.

## OPERATION EFFECTS

During the past years several operation effects have been discussed which could reduce leachate emissions and/or enhance gas production:

1. water addition,

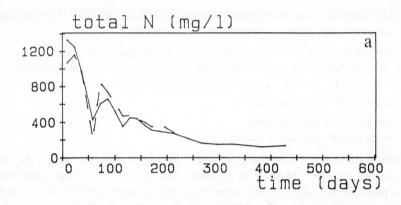

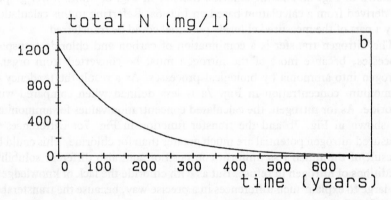

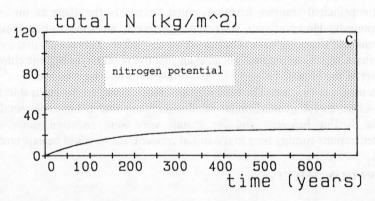

**Figure 7.** (a) Measured nitrogen slope at laboratory scale test cells, (b) estimated N concentrations of a full scale landfill (after 7a), (c) nitrogen transfer and potential (after 7b); (refuse column: area = 1 m², height = 20 m).

**Table 3.** Metal contents of refuse and transfer rates.

| Parameter | Contents mg/t dry refuse | Transfer rate* mg/(t dry refuse/year) |
|---|---|---|
| Ni | ≈15 000 | 2.7 |
| Cr | 5000–100 000 | 4.0 |
| Cu | 238 000 | 1.1 |
| Pb | 399 000 | 1.2 |
| Zn | 521 000 | 8.0 |
| Cd | 3480 | 0.08 |
| Hg | 640 | 0.13 |

* Refuse column: area $= 1\,m^2$; high $= 20\,m$; leachate production $=$ 20% of precipitation.

2. leachate recirculation,
3. seeding with digested sewage sludge,
4. lime addition,
5. uncontrolled refuse composting,
6. refuse compaction,
7. thickness of dumped layers,
8. landfill build up speed.

Modifications such as water or lime addition are only used in laboratory and pilot scale experiments. The results are not so excellent that they could compensate for other disadvantages and costs.

Seeding with digested sewage sludge shows highly discrepant results. In a full scale landfill experience in Switzerland a sharp decrease of organics over a short time was observed. On the contrary, pilot scale results show both increases and decreases of leachate organics, depending on the distribution of sludge in landfills.

Leachate recirculation is a common practice of landfill technique in several experimental scale landfills. The published results of many small scale landfills show a sharp decrease of leachate organics over a short time estimated as a result of increasing moisture content. Similar values were observed at many operated landfills in the northern part of West Germany. In contrast to small scale experiments the rate of recirculated leachate is much smaller, often less than one-hundredth. Additionally at some landfills with high build up speed or layers higher than 2 m per year, no effects of recirculation could be observed. In general, it could be estimated that an increasing moisture effect is not the main factor of leachate recirculation. Figure 8 shows $BOD_5$ and COD-values from landfills with different operation techniques. Beside leachate recircu-

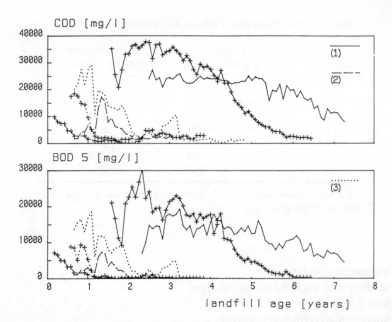

**Figure 8.** Operation effects on leachate organics; (1) 2 m layers, (2) thin layers, (3) 2 m layers and leachate recirculation (2 landfills for each case–with and without crosses).

lation dumping in thin layers (onion skin) shows also shorter times before the methanogenic phase with low organics is reached.

An explanation for this could be the penetration of oxygen into the landfill surface up to 1 m depending on density. Figure 9 shows calculated degradation rates of refuse organics versus build up speed and layer thickness. This calculation also takes into account the lower oxygen penetration depth with decreasing layer thickness.

This uncontrolled composting effect could be an explanation for low leachate organics at landfills with only a small annual refuse amount. Using leachate recirculation could increase oxygen effects, because leachate could be degraded in the topmost meter of refuse. With 0.5 l/m²d leachate, 15 000 mg $BOD_5$/l, 10% recirculation area, and an aerobic depth of 0.5 m the loading rate is 0.15 kg $BOD_5$/m³d. With such a loading rate and the overall short breaking-in period low $BOD_5$–effluents could be produced without oxidation of ammonium. The result of infiltrated recirculated leachate is a low loaded but high buffered water, which could positively influence biological

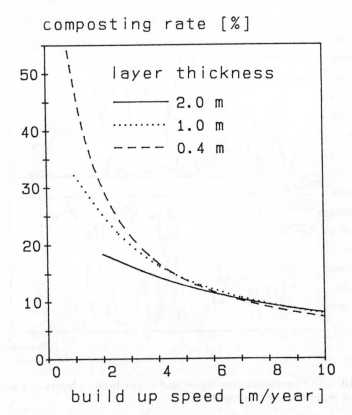

**Figure 9.** Composting rates of solid waste depending on layer thickness and build up speed.

processes. The described aerobic effects could also be observed at landfills where uncontrolled composted waste (low compacted, 2–3 m per year) was dumped. In general, most of the employed modified landfill operation techniques based on aerobic waste pretreatment are designed to prevent the violent production of acetic acids. Measurements below each layer of test cells have shown that such modifications are only necessary for the first layer. Leachate of the following layers could there be extensively anaerobically pretreated (Fig. 10). Taking this fact into account also effective and expensive modifications for the first layer are possible. The above described modifications have an important effect on leachate organics and some other components similar to values in Table 1 with a reduction of leachate treatment

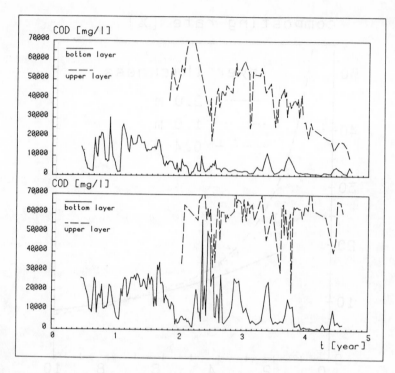

**Figure 10.** COD concentrations above and below bottom layers of 2 test cells (effect of anaerobic pretreatment).

costs. But the overall effect on long time leachate emissions must be estimated near zero.

## CONCLUSIONS

Sanitary landfill leachate is a complex polluted wastewater with high concentrations of several organic and inorganic pollutants. The main leachate parameters, including metals, are well known, but the knowledge of hazardous micropollutions in leachates and also in solid wastes is very small. Modifications of landfill operation only shorten the time for reaching the methanogenic phase, but no effect on long time emissions has been estimated. With the exception of carbon and thereafter $BOD_5$ and COD, emission times could be estimated.

## ACKNOWLEDGEMENTS

The experimental data for this chapter were developed during the realization of many research projects at the Institut für Siedlungswasserwirtschaft at the Technical University of Braunschweig, West Germany.

## REFERENCES

Ehrig, H.J. (1980). 'Beitrag zum quantitativen und qualitativen Wasserhaushalt von Mülldeponien'. Veröffentlichungen des Institut für Stadtbauwesen, TU Braunschweig, 2nd edition.
Ehrig, H.J. (1988). 'Water and element balances of landfills'. Swiss Workshop on Land Disposal, March 14–18, 1988, Gerzensee, Switzerland.
Ehrig, H.J. 'Sickerwasser aus Hausmülldeponien'. Menge und Zusammensetzung, Müllhandbuch (in press).
Greiner, B. (1983). 'Chemisch-Physikalische Analyse von Hausmüll, Berichte 7/83, Erich Schmidt Verlag.

## 4.3    Leachate Recirculation

HEIKO DOEDENS and K. CORD-LANDWEHR

*University of Hannover, Welfengarten 1, D-3000 Hannover 1,
West Germany; FH-NO-Niedersachen, D-3113 Suderburg,
West Germany*

### INTRODUCTION

Recirculation of leachate through a landfill combines anaerobic pretreatment within the landfill, regarded as a fixed-bed reactor (FBR), with evaporation occurring during each cycle.

Due to these features recirculation can contribute towards the minimizing of problems arising from the high variations in the quality and quantity of leachate. These variations exert a negative effect on the treatment of leachate either together with sewage in public treatment plants or in special plants for leachate alone.

The highest concentrations of BOD, COD and even heavy metals caused by low pH values are to be found shortly after dumping of the waste, i.e. in the newest layers or sections of the landfill. If external treatment of this highly concentrated leachate can be avoided by internal pretreatment then the working capacity of the treatment plant can be reduced to a small fraction of the load involved if no such measure is taken.

Since high amounts of COD and BOD are caused mainly by organic acids in the leachate, a suitable measure to take is that of enhancement of methane production through improvement of such conditions as water content and transport. This can be achieved by recycling leachate also by means of partial evaporation in the case of spray irrigation over the landfill. Besides spray irrigation, further methods are possible whereby water collected from the bottom seal of the landfill is redispersed onto the surface or infiltrated into the landfill.

The possible advantages in the recirculation of leachate are, however, accompanied by several disadvantages and risks:

1. Infiltration of additional water involves an additional risk of leachate infiltration with pollution of the soil and ground water in the case of non-existent or damaged bottom sealing. Thus an increased amount of infiltration does not correspond with the general effort of reducing the water content of landfills.
2. Multiple elutriation by leachate recycling could result in higher concentrations of salts and heavy metals in the leachate.

In order to further understanding of leachate recycling, a research project financed by the Ministry for Research of the Federal Republic of Germany was carried out, over a five-year period (Cord-Landwehr *et al.*, 1986). Investigations were carried out on three scales: test cells and landfill sections, both with and without leachate recycling, as well as large-scale landfills equipped with recycling techniques.

## LEACHATE RECYCLING IN LARGE-SCALE LANDFILLS

In 1981 13 large-scale landfills, mainly in the north of the Federal Republic of Germany were practising leachate recycling (LR). Table 1 reports important data characteristic of these landfills. Further results obtained are as follows.

Firstly, tanker-lorries or water dispersal techniques either on or below the surface of the landfill were used in the distribution of leachate spray irrigation systems.

Secondly, the specific amounts of waste dumped varied greatly from 2800 to 80 000 t/ha y, corresponding to an elevational speed of 0.3–10 m/y. Leachate occurs even at the points of highest elevational speed, although the additional storable quantities of water caused by the homogeneous distribution of humidity amount to almost ten times that due to seepage or percolation.

Thirdly, the quantity of leachate obtained from the bottom of landfills cannot be used as a process parameter to describe LR-landfills either in the dimension $m^3$/ha y or in the percentage of precipitate, on account of possible short circuiting due to multiple back-pumping. A suitable process parameter is represented by that obtained from drained-off quantities of leachate. According to Table 1 these values varied from 0 at four sites to 1.459 $m^3$/ha y in comparison to 1.825 $m^3$/ha d based on an average of 4.5 $m^3$/ha d. The evaporation potential can be most effective if the single spray irrigation dose is maintained below 5 mm. This has not as yet been realized on many landfills.

Fourthly, in situations such as that observed in Germany, where the potential of evaporation is higher than the level of precipitation only from April to October and where the majority of excess leachate is produced during the remaining period, a large storage capacity should be provided, of up to 1500–2000 m³/ha in order to take maximum advantage of evaporation potential during summer time. In reality, four of the landfills visited had no buffer capacity at all whilst at the other nine landfills the storage volume was only 4–400 m³/ha landfill area.

Fifthly, in German climatic conditions, a complete evaporation of leachate by LR would not be possible even in conditions of sufficient storage capacity. Thus, an external system for the treatment of excess leachate has to be provided but, due to the LR, the excess amounts can be calculated according the following figures: 0.5 m³/ha d (average) with spray irrigation; 2 m³/ha d (average) with surface percolation; as opposed to 4–5 m³/ha d without LR.

Finally, in order to avoid odour problems, leachate should not be irrigated unless the ratio BOD : COD is below 0.1 or BOD is below 1000 mg/l.

In Figs. 1 and 2, levels of BOD and COD from LR landfills are compared with those obtained in another research study (Ehrig, 1980). Generally statements relating to such a time-dependent course are rendered more difficult in that the landfills are often equipped with an undivided drainage network where more and less polluted leachates from various old sections of the landfill combine (the age of the landfill relates to the oldest section thereof). Those landfills where LR has been practised since commencement of operation of the landfill demonstrate a faster reduction of BOD and COD than that represented in Ehrig's enveloping graph: after a maximum of four years from starting the landfill, all LR sites investigated had decreased to less than 1000 mg BOD/l and 10 000 mg COD/l. At the beginning both landfills in Hannover and Braunschweig-Wolfenbüttel show extremely high concentrations with high elevational speeds and 2–3 m thick layers. Leachate is present in very low concentrations at the two 'thin-layer' sites and that proves, in accordance with Ehrig (1980), that the combination of 'thin-layer' sites and LR is indeed most advantageous.

Thin layer operation can be equated with natural ventilation and brief aerobic decomposition which reacts favourably with leachate concentrations and the onset of methane production (Jourdan, 1981; Stegmann and Spendlin, 1986). No increase in the concentrations of salts or heavy metals which could be attributed to LR was observed.

## TEST CELL EXPERIMENTS

On the smallest scale, four air-tight, temperature-regulated, steel cylindrical test cells ($D = 1.5$ m, height of waste layer $= 1.35$ m, waste

234 Doedens and Cord-Landwehr

**Table 1.** Results of questioning at landfills with leachate recycling.

| Name of landfill (part number) | Stapel-feld | Flechum | Dörpen | Venneberg I | II | III | Morgen-stern | Blanken-hagen |
|---|---|---|---|---|---|---|---|---|
| *Subject of question* | | | | | | | | |
| Area of landfill (ha) | 10 | 2.4 | 2.7 | 2 | 2 | 2 | 4.8 | 12 |
| Elevation of waste[a] (m) | 6 | 8 | 9 | 9 | 9 | 3 | 35 | 12 |
| Area connected to the landfill (km²) | 931 | 314 | 1196 | 734 | | | — | 1300 |
| (inhab.) | 72 000 | 18 000 | 80 000 | 82 100 | | | 143 000 | 160 000 |
| Amount of waste (t/y) | 28 000 | — | — | — | | | 124 000 | 188 000 |
| Compacted (m³/y) | — | 17 000 | 70 000 | 65 000 | | | — | 180 000 |
| Average precipitation (mm/y) | 750 | 812 | 750 | 790 | | | 750 | 650 |
| Amount of leachate (m³/y) | — | 1000 | 570 | — | | | — | — |
| (m³/ha · y) | — | — | — | 5900 | | | — | — |
| Excess leachate since beginning of LR[a] (m³) | 2000 | 0 | 0 | 7000 | | | 0 | — |
| Storing capacity of leachate (m³) | 1650 | 500 | 800 | 2100 | | | 0 | 50 |
| Distribution system[a] | | | | | | | | |
| irrigation | × | | | × | × | × | | |
| tank lorry | | × | | | | | | |
| others | | | | | | | ×[b] | × |
| Leachate quality (mg/l) | | | | | | | | |
| BOD | 820 | 2900 | 10 | 2395 | | | 676 | 1400 |
| COD | 1680 | 4100 | 500 | 4700 | | | 5456 | 2900 |
| pH | 7.4 | 6.0 | 7.6 | 7.7 | | | 8.3 | 7.65 |
| Starting landfill in | 1973 | 1975 | 1979 | 1976 | 1977 | 1981 | 8/77 | 1976 |
| Starting LR in | 10/81 | ×[d] | ×[d] | 9/77 | | | ×[d] | 1978 |
| Date of questioning | 3/82 | 7/82 | 7/82 | 7/82 | | | 1/83 | 8/82 |

[a] At the date of questioning.
[b] Distribution pipes.
[c] Percolation on top by tubes.
[d] After first leachate in the drains.
[e] Infiltration below the surface.

| Nauroth | Reinstetten I | Reinstetten II | Belters-rot | Süpplingen I | Süpplingen II | Watenbüttel I | Watenbüttel II | Bornhausen I | Bornhausen II | Bornhausen IIIa |
|---|---|---|---|---|---|---|---|---|---|---|
| 7 | 2.9 | 2.5 | 3 | 2.5 | 2.5 | 10 | 5 | 1.0 | 1.3 | 0.6 |
| 24 | 15 | 10 | 12 | 8 | 11 | 32 | 7 | 13 | 12 | 6 |
| 642 | 941 | | — | — | | — | | | 132 | |
| 122 000 | 111 400 | | 85 000 | 98 400 | | 257 800 | | | 24 600 | |
| — | | 62 000 | 51 000 | 40 000 | | 250 000 | | | 36 800 | |
| 120 000 | — | | — | — | | — | | | — | |
| 1100 | 800 | | 650 | 535 | | 639 | | | 710 | |
| — | | 8500 | — | 9000 | — | — | | 2281 | 5792 | 54 092 |
| | — | | — | — | | 140 | — | | | |
| 0 | | 470 | 8333 | 1800 | | 0 | — | — | 7800 | |
| 2250 | | 110 | 300 | 1000 | | 0 | 0 | 0 | 0 | 0 |
| × | 0 | × | | 0 | × | 0 | × | | | |
| | | | ×[b] | | | | | ×[b] | ×[b] | ×[c] |
| 202 | 6000 | | 140 | 525 | 850 | 368 | 23 000 | 44 | 58 | 135 |
| — | — | | 4800 | 2250 | 2580 | 2285 | 46 500 | 959 | 1443 | 2680 |
| 7.4 | 7.4 | | 8.1 | 7.5 | 7.5 | 7.7 | 6.1 | 8.3 | 8.3 | 8.0 |
| 1973 | 1975 | | 7/80 | 1/75 | 1/78 | 7/67 | 2/82 | 11/74 | 7/77 | 10/81 |
| 1975 | 0 | 5/82 | 1980 | ×[d] | | ×[d] | | ×[d] | | |
| 8/82 | 10/82 | | 10/82 | 1/83 | | 1/83 | | 1/83 | | |

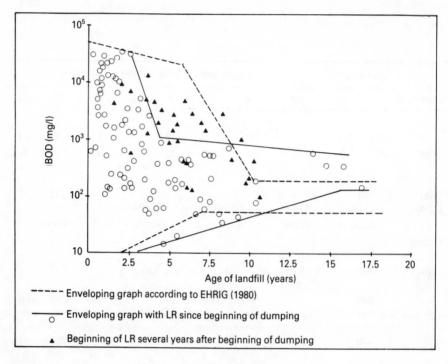

**Figure 1.** BOD concentration vs. age of landfill.

volume = 2.4 m³; $T$ = 35°C) were installed (Fig. 3). Waste which had first been shredded by a shredding mill and then stored for 14 days covered in foil was packed into the test cells with a hydraulic hammer until a density of 0.77–0.86 t/m³ (inclusive of water) was reached, which is comparable to densities reached by compactors. The original water content amounted to 24–31%.

Test cell 1 was set up with 660 mm/y rainwater. All leachate released was fed back into the cell. The remaining three test cells were set up with 50% of the years precipitation; however various water compositions were chosen. Test cell 2 received rain water only, i.e. it was not to be subjected to recycling. Test cell 3 was irrigated with rainwater and leachate. As leachate left the cell, it was fed straight back in and topped up with rain water to a total water volume of 330 mm/y, whereby the proportion of leachate only ever reached a maximum of 40% of the total amount. Test cell 4 was brought to the saturation point with the leachate from a stabilized landfill, i.e. to the point where the amount of water entering the cell daily equalled approximately that leaving it. Thereafter

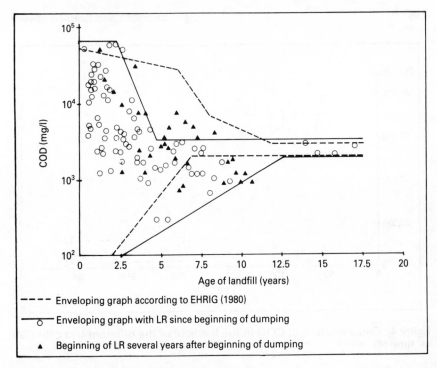

**Figure 2.** COD concentration vs. age of landfill.

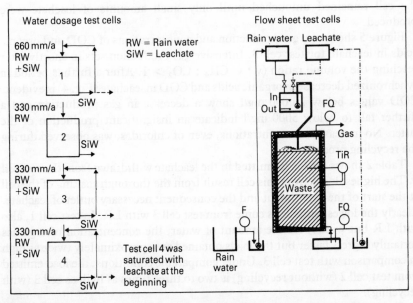

**Figure 3.** Flow sheet and water dosage scheme for the four test cells.

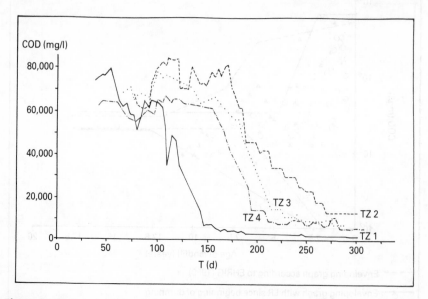

**Figure 4.** Concentration of COD in the leachate of the different test cells (TZ) vs. time (*T*).

the cell remained untouched until only small amounts of leachate were produced.

Figure 5 shows the gas production and concentrations of COD and organic acids in leachate for test cell 3. Intensive gas production sets in shortly after reaching the volume ration (v : v) $CH_4 : CO_2 > 1$. After a further period, a synchronized decrease in organic acids and COD in leachate (Fig. 4) is evident. COD values below 20 000 mg/l show a decrease in gas production and a further fall to below 5000 mg/l indicate an insignificant production of the latter. No increase in concentrations, even of chlorides, was observed during the recycling process.

Table 2 shows the loads emitted in the leachate withdrawn from the test cell 4. The highest loads from this cell result from the thorough soaking of the cell at the start of the experiment and the consequent necessary outlet of leachate. Clearly the lowest emissions come from test cell 3 with LR. In test cell 1, also with LR but with twice the amount of water, the concentration of organics certainly regress faster but the loads obtained are approximately twice as high in comparison with test cell 3. Under comparable conditions, the load emitted from test cell 2 (without recycling) is two to three times as high as cell 3 (with recycling).

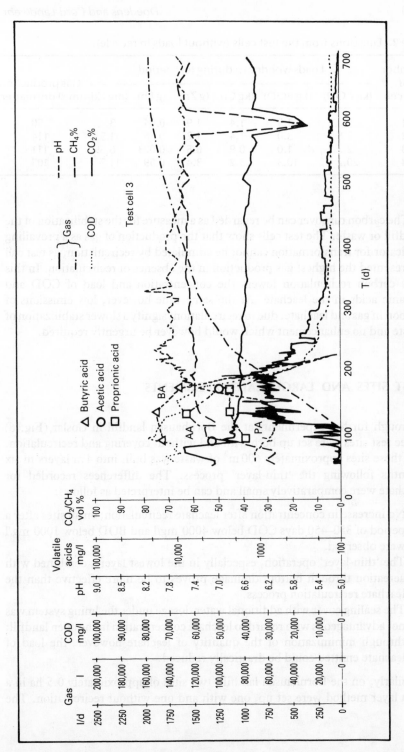

**Figure 5.** Relation between leachate and gas production in test cell 3 vs. time (*T*).

**Table 2.** Emissions from the test cells (without loads in recycle).

| Number of test cell | Loads withdrawn during test period | | | | | | Gas production (m³/t dry matter) |
|---|---|---|---|---|---|---|---|
| | (kg COD) | (kg BOD) | (kg Cl) | (g Zn) | (g Pb) | (mg Cd) | |
| 1 | 5 | 1.5 | 1.4 | 1.9 | 0.04 | 3 | 50 |
| 2 | 7.8 | 3.0 | 1.4 | 1.2 | 0.38 | 1.5 | 114 |
| 3 | 2.7 | 1.0 | 0.9 | 0.5 | 0.023 | 0.55 | 111 |
| 4 | 20.2 | 10.3 | 3.2 | 3.4 | 0.59 | 11.7 | 105 |

The carbon turnover can be regarded as a measure for the stabilization of the landfill or waste. The test cells show that the production of gas as a prevailing indicator for transformation cannot be stimulated by recirculation, as test cell 2 presented the highest gas production in the absence of recirculation. In this way carbon recirculation lowers the concentration and load of COD and organic acids in the leachate. At the same time however, low emissions of carbon in gas and leachate, due to recirculation, signify a slower stabilization of waste and no enhancement which would however be urgently required.

## TEST SITES AND LARGE-SCALE EXPERIMENTS

Through further experiments at the Bornhausen landfill in Goslar (Fig. 6) three test sites were set up both with and without covering and recirculation. On these sites approximately 600 m³ of waste was built into 4 m layers in six months following the 'thin-layer' process. The differences recorded for leachate were comparatively small and can be interpreted as follows:

1. No increase in concentration after leachate recirculation; on all sites after a period of 350–450 days COD below 4000 mg/l and BOD below 1000 mg/l were observed.
2. The 'thin-layer' operation, especially in the lowest layers, combined with aereation through bottom drainage proves to be more effective than the leachate recirculation process.
3. The sealing cover with additional water dosage under the lining system was not advantageous with regard to leachate concentration in the open landfill; through minimization of the quantity of leachate however, the load of leachate emitted could be drastically reduced.

Similarly, on the Bornhausen landfill, two sites of approximately 0.5 ha in a 2 m layer method were set up, one with and one without recirculation. The

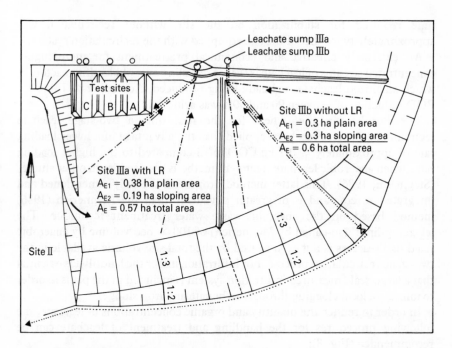

**Figure 6.** Plan of landfill and test sites at Bornhausen (County of Goslar).

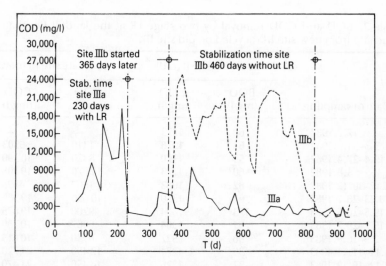

**Figure 7.** Concentration of COD in leachate from sites IIIa and IIIb (see Fig. 6) vs. time (*T*).

time required for stabilization on the site without recirculation was approximately twice that of the site equipped with the recirculation system.

An extremely effective and economical pretreatment for more highly concentrated streams of leachate is that of recycling over old landfill reactors in which stabilized leachate is already being produced. According to Table 3 a BOD reduction of between 90 and 99% was achieved.

The Bornhausen site, where for the first time a two-stage leachate recirculation process has been applied, has the advantage of a lower loading rate of approximately $\leqslant 0.1$ kg $COD/m^3$ d compared to the higher loading rates when anaerobic leachate treatment methods are used, as in other studies (Stegmann, 1986). This latter method proposes a specially dimensioned and integrated reactor and is of greater advantage than the Mennerich (1986) method based on the anaerobic-waste water treatment technique. The 'clogging-limit' (so-called by Mennerich) of 100 × bed volume by anaerobic fixed bed reactors is not even neared by the total site volume available in the two-stage recirculation process. For infiltration under the landfill it is essential that a large-scale measured infiltration system is included in the plans in order to minimize local clogging through carbonate and iron sludge.

In order to reduce the quantity and organic concentrations of leachate, the following procedures for the handling and treatment of leachate can be recommended (Fig. 8):

1. preparation (as early as possible) of an anaerobic fixed-bed landfill reactor

Table 3. BOD and COD removal by two stage LR at the landfill Bornhausen (leachate from new site III recycled on old site II).

| Date of sampling | Leachate site II (old) | | Leachate site III (new) | |
|---|---|---|---|---|
| | BOD (mg/l) | COD (mg/l) | BOD (mg/l) | COD (mg/l) |
| 17.2.1982 | 60 | 1473 | — | — |
| 3.3–31.3.1982 | 64 | 1278 | 1310 | 5303 |
| 15.4–27.4.1982 | 59 | 1370 | 5320 | 10 390 |
| 3.8.1982 | 60 | 1273 | 11 970 | 19 308 |
| 2.12–30.12.1982 | 82 | 1083 | 1807 | 4898 |
| 10.23–28.2.1983 | 61 | 1350 | 10 650 | 19 385 |
| 19.5–26.5.1983 | 173 | 1604 | 9200 | 19 675 |
| 15.7–29.7.1983 | 91 | 1364 | 6387 | 10 780 |
| 2.9–16.9.1983 | 96 | 927 | 8750 | 10 615 |
| 4.11–11.11.1983 | 39 | 1271 | 12 450 | 21 720 |
| 2.12–16.12.1983 | 37 | 1226 | 14 450 | 21 470 |
| 20.1–27.1.1984 | 75 | 1725 | 6450 | 16 425 |

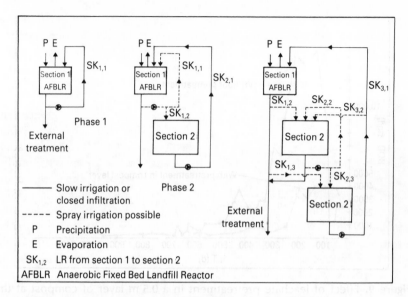

**Figure 8.** Two-stage leachate recycle scheme.

(AFBLR) by aerobic pre-stabilization of a bottom layer consisting of aereated waste or compost;

2. recycling of all leachate with a BOD level higher than 500 mg/l to the AFBLR;
3. recycling of pre-treated leachate from the AFBLR by means of spray irrigation to minimize the quantity;
4. external treatment of excess amounts of leachate.

By these means conversion of TOC in organic acids to landfill gas can be achieved instead of conversion to mainly excess sludge in the case of aerobic effluent treatment.

The first German experiences of this type of internal treatment have already been published (Damiecki, 1987). Other landfills have similar systems under construction and the results are summarized in the graphs shown in Fig. 9.

At the Northeim landfill (Federal Republic of Germany), leachate with higher concentrations of organics is recycled in an infiltration system beneath the top lining of the landfill which consists of a polythene sheet and a layer of clay: a moisture content of more than 30% is essential for satisfactory methane production but a higher water content will not mean further enhancement (Barlaz *et al.*, 1987). Therefore top sealing of landfills may be advisable if a

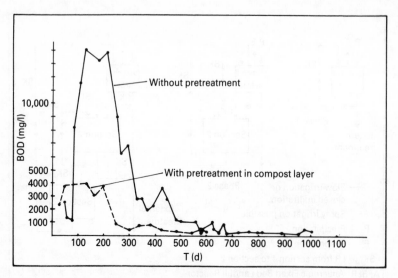

**Figure 9.** Effect of leachate pretreatment in a 0.5 m layer of compost at the bottom of the landfill (Damiecki, 1987).

sufficient water content is provided prior to sealing, although water content can be controlled after sealing by small amounts of recycled leachate.

The Bornum landfill in the Wolfenbüttel district has no connection to public sewers. The leachate therefore has either to be transported or fed back into the landfill. The collection of leachate is achieved through the use of vitrified clay pipes with a diameter of 200 mm which are installed at a 50 m distance from each other. The leachate is fed into a 2800 m³ effluent treatment plant by means of a sump. The activated sludge tank is supplemented by a 2800 m³ storage tank.

Between July 1984 and June 1987 two landfill sites, each with a capacity of 12 500 m² were operating. Up until June 1985 only domestic waste which was deposited in a loose bottom layer was introduced into site one. Since November 1985 field two has been filled in the same manner.

Figure 10 demonstrates the total amount of precipitation and waste accumulated over a three-year working period. Waste received was 161 470 t and precipitation amounts to 1.749 mm.

The development of the quality of leachate is illustrated in Fig. 11. Commencement of operation of the landfill (field 2) can be clearly observed. Prior to this, the 2800 m³ activated sludge tank had to be filled. In July 1985 leachate treatment was made possible although initially only small quantities of

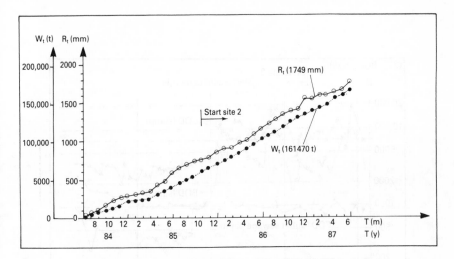

**Figure 10.** Total waste production ($W_t$) and total rain precipitation ($R_t$) vs. time at the Bornum landfill site (m = months, y = years).

leachate were removed. Due to the difficulties encountered in spray irrigation in winter the recycling of leachate did not occur until March 1986. The quantity of leachate was measured by means of a meter; the quantity transported through weighing of the tanker lorries and the quantity of sump was measured by means of an inductive flow meter. Prior to September 1986 when the latter measurements were commenced, the filling and pumping time required for the two 2800 m$^3$ storage tanks were measured (Fig. 12).

The quality of recycled leachate is equal to the sewage plant outlet (**Fig. 11**). With regard to the development of leachate quality in the Bornum landfill it is evident that the progress of the pollution load becomes positive extremely rapidly. Within approximately two years COD values of < 4000 mg/l and BOD values of < 1000 mg/l are reached. The pH value is almost invariably neutral. The reason for this positive development can be found in the loose bottom layer of domestic waste and in the leachate recycling. It must be pointed out that there are relatively high values of $NH_4$ similarly to landfills which are not equipped with leachate recycling systems. In an intended procedure of nitrification and denitrification connected with a reduction of the excessive organic load these high values may lead to a carbon deficit.

The measured concentration and the known leachate quantity result in the loads generated within and those removed from the landfill. In calculating these results an essential advantage of the recirculation of leachate is

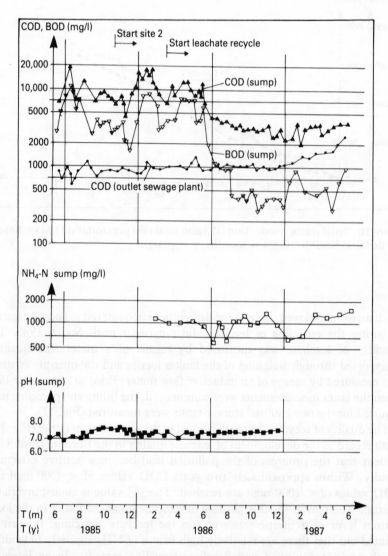

**Figure 11.** Leachate quality vs. time at the Bornum landfill site (County of Wolfenbüttel).

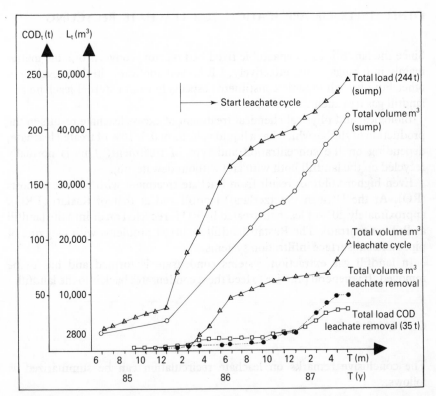

**Figure 12.** Total COD load ($COD_t$) and total volume of leachate ($L_t$) vs. time at the Bornum landfill site.

evidentiated: at high concentrations leachate can be retained inside the landfill. Even when conditions are disadvantageous, such as when partially purified leachate is not removed from the storage tank but is extracted from the sump, COD loads can be calculated as follows (Fig. 12): 244 t collected from sump; 35 t removed from landfill system.

As the aim of the Bornum landfill is to achieve high natural evaporation which is only possible by means of spray irrigation, leachate must first be partially purified. As long as the stabilized methane phase is not reached there will be a considerable odour problem due to organic acids, especially butyric acid, during the phase of acid fermentation. If leachate no longer contains organic acids spray irrigation may be carried out without pretreatment.

As 1987 was an exceptionally damp year, removal of leachate has had to be intensified; a third landfill section is in preparation, and further developments remain to be seen.

## OTHER TYPES OF APPLICATION FOR LEACHATE RECYCLING

Since the landfill, as an anaerobic fixed-bed reactor, converts organic matter mainly into biogas quite effectively, LR is even applicable in the recycling of other liquids with organic constituents especially from external leachate and landfill gas treatment.

Biological and physical/chemical treatment of excess leachate results in the production of sludge which has a liquid volume of 0.5–10% of treated leachate, depending on the concentration and type of treatment. This is normally recycled on the landfill both with and without dewatering.

Even higher volumes result from leachate treatment with reverse osmosis (RO). At the Uttigen (Switzerland) landfill and at that of Rastatt (FRG), approximately 20% of leachate treated by RO is recycled on or into the landfill as RO concentrate. The Rastatt landfill incurred problems with clogging of pipes and subsurface infiltration systems.

In landfill gas extraction systems condensate is formed and has to be removed. It is general practice to feed these condensates back into the landfill.

## CONCLUSIONS

The conclusive remarks on leachate recirculation can be summarized as follows:

1. Recirculation stimulated a rapid decrease to BOD and COD in leachate.
2. It was not proven that recirculation could enhance waste stabilization or increase the quantity of landfill gas.
3. Two-stage leachate recirculation can be used as a fairly economical and effective means of leachate pretreatment in an anaerobic fixed-bed landfill reactor. The same results can be achieved through composting or aerobically pretreating the bottom waste layer in a landfill.
4. Sludge from leachate treatment and condensate from landfill gas are further examples of material which could be recirculated through the landfill.

## REFERENCES

Barlaz, A.M., Wilke, M.W. and Ham, R.K. (1987). 'Gas production parameters in sanitary landfill simulators', *Waste Management and Research*, 5, 27–39.
Cord-Landwehr, K. and Doedens, H. *et al.* (1986). 'Stabilisierung von Mülldeponien durch Sickerwasserkreislaufführung', FE-Report, BMFT-FB-T 86-124, Karlsruhe.

Damiecki, R. (1987). 'Vorbehandlung des Sickerwassers innerhalb der Deponie', *Entsorga-Magazin* 5, 41–44.

Ehrig, H.J. (1980). 'Beitrag zum quantitativen und qualitativen wasserhaushault von Mülldeponien'. Veroffentlichungen des Institut für Stadtbauwesen, TU, Braunschweig, 2nd edition.

Jourdan, B. (1981). Reduzierung des Sickerwasseranfalles auf einer Rottedeponie', Stuttgarter Berichte zur Abfallwirtschaft 15, 265–286.

Mennerich, A. (1986). 'Untersuchungen zur anaerob-aeroben Reinigung von Sickerwassern aus Mülldeponien'. Lecture held at the symposium on Leachate Treatment, Aachen, 1986.

Stegmann, R. (1986). 'Verfahren zur kontrollierten anaeroben Sickerwasserbehandlung im Deponiekörper'. Lecture held at the symposium on Leachate Treatment, Aachen, 1986.

Stegmann, R. and Spendlin, H.H. (1986). 'Untersuchungen zur Beschleunigung der anaeroben biochemischen Umsetzungprozesse in Mülldeponien'. Internal report TU Hamburg-Harburg, FE-Project Nr. 1430267.

# 4.4    Biological Treatment

R. COSSU,* R. STEGMANN,** G. ANDREOTTOLA,***
P. CANNAS****

*Institute of Hydraulics, University of Cagliari, Piazza d'Armi,
09100 Cagliari, Italy
**Technical University of Hamburg-Harburg, Eissendorfer Str. 40,
D-2100 Hamburg 90, West Germany
***Institute of Sanitary Engineering, Polytechnic of Milan,
Via Fratelli Gorlini 1, 20151 Milano, Italy
****CISA, Environmental Sanitary Engineering Centre,
Via Marengo 34, 09123 Cagliari, Italy

## INTRODUCTION

The more important characteristics of leachate which could influence biological treatment concern the high concentration of organic and inorganic substances, irregular production depending on the amount of rainfall, variations in the biodegradable fraction of organic substance depending on age of the landfill, and the low or negligible amounts of phosphorus. Several indications regarding the degree of biological degradability of organic substances present in leachate are given by the ratio BOD:COD, values of approximately 0.5 indicate that organic components will be easily broken down by bacteria whilst lower values signify that the organic fraction present in leachate will not be overly biodegradable and therefore biological treatment systems are not advised.

These characteristics of leachate impose operational difficulties on treatment processes not normally found when treating waste water of consistent strength and volume. Different treatment strategies are therefore required to match treatment to the differing leachate volumes and strengths during the filling phase and aftercare of a landfill.

Furthermore leachate from landfills may contain substances which are able to limit biological treatment efficacy. Those compounds which may affect

251

treatability are as follows (Harrington *et al.*, 1984):

1. *Metals*: the concentrations of metals are reduced inside the landfill under the prevailing anaerobic conditions. This attenuation is a function of the pH value, carbonate and sulphide solubilities. The presence of heavy metals in leachate, which can successfully be tolerated, is dependent on the concentration and activity of the bacteria. If metal inhibition is considered a problem, simple pretreatment using lime to precipitate and remove the metals as hydroxides should be practised.

2. *Carbon compounds*: these include chlorinated solvents, cyanide compounds and phenols. Once again, attenuation of these compounds inside the landfill occurs and no pre-treatment is foreseen. As a precaution, however, a balancing lagoon should be constructed prior to the biological stage if these compounds are considered a problem. Very high concentrations of both phenols and cyanide compounds can be biologically oxidized by acclimatized biological systems. Chlorinated solvents will almost certainly be rapidly volatilized with the biological system against the effects of high concentrations of toxic components of leachate.

3. *Ammonia*: high ammoniacal nitrogen concentrations may be present in the leachate from household waste sites. Very high concentrations of ammonia in influents to biological treatment have been tolerated both by anaerobic (2700 mg/l) and aerobic (6350 mg/l) systems (Melbinger and Donnellon, 1971; Ashmore *et al.*, 1967). These values will not be exceeded in leachate.

4. *Chloride*: chloride is tolerated at relatively high concentrations by aerobic processes after acclimatization, up to 1–2% according to conditions (Ashmore *et al.*, 1967). Anaerobic processes are more sensitive and up to 10 g/l had a marked effect on gas production. However, excessive chloride concentrations above these would be rare.

5. *Sulphide*: inside the landfill complexation and precipitation of sulphides reduces the concentration of sulphide in leachate. However, high concentrations of sulphide have been found particularly at sites which have received large quantities of sulphate-containing wastes. Anaerobic digestion can tolerate up to 200 mg/l as soluble sulphide and little effect was observed up to 400 mg/l. Aerobic processes can tolerate up to 10 000 mg/l of $S^{2-}$ and satisfactorily treat 1000 mg/l without impairing performance (Dewson and Jenkins, 1950).

However the sensitivity of biological treatment processes to the effects of toxic compounds is reduced by several factors. Those microorganisms which are within the biological floc may be protected to some extent from transitory shock discharges of toxic substances. Further, even if the population is severely depleted by poisoning, this can be rapidly retrieved because the regeneration time of the microorganisms is short. Finally, bacterial populations have remarkable capacities to acclimatize to the point of actually utilizing many toxic compounds as substrates. Examples are phenols, formaldehyde, pesticides and organic solvents.

Anaerobic biological treatment is generally the more susceptible to poisoning by toxic substances, especially heavy metals, phenols and chlorinated organic compounds (Swanwick *et al.*, 1969).

Aerobic processes have the capacity to acclimatize to the presence of certain toxic organics substances (which can also oxidize in some cases) and of heavy metal ions, although in this last case the toxic substances are not biologically oxidized but are absorbed by the biological floc.

Even though biological treatment processes are tolerant and effective in removing a wide range of toxic substances a pretreatment stage may be necessary if inhibition of the process is observed, for example by a reduction in the respiration rate of the biological floc or if overall performance is anyway reduced.

Finally, systems proposed for disposal of leachate must be simplified, economical and require the least possible amount of energy and staff for management of the same.

## AEROBIC BIOLOGICAL TREATMENT METHODS

Several treatability studies have evaluated aerobic biological processes. It can be concluded generally from the following review, that leachates generated from recently emplaced wastes, in which the organic matter consists mainly of volatile fatty acids, can be degraded readily by biological means. However, the few fullscale experiences have often shown some operating problems due to load fluctuations, low temperatures, foam formation and clogging.

In leachate from older more stabilized landfills a greater proportion of the organic fraction consists of refractory material, and therefore such leachates are less amenable to biological treatment.

In the following paragraphs several methods of aerobic treatment of leachate are considered and the main experiences on laboratory, pilot and fullscale described.

### Aerated Lagooning

Introduction

When the production of leachate is relatively low, aereated lagoons may represent an efficient system for leachate treatment. Basically the idea is that the period of hydraulic retention is long enough to allow development of a high number of bacteria during the same, in such a manner that the balance of bacteria formed inside the lagoon will be on a par with those removed in the discharge.

Aereated lagooning is performed using particular devices which enhance the aerobic processes of degradation of organic substances and which act over the entire depth of the tank. This is obtained by continuously agitating the leachate by means of aereation systems (surface aerators or air pumps) in such a way that the mixture obtained hinders the formation of layers in anaerobic conditions and facilitates the introduction of oxygen from the external atmosphere. The processes of degradation of organic substances are carried out aerobically due to the presence of dissolved oxygen and are therefore faster with respect to anaerobic processes. On the other hand a more complex plant is required due to the presence of electromechanical components (aerators or pumps). The shape and depth of the tank used for aereated lagooning should be such that a satisfactory mixing is obtained by means of the aerators and that the turbulence generated ensures the presence of a sufficient amount of dissolved oxygen in the sewage.

## Application to Leachate Treatment

Knoch (1972) reports on laboratory and half-technical scale investigations; as a result, kinetic characteristics have been determined for two different leachate qualitites. The calculated degradation rates are as follows:

$k = 0.075$ (first leachate)

$k = 0.08$ (second leachate)

Leachates with $BOD_5$–concentrations $< 1000$ (mg/l) were degradated down to values around 25 mg $BOD_5$/l in about 14 days at room temperature; with longer detention times (28 days) $BOD_5$–effluent concentrations $< 25$ mg/l were monitored. Leachates with an organic content between 4000 and 10 000 mg/l could not be degradated down to values of 25 mg/l in a period of 80 days. In these experiments no phosphorus was added to the leachate.

Boyle and Ham (1972, 1974) obtained substantial oxidation of organic matter over a period of several months by aerating leachate from fresh refuse in one-litre glass bottles on a fill-and-draw basis, with a mean retention period of 5 days. It was shown that 80–93% removal of COD could be accomplished at loadings of between 0.54 and 1.04 kg COD/$m^3$ per day. However, an increase in organic loading to 1.74 kg COD/$m^3$ per day (corresponding to 1.39 kg BOD/$m^3$ per day) resulted in considerable reduction in process efficiency. In all experiments it was found necessary to add small amounts of an antifoam solution in order to reduce the foaming problems encountered. Also, the units required relatively large amounts of power, and Boyle and Ham considered that more effort should be directed towards investigation of anaerobic processes.

Cook and Foree (1974), in similar laboratory-scale experiments, added combinations of lime and nutrients to some units in order to adjust the pH value and lower the BOD : N : P ratios to less than the value of 100 : 5 : 1 widely recommended for sewage treatment (from an original value of 432 : 10 : 1). Results indicated that the best operational conditions were a retention time of 10 days for completely mixed units with no recycling of sludge. Under these conditions COD removal was better than 97% giving an effluent BOD of less than 26 mg/l from all units. It was concluded that addition of lime and/or nutrients did not significantly improve removal efficiencies. Treatment units with retention periods of 2 and 5 days failed, as indicated by high effluent COD and decreasing MLVSS concentrations. Uloth and Mavinic (1977) investigated aerobic treatment of a high-strength leachate (COD 48 000, BOD 36 000 mg/l) in similar fill-and-draw reactor vessels. They obtained 96.8% COD removal after a retention period of 10 days and greater than 98.7% COD removal after retention periods of more than 20 days. BOD values in settled effluents were 129 mg/l after 10 days retention and < 33 mg/l after 20 days or more. Additions of N and P were necessary and foaming problems were avoided by use of an antifoaming agent.

Chian and DeWalle (1977a, b) determined from laboratory experiments that treatment in an aerated lagoon at retention periods ranging from as low as 7 days to as high as 85 days could remove between 93 and 96.8% of the organic matter from a leachate having a COD of almost 58 000 mg/l. Pretreatment of the influent leachate was not necessary, but extensive evaluation of phosphate requirements showed that the COD : P ratio in the influent of a 30-day unit should be no greater than 300 : 1. With longer retention greater ratios were satisfactory (e.g. 85 days, COD : P of 1540 : 1). This was considered to be due to recycling of phosphate by bacteria. The growth velocity of the microorganism from these test series is the following:

$$Y = 0.42 \text{ mg MLVSS/mg COD}$$
$$b = 0.025 \text{ d}^{-1}$$

It was concluded that aerobic biological treatment of leachate would not be successful at high organic loading and low retention periods without addition of nutrients, as shown by failure of such experimental units when nutrient additions were decreased and stopped. In general, Chian and DeWalle showed that most of the organic matter in effluents from aerated lagoons consisted of stable refractory materials, often with a high molecular weight. On the basis of molecular-weight characterization and other analyses, they concluded that most of these refractory compounds were present as material such as fulvic and humic like acid and they considered that these effluents were similar to leachates from older, relatively stabilized fills, from the point of view of treatment.

Other investigations at the laboratory-scale were done by Robinson and Maris (1985). The test facilities (simulated aerated lagoons) had a volume of 20 l. The room temperature was kept constant at 10°C. The mean influent concentrations were 4805 mg COD/l and 2845 mg $BOD_5$/l. The study showed that sludge age is equal to or higher than 10 days (= organic load of 0.28 kg $BOD_5$/m$^3$ per day) a relatively constant degradation process with effluent $BOD_5$–concentrations < 25 mg/l could be achieved. Changing the temperature from 10°C to 5°C resulted in higher $BOD_5$–concentrations if the detention time was shorter than 15 days (= organic load < 0.19 kg $BOD_5$/m$^3$ per day). In another test ammonia was added in order to find out in how far nitrification processes took place. If the detention time is 10 days no nitrification process could be observed; setting a detention time of 20 days after a test run of 70 days, nitrification processes could be observed.

At the Technical University of Braunschweig, West Germany, extensive studies including laboratory and pilot scale experiments were performed (Stegmann and Ehrig, 1980). The laboratory tests were run using different leachates from 20 landfills over a period of altogether more than 300 months. During the experiments the detention times varied from 10–70 days. As a result of the varying leachate concentrations (27.0–30 000 mg $BOD_5$/l) the organic loadings were in the range of 0.0005 until 1.128 kg $BOD_5$/m$^3$ per day. By means of self-regulating processes the sludge load (F/M ratio) was as low as < 0.001 and as high as 1.8 kg $BOD_5$/kg MLSS per day.

Results from laboratory-scale aeration of leachates, reported by other workers (Schoenberger *et al.*, 1970; Palit and Qasim, 1977) confirm that leachates from recently emplaced refuse may be treated satisfactorily by aerobic processes on a small scale, although in some cases nutrients (especially P) may be deficient and problems with foaming and poor settling characteristics may have to be overcome.

In no studies has it been shown that removal of metals is necessary before aerobic biological treatment in order to reduce potential toxic effects. Both Chian and DeWalle (1977a, b) and Uloth and Mavinic (1977) reported high percentage removal of some metals in the settled biological floc, especially iron (>98%), zinc (>99%), calcium (>93%), manganese (>95%), cadmium (>96%), lead (>79%), and magnesium (>54%).

Application of these results to the design of full-scale aeration systems has been very limited. Scaling-up is important for taking account of fluctuating flows and temperatures and of changes in strength and composition of leachate over a period of time. Variations in composition will occur in the long term, and the volume of leachate will also change in the short term. This has not been taken into account in most reported laboratory experiments, where relatively constant leachate compositions and flows have been used.

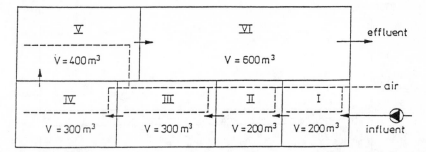

**Figure 1.** Groundplan of the aerated lagoons in Lingen, West Germany (Stegmann, 1982).

Full-scale aerated lagoons for leachate treatment were investigated over a period of several years. Figure 1 shows as an example the schematic ground plan of the plant (Stegmann, 1982).

The treatment system consists of 6 lagoons where 5 lagoons are artificially aerated by means of a coarse bubble aeration. Lagoon No. 6 is a settling tank and is also used as a reservoir. The influent concentrations were measured in the following ranges:

$BOD_5$: 51–4670 mg/l
COD: 1753–9210 mg/l
$NH_4$–N: 473–773 mg/l

Figure 2 shows the $BOD_5$–effluent concentrations relative to the organic load and different groups of temperature. The results indicate that temperatures < 5°C result in significantly higher effluent concentrations. Short detention times (below 20–50 days) are only of influence if the organic load is > 0.1 kg $BOD_5$/m$^3$ per day.

The experiments with laboratory, pilot and full scale aerated lagoons provide the basic data for the design of biological leachate treatment. All the experimental data can be used to obtain summarizing graphics (Andreottola et al., 1989). COD removals and effluent values depending on detention time are shown in Figs 3 and 4. The results are divided in three groups according to temperature operation ranges. The influence of retention time is very strong both on $BOD_5$ (not shown in figures) and COD removal efficiencies and final effluent concentrations. Experimental results are in good agreement with the theoretical relationship between hydraulic detention time $(t)$ and effluent substrate concentrations for completely mixed, no-recycle systems which is as follows:

$$1/t = \frac{Y \times K \times S}{K_s + S} - b \tag{1}$$

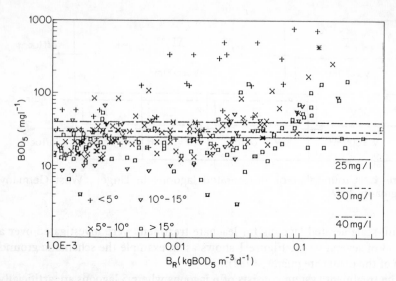

**Figure 2.** BOD₅-effluent concentrations relative to the organic load (B$_R$) and four temperature ranges, resulting from full scale aerated lagoons at Lingen, West Germany (Stegmann, 1982).

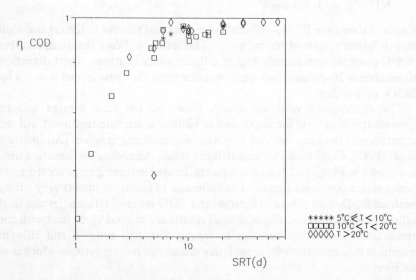

**Figure 3.** COD removal efficiencies versus sludge retention time (SRT) in aerated lagoon treatment plants on laboratory, pilot and full scale, obtained by various authors (Andreottola *et al.*, 1989).

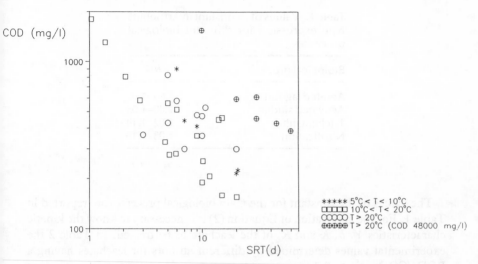

COD (mg/l)

SRT(d)

Legend:
- ***** $5°C < T < 10°C$
- ▭▭▭▭▭ $10°C < T < 20°C$
- ○○○○○ $T > 20°C$
- ⊕⊕⊕⊕⊕ $T > 20°C$ (COD 48000 mg/l)

**Figure 4.** COD effluent concentration versus sludge retention time (SRT) in aerated lagoon treatment plants on laboratory, pilot and full scale, obtained by various authors (Andreottola *et al.*, 1989).

where:
$t$ = hydraulic detention time;
$Y$ = growth yield coefficient;
$K$ = maximum rate of substrate utilization per unit weight of micro-organisms;
$S$ = concentration of substrate surrounding the microorganisms;
$b$ = microbial decay coefficient;
$K_s$ = substrate concentration at one-half the maximum growth rate.

From Equation (1) it is possible to define the detention time necessary to obtain a certain effluent substrate concentration:

$$S_e = \frac{K_s(1 + bt)}{t(YK - b) - 1} \qquad (2)$$

$K$ is influenced by temperature on the basis of the Arrhenius type expression:

$$K = K_{20}\theta^{(T-20)}$$

where:
$K_{20}$ = maximum rate of substrate utilization per unit weight of micro-organisms at 20°C;
$\theta$ = constant;
$T$ = temperature (°C);

**Table 1.** Values of $\theta$ constant in Arrhenius type expression for different biological processes.

| Biological process | $\theta$ |
|---|---|
| Aerated lagoon | 1.03–1.09 |
| Activated sludge | 1.00–1.03 |
| Trickling filter | 1.02–1.14 |
| Nitrification | 1.08–1.10 |

The values of the constant for the main biological processes are reported in Table 1. For the application of Equation (2) it is necessary to know the kinetic characteristics $Y$, $b$, $K$ and $K_s$ of the leachate to be treated. In Table 2 the experimental values determined by different authors for leachates having a $BOD_5/COD$ ratio $> 0.4$ are reported.

An aerated lagoon is a self-regulating system where the mixed liquor microbial mass concentration, $X$, is a function of substrate load and leachate characteristics:

$$X = \frac{Y(S_0 - S_e)}{1 + bt}$$ (3)

where $S_0$ is the substrate influent concentration.

The failure of leachate aerated lagooning reported in literature is due mainly to too short detention times or to low temperatures. The theoretical detention time, under which the system fails varies from 5 days at 20°C to 10 days at 5°C.

For detention times superior to 10 days removal efficiencies higher than 90% of $BOD_5$ and of COD can be obtained for $BOD_5/COD$ ratios $> 0.4$. The influence of temperature is quite strong, particularly if a complete treatment of leachate is expected.

**Table 2.** Values of leachate kinetics constants calculated by different authors.

| Author | Coeff. basis | $Y$ | $b$ ($d^{-1}$) | $K$ ($d^{-1}$) | $K_s$ (mg/l) |
|---|---|---|---|---|---|
| Cook and Foree (1974) | BOD | 0.332 | 0.0025 | 0.75 | 21.4 |
| Uloth and Mavinic (1977) | COD | 0.4 | 0.05 | 0.6 | 175 |
| Palit and Qasim (1977) | COD | 0.59* | 0.115 | 1.8* | 182 |
| Domestic sewage | COD | 0.67* | 0.07 | 5.6* | 22 |

* Calculated using SS, the others are calculated using VSS.

Stegmann (1982) reported that temperatures $< 5°C$ result in significantly higher effluent concentrations. His results show also that short detention times (10–20 days) are only of influence if the organic load is $> 0.1$ kg $BOD_5/m^3$ per day. This fact is confirmed by laboratory and pilot plant experiences where an optimum detention time of 10 days resulted for organic loads ranging from $0.14$–$3.6$ kg $BOD_5/m^3$ per day. Stegmann suggests the following design data for getting a final effluent of 25 mg $BOD_5/l$ (filtered) with a $BOD_5/COD$ ratio $> 0.4$:

$$0.025 < \text{BOD loading} < 0.05 \text{ kg } BOD_5/m^3 \text{ per day}$$

Because of low winter temperature and probably of the high fluctuation of influent substrate concentration (51–4670 mg $BOD_5/l$), detention times necessary to get a substrate concentration of 25 mg $BOD_5/l$ vary from 90 to more than 500 days. These detention times are in good agreement with theoretical evaluations if temperatures $< 5°C$ and such low effluent $BOD_5$ concentrations are considered. In fact, at low temperatures a very long detention time is necessary to increase removal efficiencies from levels as high as 95% to 99.95%.

For $BOD_5/COD$ ratios $< 0.4$ no data are available about leachate kinetic characteristics. In Fig. 5 is shown the relationship between $BOD_5/COD$ ratio and removal efficiencies, derived from laboratory and pilot plant literature experiences (Andreottola *et al.*, 1989).

Optimum efficiencies are obtained for a $BOD_5/COD$ ratio $> 0.4$. The only design data derive again from Stegmann's full scale experience. The following organic loads are suggested for getting complete treatment (25 mg $BOD_5/l$):

BOD loading $(0.3 < BOD_5/COD < 0.4) = 0.01/0.025 \text{ KgBOD}/m^3$ per day
BOD loading $(0.05 < BOD_5/COD < 0.3) = 0.001/0.01 \text{ KgBOD}/m^3$ per day

As far as COD is concerned, removal efficiencies decrease strongly for $BOD_5/COD$ ratios $< 0.4$; for $BOD_5/COD$ ratios $< 0.1$ effluent COD concentration is practically equal to the influent one (Stegmann and Ehrig, 1980).

Nearly complete removal of ammonia was achieved in most laboratory and pilot scale studies with SRT values of only 20 days, for leachates having a $BOD_5/COD$ ratio $> 0.4$. This is mainly due to conversion to organic nitrogen (biomass). Longer SRTs allow a gradual conversion of organic nitrogen to nitrate. Nitrification generally requires SRTs of greater than 60 days. Successful nitrification of ammonia in leachates on full scale clearly requires a high degree of control, particularly at low temperatures of 10°C and below (Fig. 6).

Treatment units on laboratory and pilot scale where SRT was 10 days or more removed most of metal ions (iron, manganese, zinc, etc.). The high

262

Cossu et al.

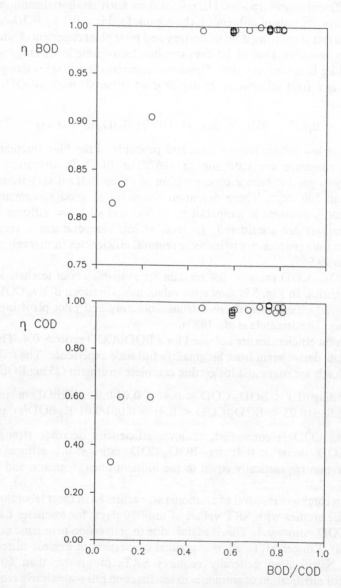

**Figure 5.** BOD₅ and COD removal efficiencies relative to BOD₅/COD ratio in aerated lagoon treatment plants on laboratory, pilot and full scale, obtained by various authors at optimal conditions (Andreottola *et al.*, 1989).

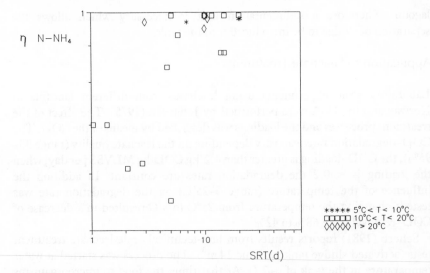

SRT(d)

**Figure 6.** $N–NH_4$ removal efficiencies versus sludge retention time (SRT) in aerated lagoon treatment plants on laboratory, pilot and full scale, obtained by various authors (Andreottola *et al.*, 1989).

reduction of most metal ions may be attributed to precipitation of metal hydroxides with subsequent entrapment in biological flocs, sorption by organic solids and consumption by biomass. The metal reduction resulted independent of temperature and loading rates. This indicates the possibility that important removal mechanisms are not controlled by biological activity, but rather by relatively constant parameters such as leachate alkalinity, mixed-liquor pH and mixed-liquor solids concentrations. Metal removal has been found low only in those units where effluents were so poorly clarified that high concentrations of metals remained in suspension (this was particularly true for failing units with SRT < 5–10 days).

## Activated Sludge Plants

### Introduction

These differ from aerated lagoons as the discharged sludge is recirculated. The retention period can be considered shorter than that necessary for an aerated lagoon as the sludge content can be carefully controlled by means of a recirculation system and is therefore equal to 3–5 times that of an aerated lagoon. Therefore a sedimentation unit is necessary which allows the separation of sludge to be introduced into the tank.

## Application to Leachate Treatment

Laboratory scale experiments using leachates from different landfills in Norway and the USA were performed by Johansen (1975). The effect of the treatment processes and the loadings was described by means of the COD. The COD–degradation rates can vary depending on the leachate quality (range 35–95%), the COD–loading is greater than 0.2 kg COD/kg MLVSS per day; when the loading is < 0.2 the degradation rates are constant. In addition the influence of the temperature (range 5–25°C) on the degradation rate was tested; changing the temperature from 25°C to 5°C resulted in a decrease of COD–removal from 68% to 42%.

Scherb (1981) reports results from half-technical scale leachate treatment tests (activated sludge tank: volume 14 m$^3$). The process was started at water temperature in the tank of 6–7°C. At that time, the food to microorganisms ratio, F/M, was 0.07 kg BOD$_5$/kg MLSS per day (mean BOD$_5$–influent: 3578 mg/l) and the effluent concentrations were in the range of 71–384 mg BOD$_5$/l. These results show the negative effect of starting the activated sludge process at low temperatures. The results improved greatly when the plant was in stable operating conditions:

$$\text{BOD loading} = 0.57 \text{ kg BOD}_5/\text{m}^3 \text{ per day}$$
$$\text{F/M} = 0.05 \text{ kg BOD}_5/\text{kg MLSS per day}$$

Removal rates of 99.6% BOD$_5$ and 96.5% COD were obtained. During this operation complete nitrification was achieved. The activated sludge had good settling properties; approximately 0.6 kg MLSS/kg BOD$_5$–removal was produced.

Extensive laboratory leachate treatment tests were practiced at the Technical University of Braunschweig, West Germany (Stegmann and Ehrig, 1980). During these tests it became obvious that the BOD$_5$/COD-ratio is of influence on the degradation process. During these experiments the temperatures changed but no significant influences were detected. It resulted that the BOD$_5$–concentrations are < 25 mg/l, when F/M is lower than 0.05 kg BOD$_5$/kg MLVSS per day.

Results of a full scale treatment plant are presented by Klingl (1981), where the same leachate that Scherb (1981) used in his tests is treated (Außernzell, West Germany). The following parameters have been used:

$$\text{BOD loading} = 0.21 \text{ kg BOD}_5/\text{m}^3 \text{ per day}$$
$$\text{F/M} = 0.017 \text{ kg BOD}_5/\text{kg MLSS per day}$$

obtaining 99.5% BOD$_5$ and 96.5% COD removal.

Results from a full scale leachate treatment plant in Italy are presented by Cossu (1981). The plant consists of three treatment steps:

1. anaerobic lagoon
2. activated sludge plant
3. aerated lagoon

The activated sludge plant had a load of $> 1.0$ kg $BOD_5/m^3$ per day. BOD and COD removal efficiency reached respectively 99.6 and 98.7%. Ammonia was removed up to 76.7%.

In addition to these experiences a two stage full scale plant was investigated at Waldeck Frankenberg in West Germany (Stegmann, 1982). The leachate was diluted by surface water resulting in the following influent concentrations:

$BOD_5$    413–3855 (mg/l)
COD      905–6296 (mg/l)
$NH_4$–N  161–124 (mg/l)

Each of the activated sludge tanks had a volume of 2060 $m^3$. The effluent concentrations are presented in Fig. 7, relative to the F/M ratio. The effluent

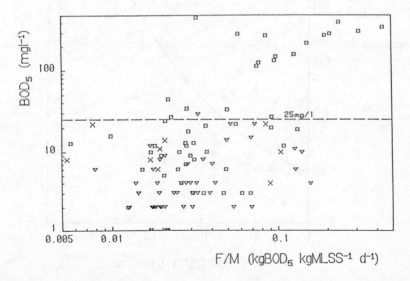

**Figure 7.** $BOD_5$–effluent concentrations relative to F/M ratio (kg $BOD_5$/kg $MLSS^{-1}/d^{-1}$), resulting from full scale activated sludge plant at Waldeck-Frankenburg, West Germany.

$BOD_5$–concentrations $> 100$ mg/l were measured in the summertime where the sludge had very poor settling properties.

Knox (1983) practiced half-technical scale leachate treatment tests with a low $BOD_5/COD$–ratio. The activated sludge tank had a volume of 8 m³, the detention time was in the range of 12 h–8 d. The average temperatures were measured between 2°C–21°C. It was the exception when higher ammonia concentrations ($> 5$ mg/l) were detected in the effluent. The ammonia removal rate was in the same range that has been observed in sewage treatment plants (Fig. 8). Since poor settling rates of the sludge were observed, high $BOD_5$-effluent concentrations were measured.

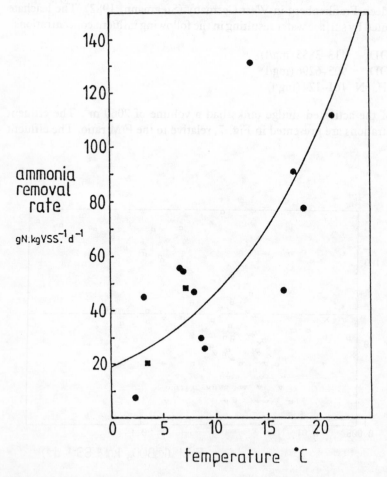

**Figure 8.** Ammonia removal rate versus temperature in an activated sludge plant with effluent concentrations ≤5 mg/l (Knox, 1983). ■ = Ammonia concentrations >5 mg/l.

Results from laboratory, pilot and full-scale experiences (Andreottola *et al.*, 1989) where $BOD_5/COD$ ratio was $> 0.4$ show that $BOD_5$ removal efficiencies $> 99\%$ can be obtained with the following loading rate:

$$F/M < 0.05 \text{ kg } BOD_5/\text{kg MLSS per day}$$

Full nitrification is generally observed at loading rates lower than 0.03 kg N/kg MLSS per day. But the longer the nitrification time the higher the possible loading rate up to values of 0.1 kg N/kg MLSS per day. With decreasing temperature a sharp decrease of removal rate is observed (Fig. 8).

When the $BOD_5/COD$ ratio is $< 0.1$ the organic load is not the right design parameter. Since the ammonia concentrations in leachate from old landfills ($BOD_5/COD < 0.1$) are of concern during the treatment process, ammonia loading (kg N/kg MLVSS per day) should be used for design purposes. High ammonia concentrations with a pH of approximately 8.5–9.0 may result in $NH_3$-toxicity in the activated sludge process. The ammonia loading should be chosen in the range between 0.07–0.1 kg N/kg MLVSS per day with such low $BOD_5/COD$ ratios.

## Rotating Biological Contactors (RBC)

### Introduction

Rotating biological contactor (RBC) could be used as aerobic system for leachate treatment. The biomass is attached to circular blades which are rotated by a mast to which they are fixed. The disks are half-immersed in a tank containing the leachate to be treated. The rotational movement means that alternately part of the disk is immersed in the sewage and subsequently comes into contact with atmospherical oxygen which is used by the biomass present on the disk to aerobically degrade the organic substances.

### Application to Leachate Treatment

Investigations in laboratory scale on the degradation of leachate in RBC were carried out by Ehrig (1983). During these experiments leachate with low $BOD_5$ and high ammonia concentrations were used. Four separate RBC were run in sequence; each unit had a surface of $0.6 \text{ m}^2$. The RBC consists of plastic media elements. The results show that about 95% of the ammonia concentrations are oxidized under optimal conditions. The problem occurs when the loading exceeds $2 \text{ g N/m}^2$ per day since the ammonia oxidation is not complete and increasing $NO_2$ concentrations are produced. In order to avoid toxic nitrate concentrations, the nitrogen loading should not exceed $2 \text{ g N/m}^2$

per day. These results could be found also in a half-technical scale plant at low temperatures.

Knox (1987) reports results obtained from a full scale leachate treatment plant using biological discs (RBC) with a surface of $30\,000\,m^2$. The main function of this plant, situated at the Pitsea landfill, is that of treating a load of approximately $150\,000\,m^3$/a of leachate in order to obtain total nitrification of ammonia present. The leachate was of low organic concentrations ($BOD_5$ 80–250; COD 850–1350 mg/l) and high ammonia concentrations ($NH_4$–N 200–600 mg/l).

The operating load was around $3\,g\,N/m^2$ per day (the design load was $4.8\,g$ $N/m^2$ per day) with a leachate temperature of 20°C. This was maintained also throughout the winter months by means of heating using biogas which was extracted from the landfill; the maintaining of this constant temperature enabled the plant dimensions to be reduced by 40% compared to that using leachate at a natural temperature. The hydraulic retention time ranged from $1.2\,d$ for loads of $100\,m^3$/d to $2.9\,h$ for $1000\,m^3$/d.

The quality of effluent with respect to BOD and $NH_4$–N was excellent. The mean concentrations were as follows:

$NH_4$–N (mg/l) 2.96
BOD (mg/l) 14

The mean influent values ranged between 100–300 mg/l $NH_4$–N and 40–130 mg/l BOD.

In conclusion rotating biological contactors show very high ammonia removal rates at loading ranging from 2–3 g $N/m^2$ per day at temperatures of 18°–22°. Higher loading rates up to 17.5 g $N/m^2$ per day lead to nitrification rates greater than 90%, but could give rise to high nitrite concentration in the effluent. Comprehensive data on the relationship between ammonia removal rates and temperature are not yet available. The following figures for sludge production can be given (Knox, 1987):

Typical solids content: 2–4%
SS yield per kg $NH_4$–N removed: 0.168 kg
VSS yield per kg $NH_4$–N removed: 0.110 kg
Typical ratio VSS/SS: 66%

Generally for full nitrification feeding of alkalinity is necessary.

## Trickling Filters

### Introduction

The trickling filter is made up of a several meter-high tanks filled with gravel or synthetic material: the wastewater is sprayed over the surface of the filling material after being pumped over the top of the filter.

The passage of the sewage through the filter is in a downwards manner, leaching from one filling element to the next so that the entire bed is never submerged and the free spaces allow the passage of air. The aerobic conditions thus constituted favour absorption of organic substances on filling elements surface by means of development of a rich bacterial population able to metabolize them.

## Application to Leachate Treatment

The only known experience is the pilot plant run by Knox (1985a, b). The leachate was of low organic concentrations ($BOD_5$ 80–250; COD 850–1350 mg/l) and high ammonia concentrations ($NH_4$–N 200–600 mg/l). The trickling filter volume (16.5 m$^3$) was filled with plastic media (surface 230 m$^2$/m$^3$). The loading was 2–7 m$^3$/d. As a result the ammonia concentrations in the effluent were lower 20 mg $NH_4$–N/l. At low temperatures the $BOD_5$–effluent concentrations exceeded 25 mg/l.

The results of this study indicate that ammonia in a stabilized leachate can be completely nitrified using a trickling filter. However the treatment of high strength leachates with trickling filters could lead to clogging and scaling due mainly to oxidation and precipitation of iron and manganese and precipitation of calcium carbonate. For this reason this process can be better employed as part of a combined plant or to treat stabilized leachates.

## Other Treatments

### Facultative Lagooning

This is carried out in tanks with limited depths (0.7–1.5 m). In the superficial layers, due to the presence of oxygen caused by photosynthesis of algae and wind-aeration, conditions are created which favour the onset of aerobic processes for degradation of organic substances. In the deeper layers, where light and oxygen do not diffuse, anaerobic processes will take place. Through facultative lagooning, a considerable nitrification of ammoniacal nitrogen can be obtained.

No data are currently available on leachate treatment in laboratory, pilot or full scale facultative lagooning.

### Aerobic Lagooning

This is characterized by the fact that degradation of organic substances is carried out aerobically without the need for any devices to transfer atmospherical oxygen into the leachate which is to be treated. Therefore, in order to obtain aerobic processes through lagooning, both the tank and the sewage used must have particular characteristics. In fact, the height of the water level must be less than 0.7 m and easily regulated so as to allow light to penetrate and oxygen to diffuse to the bottom of the tank and create favourable conditions for development of wholly-aerobic processes. The sewage used should have a low organic load and moreover should be sufficiently clear in order to enhance the production of algae and therefore photosynthetic oxygen. To this regard, the sewage should be sedimented prior to introduction into an aerobic lagoon. It is necessary to remove the suspended biomass which has formed during previous treatments.

Aerobic lagooning is characterized by an intense production of algae and the degree of removal of organic substances can be increased by inseminating the lagoon with macrophytes such as Lemnaceae.

The only known experience was made by Cossu in a full scale combined plant, already described earlier. Removal efficiencies of 72.3% $BOD_5$, 56.2% COD and 68.9% N were obtained with surface loads oscillating between 35 and 60 g $BOD_5/m^2$ per day. The short period of research with the aerobic lagoon does not appear sufficient to define a significant correlation between removal rates and organic surface loadings.

## ANAEROBIC BIOLOGICAL LEACHATE TREATMENT

Anaerobic biological processes for the treatment of leachates have several potential advantages over aerobic biological processes. These include the generation of methane gas as a by-product and much lower production of biological solids in the form of sludges or suspensions; in addition the systems have no need of aeration equipment requiring a source of power.

The advantage of anaerobic treatment systems is also evidentiated by the low organic load of the discharge: this implies that the introduction of air in a subsequent aereated treatment (aereated lagoons or active sludge plants) can be limited, thereby considerably reducing the cost of the operation.

In an anaerobic-treatment system, complex organic molecules in the influent are fermented by bacteria to volatile fatty acids, mainly acetic, propionic and butyric. These in turn are converted by methanogenic bacteria to methane and carbon dioxide, resulting in low production of biological solids requiring disposal.

The anaerobic process naturally generates hydrogen sulphide from the biological reduction of sulphate. This provides efficient precipitant for most of metals. They are almost completely precipitated and the efficiency of their removal by the anaerobic plant appears to depend only upon the efficiency of capture of suspended solids generally.

A disadvantage of anaerobic treatment processes is that the microorganisms (methanogenic bacteria in particular) are inhibited easily by acidic pH values and are also sensitive to the presence of some metals. These inhibitions can cause reduced growth rates and lead to a net washing-out of microbial cells from a completely-mixed reactor system. Such problems may be overcome by addition of buffer solution to the influent leachate. An alternative technique in recent years has been the use of the anaerobic filter. Here leachate is pumped slowly upwards through a bed containing a coarse solid substrate to which microorganisms are attached, reducing losses by wash-out. For a self supporting anaerobic process it is estimated a minimum concentration of approximately 7000 mg COD/l is required (de Bekker and Kaspers, 1981).

It seems likely that some form of heating of leachate will be required to raise temperatures above ambient winter temperatures in cold regions and it may be possible to provide at least part of this using the methane gas which is generated. For this reason anaerobic processes could be most useful for treatment of relatively strong leachates; they may be uneconomic for more dilute flows and possibly also in cases where production of leachate ceases in the summer months.

## Anaerobic Lagooning and Digesters

### Introduction

The anaerobic lagooning tank which is generally situated at the head of a treatment plant carries out not only a process of depuration but also tones down the peaks in quality of leachate in order to enable a homogeneously characterized wastewater which, as observed, could positively influence the depuration processes.

The depuration process by anaerobic lagooning is carried out by means of both physical and physicochemical processes (sedimentation, floculation and precipitation), which prevent aeration of the liquid mass, and anaerobic processes, which take place producing reduced compounds ($CH_4$, $H_2S$, $NH_3$, etc.). Generally the emission of odours is prevented by the formation of a skin over the liquid surface caused by the progressive surfacing of sludge brought up by gas bubbles.

Nitrogen is removed only by means of assimilation (cellular synthesis of new bacteria) and therefore to a minimal degree due to the low concentration of biomass. The most important effect with regard to components of this element is hydrolysis of organic nitrogen in ammonia. Thus, the latter is observed in the discharge from anaerobic lagoons in concentrations higher than those present in untreated leachate. During this phase the quantity of heavy metals is reduced due to precipitation and enmeshing in the bacterial mass.

The same depuration processes which characterize anaerobic lagooning take place in an anaerobic digester. However, the manner in which these take place differ due to several devices present in the plant and characterize treatment. In fact, the anaerobic digester is composed of a closed container, contrary to anaerobic lagooning which entails an open tank. Moreoever, the digester will be continuously agitated by re-introducing the biogas produced during degradation processes.

The mixture thus obtained enables those substances which are to undergo degradation to more easily get in contact with the biomass, so accelerating the process of degradation. The greater speed of depuration processes is the main feature which differentiates anaerobic digestion from anaerobic lagooning.

A further acceleration of depuration processes could be elicited by maintaining the anaerobic digestor at a temperature of approximately 20°C. This is obtained by heating the leachate using biogas produced in the landfill.

The improvement thus obtained in depuration processes would compensate the need for a more complex plant required for heating.

## Application to leachate treatment

Boyle and Ham (1972) showed that greater than 90% removal of organic matter from leachate, as measured by COD and BOD, was possible by storage in anaerobic conditions for 10–12 days at temperatures between 23 and 30°C. The organic loading was $1.05 \, kg \, COD/m^3$ per day. Further experiments showed that temperature was an important factor affecting the efficiency of anaerobic units in the range 11–23°C; with an organic loading of $0.67 \, kg \, COD/m^3$ per day removal efficiency dropped from 87% at 23°C to 22% at 11°C, with retention periods of 12.5 days.

Foree and Reid (1973) operated five completely mixed, fill-and-draw anaerobic digester units of capacity 1.5 l under various conditions of organic loading and temperature, with and without additions of lime and nutrients. They concluded that addition of nutrient and lime did not contribute significantly to removal of organic material (the leachate initially had a COD of 12 900 mg/l and total soluble phosphorus concentration of 12.5 mg/l). 95% COD removal was achieved at 35°C with loadings of $0.64 \, kg \, COD/m^3$ per day, but only 77% removal was obtained with a similar digester operated at 20°C.

Results for this digester could be compared with data from an anaerobic filter also with an organic loading of $1.28 \, kg \, COD/m^3$ per day at 35°C. This filter consisted of a column 1.8 m (6 ft) high, 150 mm (6 in.) diameter, containing limestone fragments through which leachate was pumped upwards; a COD removal of 96.1% was achieved when the unit had reached a steady state after 79 days.

In general, the greater efficiency of filters when compared with corresponding completely-mixed digesters is explained by the fact that micro-organisms are largely retained within a filter, whereas they may be lost in the effluent from a digester. Suspended solids in the filter effluent were approximately 40 mg/l, indicating that most of the solids were being retained within the filter and that effluent COD resulted mainly from soluble material. This compared with a suspended-solids concentration of 2754 mg/l in the effluent from the digester.

In all experimental units, including the filter and the mixed digesters, considerable increases in concentration of ammoniacal nitrogen were observed. Concentrations were raised from an initial value of 10 mg/l in the leachate to 156 mg/l in the effluent from the filter and to 417 mg/l in the corresponding digester effluent.

Chian and DeWalle (1977a, b, c) found that a completely-mixed anaerobic filter, where influent leachate was diluted with recirculated effluent, could remove more than 95% of the organic matter from a leachate with a COD of 54 000 mg/l and a pH value of 5.4. Relatively large concentrations of potentially toxic metals were present in the leachate (e.g. Fe 2200 mg/l, Zn 104 mg/l, Cr 18 mg/l, Cu 0.5 mg/l and Ni 13 mg/l), but possible toxic effects were eliminated by addition of sulphide. 99% of the COD removed by the unit was accounted for as methane production, and a solids balance indicated that only 0.012 g of volatile suspended solids were produced per gram of COD removed.

Further work by DeWalle and Chian (1977) looked in more detail at removal of toxic metals from leachate in an anaerobic filter. They concluded that percentage removal of iron, zinc, nickel, cadmium, lead and chromium increased with increasing concentrations of metals in the leachate and also with increasing hydraulic retention time. Metals were precipitated as sulphides, carbonates and hydroxides and most removal took place in the lower part of the filter.

Cameron and Koch (1980) carried out laboratory tests on anaerobic digestion at a constant temperature of 34 ± 1°C. 7 l of digested sludge from municipal water depuration plants was introduced into the reactors which had a capacity of 14 l. All reactors were kept mechanically agitated. Retention periods of 5, 10 and 20 days were observed and the percentages of BOD removal ranged between 82–99% depending on retention time and likewise gas production ranged between 0.74–0.94 m$^3$/kg BOD removed, with a CO$_2$ value of approximately 25% for all reactors.

The only fullscale experience of anaerobic lagoons reported in literature was performed by Cossu (1981). The plant consisted of an anaerobic lagoon, activated sludge oxidation and sedimentation. The anaerobic pond (900 m$^3$) was lined with PVC membranes. The raw leachate (8–10 m$^3$/h) was pumped from the collecting facilities to the anaerobic pond, designed to provide storage, equalization and removal of biodegradable organic substances.

The removal rates of organic substances achieved (BOD eff. = 50–65%) are partly attributable to the sedimentation of the suspended solids in the raw leachate, and were obtained using volumetric loads of 1–2 kg BOD/m$^3$ per day with a 15-day average retention period and a range of 10–40 days due to the discontinuous pumping of leachate.

As far as nitrogen is concerned, the main effect in the anaerobic pond is the hydrolysis of the organic nitrogen to ammoniacal nitrogen, with the result that the effluent from this unit has ammonia concentrations which are normally higher than those of the raw leachate.

Bull *et al.* (1983) investigated, on a laboratory scale, the anaerobic treatment of leachate with an average organic loading of 6000 mg BOD/l and 8000 mg COD/l. At room temperature (20–25°C), anaerobic treatment was shown to remove over 95% of the soluble BOD and essentially all the iron. In addition, over 90% of the nitrogen originally present (mainly as organic nitrogen) was converted to free ammonia. Anaerobic processing consumed about 1 mg phosphorus g$^{-1}$ BOD and yielded 100 mg of suspended solids g$^{-1}$ BOD consumed. Anaerobic processing with phosphate enrichment provided an effluent which had a BOD of 200 mg/l, an iron content of less than 2 mg/l and a free ammonia content of 700 mg/l.

Cernuschi *et al.* (1983) carried out treatability tests of anaerobic digestion on laboratory scale plant, at three different temperature values (25, 35, 55°C), using leachate with 15 400 mg/l COD and 7 600 mg/l BOD. Results showed organic removal yields decreasing with temperature, due to the toxic effect of non-ionized ammonia, present at greater concentrations and a higher temperature. At 25°C removal yields of 90.8% BOD and 81.9% COD have been reached. Specific mean production of gas (CH$_4$ = 85%) increased with temperature ranging between 147.2 and 262 l of gas per kg of removed BOD. The heat required to warm up the raw leachate fed to the anaerobic reactor cannot be fully supplied by the biogas released during the process. 95% of iron, 87% of zinc and 75% of nickel were taken up by the sludge.

Cossu (1984) and Blakey and Maris (1987) carried out laboratory studies on anaerobic lagooning. Of the 14 'batch' reactors studied, 7 were maintained in agitation and 7 were left untouched in order to evaluate the effect of agitation on the process. The trials were performed at three different temperatures, 25, 10 and 4°C. The volume of leachate placed in reactors was 10 l. Phosphorus was added to some reactors as phosphoric acid so as to guarantee a BOD$_5$/P ratio equal to 100. Moreover 1 l of digested biological sludge from domestic treatment plants was added to some reactors. A positive effect was observed only at 25°C after addition of sludge. After 28 days a COD removal of 65% was reached in reactors containing sludge whilst in those without the percentage was between 46–50%. The influent COD value was equal to 13 700 mg/l. After 42 days removal was observed to be 87.2% whilst in other reactors it was from 68–75%. After 56 days the differences between various reactors was no longer noticeable and removal was approximately 90% for all reactors at 25°C. At temperatures of 10°C and 4°C the addition of digested sludge did not affect removal. At all temperatures the addition of phosphorus was of no effect. At the temperature of 10°C removal of 80% was reached only with retention times higher than 160 days whilst at 4°C the biological processes were almost completely blocked and even with longer retention times COD-removal did not exceed 12%. Lema *et al.* (1987) studied anaerobic digestion of landfill leachates using a series of four laboratory digesters working at 37°C with HRT 5–35 days. Most of the COD of the leachates (18.0–40.0 g/l comes from VFA (60–70%) followed by protein and hydroxyaromatic compounds). HCl was regularly added in order to keep pH approximately neutral. A lack of phosphate in the media was observed and it was necessary to add nutrient phosphate for organic loading rates higher than 1.3 kg $COD/m^3$ per day.

After the Cossu study in 1981, no other full scale experience has been reported in the literature.

In order to optimize the design and operation of any such unit it will be necessary to know more about interrelationships of organic loading, retention

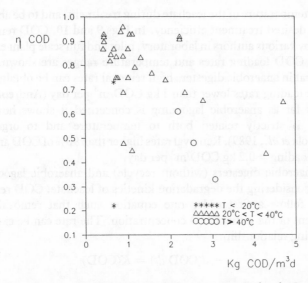

**Figure 9.** COD removal efficiencies relative to organic load ($C_v$ = kg COD/$m^3$/d) in anaerobic digesters on laboratory scale, obtained by various authors (Andreottola *et al.*, 1989).

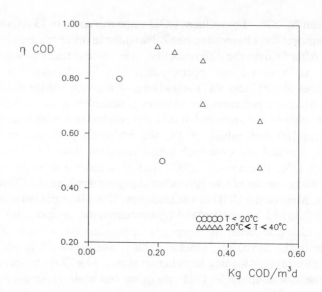

**Figure 10.** COD removal efficiencies relative to organic load ($C_v$ = kg COD/ $m^3$/d) in anaerobic lagoons on laboratory scale, obtained by various authors (Andreottola *et al.*, 1989).

time, and temperature of the leachate during treatment, and to be able to relate these to a desired treatment efficiency. In Figs 9 and 10, COD removal rates obtained by various authors in laboratory, pilot and full scale plant experiences related to COD loading rates and temperature ranges are shown. It clearly appears that in anaerobic digesters high removal rates can be obtained in most cases with loading rates lower than 1 kg COD/$m^3$ per day (Andreottola *et al.*, 1989). As far as anaerobic lagooning is concerned it shows how removal efficiency is strictly related both to temperature and to organic loads (Andreottola *et al.*, 1989). Removal rates higher than 90% of COD are obtained for COD loading = 0.2 kg COD/$m^3$ per day.

Both anaerobic digesters (without recycle) and anaerobic lagoons can be designed considering the degradation kinetics of leachate. COD removal was found to follow a first order rate equation such that removal rate was independent of the initial COD concentration. The rate can be expressed by the following relationship:

$$- d(COD/dt) = K(COD) \qquad (4)$$

where:
COD = initial COD concentration (mg/l)
$t$ = retention time (days)

$K$ = reaction constant $(d^{-1})$

Integrating Equation (4) between the initial time $t_0$ and time $t$ yields:

$$\ln (COD_t/COD_0) = - K(t) \qquad (5)$$

First order rate constants for the removal of COD have been experimentally obtained for three different temperatures (Cossu, 1984):

$$K_{25°} = 0.0317 \, (d^{-1})$$
$$K_{10°} = 0.0083 \, (d^{-1})$$
$$K_{4°} = 0.0012 \, (d^{-1})$$

The effect of temperature on the reaction constant has been determined from Arrhenius type relationship. Taking the experimental values of $K$ reported above at 25°C and 10°C the value of $\theta$ can be calculated:

$$\theta = 1.093 \, (T > 10°C)$$

For temperature lower than 10°C the value of $K_{4°}$ can be substituted obtaining the following value of $\theta$:

$$\theta = 1.17 \, (T < 10°C)$$

These values agree well with those reported in the literature where the range for $i$, under different biological conditions is 1.0–1.2. Using the relationships described above, an expression can be derived linking the mean operating temperature and the target percentage COD removal ($n_{COD}$) to the volume of a suitable anaerobic digester or lagoon:

$$n_{COD} = (COD_0 - COD_t)/COD_0 = 1 - e^{-K(t)} \qquad (6)$$

If $Q$ equals the daily leachate flow $(m^3/d)$ and $V$ is the volume of the anaerobic reactor, then substituting $V/Q$ for $t$ the following expression is obtained:

$$V = -Q[\ln(1 - n_{COD})/K]$$

This approach can provide an approximate guide to anaerobic digester and lagoon size, since performance has been measured in batch experiments (Cossu, 1984).

**Anaerobic Filters**

Introduction

Most of the experimental work carried out in the past on anaerobic biological treatment of leachate concerned completely-mixed digesters (considered more suitable for on-site treatment of leachates), although filters were found to be more efficient. However, recent advances in anaerobic techniques have revived interest in this process as a means of treating leachate. New reactors (upflow anaerobic sludge blanket, UASB) have been used to treat leachate of high strength.

Advantages claimed for UASB reactors include high loading capability and short retention period of treatment implying small plant size compared with that required for aerobic systems. Sludge production is minimal, nutrient requirements are low, and because of high energy biogas production an energy surplus may be obtained. A disadvantage is the initial high capital cost of the plant. However, this may be counter balanced by low operational costs and savings from excess biogas recovery.

## Application to Leachate Treatment

De Bekker and Kaspers (1981) have described tests run in both laboratory and half-technical scale using an upflow reactor (UASB). It resulted in removal efficiencies of 70% COD at 20°C and of 92% COD at 33°C with organic loads of 4.9 and 4.0 kg COD/m$^3$ per day respectively.

Mosey and Maris (1982) performed 'batch' laboratory tests. Four samples were inoculated with digested sludge from municipal depuration plants and a further four with 'methane granules' obtained from an anaerobic sludge bed (UASB). Results showed that the UASB granules induced an immediate gas production without any period of acclimatization and at a much higher speed than in reactors containing digested sludge. At the end of the test when gas production had ceased, both types of reactors had removed approximately 95% of organic carbon from leachate. The gas contained around 80% of methane and 20% $CO_2$ and the production of 6.4 l of gas per litre of leachate suggested that anaerobic digestion of leachate with 'methane granules' in a UASB reactor could represent a good pretreatment method. The energy obtained from gas produced would be sufficient to maintain reactors at a temperature of 28°C in winter and 35°C in summer.

Jans *et al.* (1987) reported the results obtained from a UASB reactor in Bavel. This reactor was planned for working with a COD load of 900–1200 kg/d. Removal obtained was between 80 and 85% of COD with an influent leachate where total COD ranged from 25 000–35 000 mg/l. Biogas production was from 360–480 mc/d with a $CO_2$ content of 25–30%, and a working temperature of 33–35°C. The hydraulic retention time ranged from 8–12 h.

Full-scale anaerobic treatment of leachate with UASB has not been reported in the literature. Shown in Fig. 11 are COD removal rates obtained in

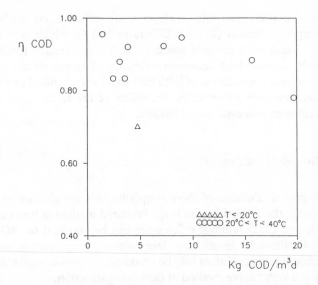

**Figure 11.** COD removal efficiencies relative to organic load (COD loading = kg COD/m³/d) in upflow anaerobic sludge blanket reactors (UASB) on laboratory and full scale, obtained by various authors (Andreottola *et al.*, 1989).

laboratory and pilot plant experiences related to COD loading rates and temperature ranges. Results show that highest removal rates can be obtained with loading rates < 10 kg COD/m³ per day. Gas production rate has an average value of 0.4 m³ $CH_4$/kg COD removed. However, further investigations are necessary to obtain reliable design data.

## CONCLUSIONS

A wide range of biological treatment options are available and include both aerobic and anaerobic processes. Long periods of retention during aeration are required to produce low strength, well clarified effluents. During aerobic treatment ammoniacal nitrogen is removed by conversion to biomass or else by nitrification at long sludge ages.

Anaerobic treatment is best used as a pretreatment method because this technique does not remove ammonia and produces a torbid effluent.

Both aerobic and anaerobic treatments are effective on leachate from recently emplaced wastes ($BOD_5$/COD ratio > 0.4), which is readily biodegradable. Leachate from aged wastes, which is less readily biodegradable and generally contains high ammonia levels, may be more effectively treated chemically or by a combination of both biological and chemical processes.

Final considerations on technical feasibility of the aerobic and anaerobic leachate treatment processes are as follows.

### Aerobic Biological Treatment

*Aerated lagoons.* Because of their simplicity and the absence of a sludge recycle facility, the aerated lagoon is the favoured method of treatment. Very high $BOD_5$ and COD removal efficiencies can be obtained for $BOD_5$/COD ratios > 0.4. Possible limits are low winter temperatures and space requirements. Floating aerators may be used and excavation within the landfill plus lining is a very simple method of lagoon construction.

*Activated sludge plants.* This treatment, even if good performances can be obtained, may be less appropriate because it demands greater operator skill and a degree of control equipment not so compatible with landfill sites.

*Trickling filters.* The use of trickling filters has been shown not to be appropriate for the treatment of strong leachates mainly because of filter blinding by inorganic deposition on the medium.

*Rotating biological contractors.* These are beginning to be considered for reducing the ammoniacal nitrogen content of leachate from aged wastes. They require very little maintenance, are simple to install and use very little power.

*Other processes.* These seem promising as part of combined treatment plants but have to be investigated more to acquire design data.

### Anaerobic Biological Treatments

*Anaerobic lagoons.* This process can be successfully employed in countries with a prevailing temperate climate. It is not a complete method of treatment but could be used for storage, balancing and as a rough pretreatment stage.

*Anaerobic digesters.* This often studied process has proved to be an effective means of removing both the organic load and heavy metals. Limits of the treatment are plant costs if high removal efficiency is required (reactor size and heating facilities).

*Anaerobic filters.* The development of upward flow anaerobic sludge blanket reactors (UASB) for the treatment of leachate has shown promise, because of high efficiencies and very low detention times. Developments in this area are currently being investigated.

## REFERENCES

Andreottola, G., Cannas, P., Cossu, R. and Stegmann, R. (1989). Overview on biological treatment of leachate. CISA, Environmental Sanitary Engineering Centre. Technical Note No. 1.

Ashmore, G. *et al.*, (1967). The biological treatment of carbonisation effluents. Investigation into treatment by the activated sludge process. *Water Research*, 1, 605–624.

Blakey, N. and Maris, P. (1987). Anaerobic treatment of leachate. Proceedings of the International Symposium on landfill, Cagliari 19–23 October.

Boyle, W.C. and Ham, R.K. (1972). Treatability of leachate from sanitary landfills, Proceedings of the 27th Industrial Waste Conference, Purdue University, Lafayette, Indiana, Engineering Extension Series 141, Part 2, 687–704.

Boyle, W.C. and Ham, R.K. (1974). Biological treatability of landfill leachate. *Journal Water Pollution Control Federation*, 46, 86•–872.

Bull, P.S. *et al.* (1983). Biological technology of the treatment of leachate from sanitary landfills. *Water Research*, 17, 1473–1481.

Cameron, R.D. and Koch, F.A. (1980). Trace metals and anaerobic digestion of leachate. *Journal Water Pollution Control Federation*, 52, 282–292.

Cernuschi, S. *et al.* (1983). Trattabilità anaerobica del percolato prodotto negli impianti di scarico controllato dei rifiuti solidi urbani. *Ingegneria Ambientale*, 12, 162–168.

Chian, E.S.K. and DeWalle, F.B. (1977a). Evaluation of leachate treatment, Vol. II. Biological and Physical-Chemical Processes. U.S. Environmental Protection Agency, Municipal Environmental Research Laboratory, USEPA, EPA-600/2-77-186/b, November, pp. 245.

Chian, E.S.K. and DeWalle, F.B. (1977b). Removal of heavy metals from a fatty acid wastewater with a completely-mixed anaerobic filter, Proceedings of the 32nd Annual Industrial Waste Conference, Purdue University, Lafayette, Indiana, pp. 920–928.

Chian, E.S.K. and DeWalle, F.B. (1977c). Treatment of high-strength acidic wastewater with a completely-mixed anaerobic filter. *Water Research*, 11, 295–304.

Cook, E.N. and Foree, E.G. (1974). Aerobic biostabilisation of sanitary landfill leachate. *Journal Water Pollution Control Federation*, 46, 380–392.

Cossu, R. (1984). Indagine sperimentale sul trattamento del percolato mediante lagunaggio anaerobico. *Ingegneria Ambientale*, 13, 226–236.

Cossu, R. (1981). Biological treatment of leachate in a full scale plant, Proceedings of I.S.W.A. Symposium, Munich, 23–25 June, pp. 35–50.

De Bekker, P. and Kaspers, H. (1981). Anaerobic treatment of leachate from controlled tips of municipal solid waste. Proceedings of ISWA Symposium, Munich, 23–25 June, pp. 51–70.

DeWalle, F.B. Chian, E.S.K. (1977). Leachate treatment by biological and physical-chemical methods — summary of laboratory experiments, Management of Gas and Leachate in Landfills (ed. S.K. Banerji). Proceedings of the 3rd Annual Solid Waste Research Symposium, St. Louis, Missouri, 14–16 March, U.S. Environmental Protection Agency, EPA-600/9-77-026, September, pp. 177–186.

Dewson, S. and Jenkins, H. (1950). The oxygen requirements of activate sludge determined by manometric methods. II. Chemical factors affecting oxygen uptake. *Sew. Ind. Wastes*, 22, 490–507.

Ehrig, H.J. (1983). Biological oxidation of sanitary landfill leachates with high ammonium concentration, International Seminar on Rotating Biological Discs, Fellbach near Stuttgart, W. Germany.

Foree, E.G. and Reid, V.M. (1973). Anaerobic biological stabilisation of sanitary landfill leachate, University of Kentucky, Office of Research and Engineering Services, Report No. UKY TR65-73-CE 17, January, pp. 51.

Harrington, D.W. *et al.* (1984). Treatment of leachate, including leachates from co-disposal sites, WRC unpublished report.

Jans, J.M. *et al.* (1987). A treatment concept for leachate from sanitary landfills, Proceedings of International Symposium 'Process, technology and environmental impact of sanitary landfill', Cagliari, 19–23 October.

Johansen, O.J. (1975). Treatment of leachate from sanitary landfill, Norwegian Institute for Water Research, Oslo; from Stegmann, 1982.

Klingl, H. (1981). Erfahrungen über die Reinigung von Müllsickerwasser an einer Deponiekläranlage im bayerischen Raum, Münchener Beiträge zur Abwasser-, Fischerei und Flussbiologie, Bd. 33, 322; from Stegmann, 1982.

Knoch, J. (1972). Reinigung von Müllsickerwasser in belüfteten Teichen, *Müll und Abfall*, 4, 123.

Knox, K. (1983). Treatability studies on leachate from codisposal landfill, *Environmental Pollution* (Series B), pp. 157; from Stegmann, 1982.

Knox, K. (1987). Design and operation of a full-scale leachate treatment plant for nitrification of ammonia, Proceedings of International Symposium 'Process, technology and environmental impact of sanitary landfill', Cagliari, 19–23 October.

Knox, K. (1985a). Leachate treatment with nitrification of ammonia. *Water Research*, 19, 895–904.

Knox, K. (1985b). Leachate treatment with nitrification of ammonia, *Water Research*, 19, 733–739.

Lema, J.M. *et al.* (1987). Anaerobic treatment of landfills leachates: Kinetics and stoichiometry. *Environmental Technology Letters*, 8, 555–564.

Melbinger, R. and Donnellon, J. (1971). Toxic effects of ammonia nitrogen in high rate digestion. *Journal Water Pollution Control Federation*, 43, 1658–1670.

Mosey, F. and Maris, P. (1982). Batch digestion tests on the leachate from New Park landfill, Water Research Centre, Laboratory Record, pp. 326.

Palit, T. and Qasim, S.R. (1977). Biological treatment kinetics of landfill leachate. *Journal Environmental Engineering Division ASCE*, 103, 353–366.

Robinson, H.D. and Maris, P.J. (1985). The treatment of leachates from domestic wastes in landfill. *Water Research*, 17, 1537–1548.

Schoenberger, R.J. *et al.* (1970). Treatability of leachate from sanitary landfills, Proceedings of the 4th Mid-Atlantic Industrial Wastes Conference, University of Delaware, Newark, Delaware, pp. 411–421.

Scherb, K. (1981). Ergebnisse von versuchen über die Reinigung von Müllsickerwasser. Münchener Beiträge zur Abwasser-, Fischerei und Flussbiologie, Bd. 33, 307; from Stegmann, 1982.

Stegmann, R. and Ehrig, H.J. (1980). Operation and design of biological leachate treatment plants. *Water Science & Technology*, **13**, 919–947.

Stegmann, R. (1982). Leachate Treatment, University of Hamburg-Harburg. Unpublished report.

Swanwick, J. *et al.*, (1969). A survey of the performance of sewage sludge digesters in Great Britain. *Water Pollution Control*, **68**.

Uloth, V.C., and Mavenic, D.S. (1977). Aerobic biotreatment of a high-strength leachate. *Journal Environmental Engineering Division ASCE*, **103**, 647–661.

# 4.5 Physicochemical Treatment

HANS-JURGEN EHRIG

*ITW-Ingenieurberatung GmbH, Erich-Nörrenberg Strasse 5,
D-5860 Iserlohn, West Germany*

## INTRODUCTION

After biological degradation (within the landfill or in a treatment plant) leachate still contains a wide range of substances (heavy metals, biorefractory and halogenated organics, etc.) which could prevent discharge to surface water, according to the law requirements existing in different countries. Studies on biological treatment of leachate gave the following results:

1. effluent still characterized by relatively high COD values (400–1500 mg/l) with unknown composition;
2. small reduction of AOX (Adsorbable Organic Halogens), approximately 10%;
3. nitrogen concentration at levels of 20–50 mg/l $NH_4^+-N$.

Comparing available effluents from biological treatment with actual and future requirements, it can be easily understood why the scientists' interests are increasingly focused on the development of advanced leachate treatment technologies based on physicochemical processes, combined or not with biological treatment.

This chapter presents an overview of the physicochemical treatment processes and an example of combined treatment.

## FLOCCULATION–PRECIPITATION PROCESS

Flocculation–precipitation of sanitary landfill leachate has been tested by many researchers (Chian and De Walle, 1976; Ho *et al.*, 1974; Thornton and

Blanc, 1973; Johansen, 1975; Stoll, 1979; Slater *et al.*, 1983; Knox, 1983; Spencer and Farquhar, 1975; Bjorkman, 1979; Wong, 1980; Graham, 1981; Edelhoff, 1978). In all cases the flocculation effects on COD-reduction using lime, iron and aluminium salts were poor, essentially with high-strength leachate. Most organics in such leachates are volatile fatty acids with low molecular weight which could not be precipitated (Narkis *et al.*, 1980). Our experiments showed a relationship between $BOD_5$/COD-ratio and flocculation effect (Fig. 1). Only with leachates from methanogenic phase (BOD/COD $\ll 0.1$) the COD-removal was in the range of 40–60%. A similar removal could be observed for AOX too. Although the $BOD_5$ of leachate was between 100 and 300 mg/l the elimination was poor and there was no effect on ammonium. The conclusions drawn about the flocculation process are as follows:

1. Flocculation is not a suitable treatment method for high strength leachate.
2. Flocculation, applied to leachate from methanogenic phase, reduces COD and AOX concentrations but not to above given standards. The flocculation effect is as higher as larger the organic molecules of leachate. An additional treatment step is necessary to remove $BOD_5$ and ammonium.

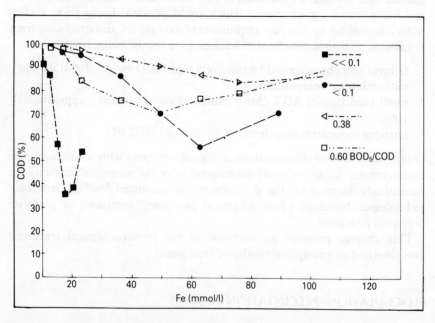

**Figure 1.** COD in the supernatant after leachate flocculation vs. flocculation agent dosage, according to different BOD/COD ratios (1 mmole = 56 mg Fe).

## ADSORPTION

Similar to flocculation many adsorption experiments with landfill leachate have been reported in the literature (Chian and De Walle, 1976; Ho *et al.*, 1974; Spencer and Farquhar, 1975; Knox, 1983; Johansen, 1975; Edelhoff, 1978).

An important parameter for the evaluation of the adsorption effect is the adsorption capacity $(X/M)$ which represents the ratio between the absorbed COD and the adsorbing activated carbon. The adsorption capacity is a function of the effluent COD values $(C_e)$. With given influent and required effluent value the activated carbon dosage increases with an increasing slope of the isotherms (Figs. 2 and 3). Figure 2 presents adsorption isotherms with acetic and humic acid and with raw leachate in acetic phase. Figure 3 shows adsorption isotherms with humic acid and methanogenic phase leachate. The isotherms for acetic acid leachate, acetic acid and also humic acid are too steep

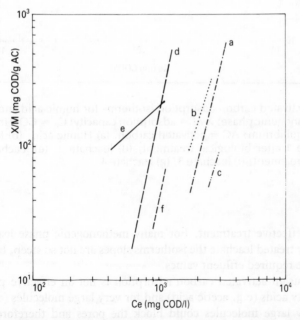

**Figure 2.** Activated carbon adsorption isotherms for different substances: $X/M$ = adsorption capacity; $C_e$ = COD in the liquid phase at equilibrium; AC = activated carbon. (a) Acetic acid; (b) humic acid; (c) raw leachate diluted 1 : 10; (d) raw leachate diluted 1 : 25; (e,f) leachate after biological treatment, with different AC dosage.

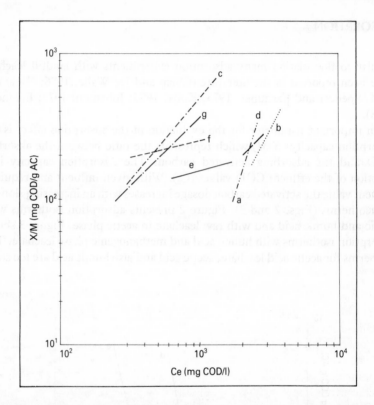

**Figure 3.** Activated carbon adsorption isotherms for humic acid and leachates from methanogenic phase: $X/M$ = adsorption capacity; $C_e$ = COD in the liquid phase at equilibrium; AC = activated carbon. (a) Humic acid; (b) leachate 1; (c) leachate 1 after biological treatment; (d) leachate 2; (e) leachate 2 after biological treatment; (f) leachate 3; (g) leachate 4.

for a cost-effective treatment. For many methanogenic phase leachate and biologically treated leachate the isotherms slopes are not so steep, but still too high for the required effluent values.

In conclusion activated carbon adsorption is not an effective process for volatile fatty acids (e.g. acetic acid) and for very large molecules (e.g. humic acid). Very large molecules could block the pores and therefore decrease adsorption capacity of other molecules.

The best effect of adsorption could be observed after biological treatment (elimination of volatile fatty acids) and flocculation (reduction of large molecules as humic like substances).

## REVERSE OSMOSIS

One of the new developments of leachate treatment is the reverse osmosis (RO). In contrast to the other techniques this is a separation process involving two liquid streams: a low-polluted permeate stream and a high-polluted concentrate stream. With new developments for membrane material it is possible to get very low-polluted permeats (Table 1) with the exception of acetic phase leachate. Some very small organic molecules present in this kind of leachate are not kept back. In such cases a biological pretreatment is necessary. In many cases also the separation of ammonium is poor or must be improved by chemical addition to adjust pH. But a very important disadvantage of reverse osmosis is the liquid concentrate.

The current technique of back-passing the non-biodegradable concentrate stream to landfill is not a solution, it is only a means of putting off the real treatment solution.

It may be concluded that:

1. The separation by RO leads in most cases to a very low loaded permeate. But the lower the pollutant reduction in the leachate the higher the concentration in the permeate.
2. For acetic phase leachate a biological pretreatment could be necessary.
3. Often the separation of ammonium is poor but could be improved by pH-adjusting. Ammonium still remains in a concentrate stream. A separate elimination process could completely eliminate ammonium.
4. The handling of the concentrate has not been solved yet.
5. At present combinations of reverse osmosis and vaporization are being discussed.
6. Other combinations, e.g. with flocculation and/or adsorption, could be possible.

## VAPORIZATION

For some years processes of vaporization of the leachate have been discussed in the FRG as a last resource: vaporizing means separating leachate in a clean water stream and in a solid phase containing all polluting materials. In reality the separation is not so strong and accurate. Normally the solids are pulpy and the condensated vapours contain volatile components (Table 2). The predominant components in the effluent from vaporization plants are volatile, sometimes chlorinated organics and ammonium with the necessity of additional treatment steps. Full-scale operation was only tested with

**Table 1.** Results of different reverse osmosis experiences on full-scale and pilot plants.

| Parameters (mg/l) | a In | a Out | b In | b Out | c In | c Out | d In | d Out | e In | e Out |
|---|---|---|---|---|---|---|---|---|---|---|
| COD | 1690 | 18–25 | — | — | 1630 | 80 | 270 | 3 | 4180 | 29.4 |
| BOD$_5$ | — | — | — | — | 140 | 30 | — | — | — | — |
| TOC | — | — | 1200 | 24 | — | — | — | — | — | — |
| NH$_4$–N | 308 | 20–34 | 1370* | 468* | 801 | 116 | 82* | 5* | 665* | 30* |
| AOX | — | — | — | — | 1.2 | 0.07 | — | — | — | — |

[a] Holzbeck et al. (1987).
[b] Ryser (1984).
[c] Marquardt (1986).
[d,e] van der Schroeff (1985).
* Values expressed as NH$_4$.

**Table 2.** Results of different vaporization experiences on full-scale and pilot plants.

| Parameters (mg/l) | a In | a Out | b In | b Out | c In | c Out | d In | d Out |
|---|---|---|---|---|---|---|---|---|
| COD | 4200–5510 | 227–508 | 4000 / 8160* | 800 / 222* | 1150–3000 | 45–460 | 5300–32 600 | 500–2500 |
| NH$_4$–N | 475–1400 | 1–219 | 2174* / 1310* | 1242 / 20* | 220–1800 | 160–1300 | 1200–3200 | 480–1000 |

[a] Amsoneit (1987).
[b] Rautenbach et al. (1986).
[c,d] Braun et al. (1985).
* Vaporization combined with stripping.

hazardous waste leachate. Small scale experiments with municipal waste leachate show problems of incrustation, foaming and corrosion. The conclusions for vaporization processes are:

1. Vaporization of domestic waste leachate is not a proven technology (unsolved problems of incrustation, foaming and corrosion).
2. Effluent polishing by adsorption and ammonium elimination could be necessary.
3. An additional solidification of the solid phase could be necessary.
4. A combination with reverse osmosis could be helpful to cut throughput of vaporization and therefore operation costs.

## COMBINATION OF DIFFERENT TREATMENT TECHNOLOGIES

All discussed and tested treatment technologies seem to present no suitable substitute for biological leachate treatment. For today's, and much more for future effluent standards, every treatment technology including biological treatment shows some deficiencies. The only solution could be an optimum combination of two or more methods. Future developments could be estimated based on the present discussion about improved leachate treatment methods. The most important aspect is the increasing of safety in service with reduced maintenance.

At the Technical University of Braunschweig a combination of biological treatment, flocculation and activated carbon has been tested at laboratory and full scale.

As already shown, flocculation and adsorption could not be used as independent treatment processes: at first the two physicochemical processes were simply combined. No adsorption filter was used, but activated carbon was dosed in a mixing chamber. During these tests it was possible to reach operation conditions for both methods.

### Flocculation

Operation conditions differ from those of phosphorus or suspended matter elimination but follow those of colour removing of potable water.

Mixing and dosing conditions could vary in a wide range without effect on flocculation efficiency. Mixing times between 10 and 600 minutes show the same results, with equal effluent values.

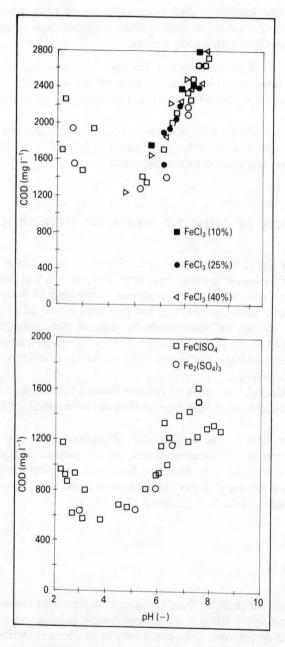

**Figure 4.** Effluent COD vs. pH during flocculation, using various solutions of iron salts.

Most effective dosing agents are liquid iron solutions such as $FeCl_3$ and $FeClSO_4$. Flocculation by aluminium solutions results in higher strength effluents. Different leachates however behaved differently.

For every leachate we could measure a special minimum dosage requirement.

The most important operation factor is the optimum pH-value (Fig. 4). In most cases the value was between pH 4.5 and 4.8 but could differ in some cases to pH 3.8 to 5.0.

To operate flocculation plants it is necessary to combine both the following requirements: firstly, minimum dosage requirement (as a rule 250 to 500 g $Fe^{3+}/m^3$); secondly, optimum pH-values = 4.5–4.8 (iron solutions are

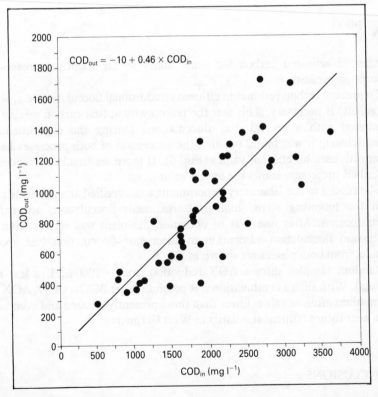

**Figure 5.** Effluent COD vs. influent COD, observed during flocculation experiments.

acid $= 2.5–2.7$ g $Fe^{3+}$ destroy 1 g alkalinity as $CaCO_3$. For practical purposes three cases may be distinguished:

1. Second requirement $\geq$ first requirement: flocculation with an iron solution alone.
2. Second requirement $\geq$ first requirement: with high leachate alkalinity, reduction of iron sludge production by minimum dosage of an iron solution, then adjustment with acid.
3. Second requirement $\leq$ first requirement: after biological treatment including nitrification, the alkalinity of leachate is often very low; thus a minimum dosage of an iron solution is applied, and then pH adjustment with lime.

Using the given operation conditions a COD-reduction of approximately 50% is possible (Fig. 5).

### Adsorption

Mixing of activated carbon for some hours by air slightly increases the adsorption capacity.

To prevent carbon residues in effluent an additional flocculation (e.g. with a coagulant) is necessary. This was the reason why at first carbon mixing was combined with a second iron flocculation. During the experiments in Braunschweig it was found out that the succession of both processes has no great influence on effluent values (Fig. 6). If there are small differences the described succession shows the lowest values.

The results of the laboratory experiments were verified at a full scale plant with the following steps: biological treatment; flocculation; adsorption; neutralization. After one year of operation the plant was rebuilt from the succession flocculation–adsorption to adsorption–flocculation (see above). Results from both phases are shown in Fig. 6.

Random samples show a AOX reduction from $\approx 1000$ $\mu g/l$ to less than $250 \mu g/l$. With such a combination it is possible to get $BOD_5$, COD, AOX and ammonium effluent values lower than those presently reached and even lower than near future effluent standards in West Germany.

## CONCLUSIONS

The common leachate treatment processes with BOD removal and sometimes

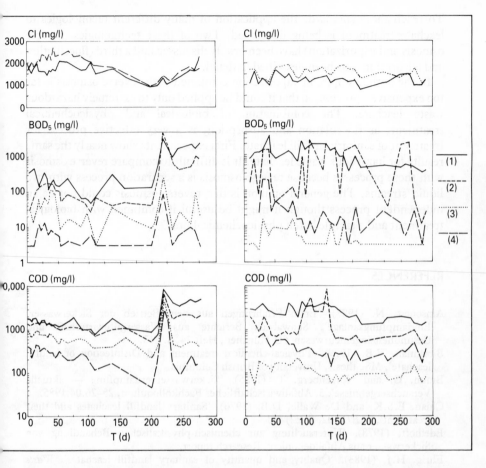

**Figure 6.** Influent and effluent quality values vs. time observed during full scale treatment tests, using different sequences of operation units: (A) biological treatment + flocculation + adsorption; (B) biological treatment + adsorption + flocculation; (1) influent; (2) biological treatment effluent; (3) flocculation unit effluent; (4) adsorption unit effluent.

nitrification may not be sufficient to respect the effluent standards prescribed by law in different countries. Today requirements are estimated as follows:

1. constant BOD$_5$-effluents $\leqslant$ 25 mg/l;
2. permanent and full nitrification;
3. reduction of the effluent non-biodegradable COD;
4. reduction of organic halides.

To reach these standards the application of many different technologies to leachate treatment is being developed. Two of these technologies (reverse osmosis and vaporization) have been briefly discussed and a third (flocculation and adsorption) has been described in detail.

In the author's opinion vaporization combined with reverse osmosis is far too expensive a process, so that it could be applied only to extremely hazardous waste leachate. The combination of biological and physicochemical treatment, as flocculation and adsorption, is a very effective process for treatment of sanitary landfill leachate. First experiments show nearly the same results for hazardous waste leachate. It is difficult to compare reverse osmosis with these processes because reverse osmosis is a separation process with two liquid streams. The general opinion is that reverse osmosis could not be an independent process though it might be useful if combined with biological treatment and a simple batch physicochemical unit.


# REFERENCES

Amsoneit, N. (1987). 'Erste Erfahrungen aus dem Betrieb der Sickerwasser-Eindampfungsanlage', Schwabach Berichte aus Wassergütewirtschaft und Gesundheitsingenieurwesen TU München, Heft 74, S. 213.

Bjorkman, V.B. (1979). 'Physical-chemical treatment and Disinfection of landfill leachate'. MSc thesis, University of British Columbia.

Braun, G. and Norrenberg, T. (1985). 'Sickerwasserverdampfung — aktuelle Versuchsergebnisse'. 5. Abfallwirtschaftliches Fachkolloquium, 25–26.04.1985.

Chian, E.S.K. and De Walle, D.B. (1976). 'Sanitary landfill leachates and their treatment', *JSED* **102**, (EE2) 411.

Edelhoff, (1978). 'Untersuchung zur chemisch-physikalischen Behandlung von Sickerwasser aus Abfalldeponien'. Research report.

Ehrig, H.J. (1983). 'Quality and quantity of sanitary landfill leachates', *Waste Management and Research* **1**, 53–68.

Ehrig, H.J. (1984). 'Treatment of sanitary landfill leachate: Biological Treatment', *Waste Management and Research* **2**, 131.

Ehrig, H.J. (1987). 'Weitergehende Reinigung von Sickerwässer aus Abfalldeponien', *Veröffentlichungen des Inst. f. Stadtbauwesen TU Braunschweig* **42**.

Graham, D.W. (1981). 'Biological-chemical treatment of landfill leachate'. MSc thesis, University of British Columbia.

Ho, S., Boyle, W.C. and Ham, S.K. (1974). 'Chemical treatment of leachates from sanitary landfills', *JWPCF* **46** (7), 1776.

Holzbeck, L. and Ventulett, G. (1987). 'Untersuchung von Behandlungsmöglichkeiten für das Sickerwasser der Zentraldeponie in Düsseldorf', *Müll und Abfall* **5**, S206.

Johansen, O.F. (1975). 'Treatment of Leachates from Sanitary Landfills', Oslo.

Knox, K. (1983). 'Treatability studies on leachate from a co-disposal landfill'. *Environment Pollution (series B)* **5**(3), 157.

Marquardt, K. (1986). 'Aufbereitung von hochbelasteten Abwässern durch Umkehrosmose und Ultrafiltration', Teil 2, *Entsorga-Magazin*, 4, S29.

Narkis, N., Henefeld-Fourrier, S. and Rebhun, M. (1980). 'Volatile organic acids in raw wastewater and in physico-chemical treatment'. *Water Research* 14, 1215.

Rautenbach, R., Kollbach, J. and Kopp, W. (1986). 'Aufbereitung von Deponiesickerwasser'. *Wasser, Luft und Betrieb* 6, S61.

Ryser, W. (1984). 'Sickerwasserreinigung am Beispiel der Umkehrosmose'. 4, Abfallwirtschaftliches Fachkolloqium, Saarbrücken.

Van der Schroeff (1985). 'Percolatiewater van viulstortplaatsen Behandeling door middel van hyperfiltratie', $H_2O$ 18, S328.

Slater, C.S., Uchrin, C.G. and Ahlert, R.C. (1983). 'Physico-chemical pre-treatment of landfill leachates using coagulation', *Journal Environment Science Health* A18(1), 125.

Spencer, C.S. and Farquhar, C.F. (1975). 'Biological and physical-chemical treatment of leachate from sanitary landfills', *Proc. 10th Canadian Symposium*, 180.

Stoll, B.J. (1979). 'Leachate Treatment Demonstration', EPA 600/(9)-79-023a, 313.

Thornton, R.J. and Blanc, R.C. (1973). 'Leachate treatment by coagulation and precipitation', *JEED* 99(EE4), 535.

Wong, P.T. (1980). 'Two-stage treatment of a landfill leachate. Aerobic biostabilization with lime-magnesium polishing'. MSc thesis, University of British Columbia.

# 5    LINING AND DRAINAGE

# 5.1    Design and Construction of Liner Systems

J.P. WORKMAN* and R.L. KEEBLE**

*Browning-Ferris Industries, P.O. Box 3151, Houston, Texas 77253, USA
**Browning-Ferris Overseas Inc., UK

## INTRODUCTION

The liner system in a landfill is the main line of defence against external migration of leachate and methane gas. A performance standard criteria is a rational approach to developing a liner system. A performance standard describes the expected performance of a lining system and specific design criteria are developed on a site specific basis.

The liner system must not only be designed properly, but must be constructed properly. Good specifications, proper equipment and an adequate quality control programme must be implemented.

## LINER SYSTEM DESIGN

There are two types of design methods for the design of landfills in the USA: the design standard and the performance standard. Design standards specify design applicable to all structures such as the number, types and thickness of liners. Design standards are easy to interpret and provide uniform designs for all landfill operators.

The performance standard describes the expected performance of a landfill and the design is prepared to meet this criteria. The performance standard increases flexibility for the landfill designer because it allows site specific information to influence the design and can result in substantial savings in construction costs.

301

The performance standard approach is a more rational approach to landfill design. Detailed site information needs to be obtained and liner performance criteria need to be developed to properly design a liner system.

## Site Investigation

A liner system design based on performance standards requires a detailed site investigation. The purpose of the site investigation is to develop a thorough understanding of the environment that will be used to evaluate the type of liner system required to protect the natural resources. The soil stratigraphy, hydrogeology and climate are particular areas that must be well defined.

Low permeability soils, such as clay, provide natural barriers to contaminant migration. Significant thicknesses of clay between the bottom of the landfill and the ground water will greatly reduce the liner requirements. A sufficient number of borings should be drilled to define soil stratigraphy. Soil samples should be preserved for laboratory testing that includes Atterberg limits, grain size analyses and permeability.

The purpose of a hydrogeological investigation is to determine the groundwater regime including the location, quality, movement and seasonal variation. Climate is another important consideration when selecting a liner system. Semi-arid and arid climates have more evapotranspiration than precipitation, and landfills in these areas can be designed and operated to virtually eliminate leachate.

## Liner Design

The following is a suggested list of criteria that should be considered for every site:

1. efficiency;
2. damage resistance;
3. long-term performance;
4. availability.

The efficiency of a liner system refers to the ability to resist the seepage forces of leachate generated within the landfill. Efficiency can be improved by controlling leachate generation and/or preventing significant levels of leachate ponding on the liner. A commonly used performance requirement for a leachate collection system is to maintain less than 0.3 m of leachate head on the liner. To maintain less than 0.3 m of head, drainage materials should have permeabilities greater than $1 \times 10^{-3}$ cm/s and networks of properly spaced drainage pipe should be installed. Liner permeability, and slope of a landfill base are also considerations that affect efficiency.

Damage is most likely to occur during construction and landfill operations. Synthetic liners, which have excellent permeability properties, are easily damaged. Clay barriers have much better resistance to damage and have self-healing properties.

Long-term performance of the liner system is of foremost importance. Criteria for determining the long-term performance are the permeability, leakage resistance and chemical resistance.

The permeability of the liner is often the most important factor in determining the long-term performance. Synthetic liner materials have very low permeabilities of $1 \times 10^{-12}$ to $1 \times 10^{-14}$ m/s. A synthetic liner provides an effective low permeability barrier that will greatly enhance the efficiency of any liner system.

The permeability of soils will vary greatly. Typical conductivity of soils classified as clays by the Unified Soil Classification System (Anonymous, 1987) will range from $10^{-7}$ to $10^{-11}$ m/s. The permeability will impact the breakthrough and leachate rate of a clay liner. Breakthrough time can be determined by the following equation:

$$t = \frac{d^2 n}{k(d + h)} \tag{1}$$

where:
$t$ = breakthrough time in years
$d$ = liner thickness in metres
$h$ = hydraulic head in metres
$k$ = permeability in m/year
$n$ = effective porosity.

Figure 1 presents the calculated breakthrough time for a 1 m thick clay liner and 0.3 m of head acting on the liner. The effective porosity was assumed to range from 0.2 to 0.3. As can be seen, permeabilities should be low in order to contain contaminants. Clays with permeabilities less than $10^{-9}$ m/s are commonly considered adequate to provide long-term protection of the environment.

The selection of synthetic liner materials should be based on the waste stream expected for the facility. Table 1 presents a summary of the effects of common chemical constituents on various synthetic liner materials prepared by Koerner (1982); see also Chapter 5.3, this volume.

The United States Environmental Protection Agency has sponsored some work to study the performance of synthetic and soil liners with leachate from

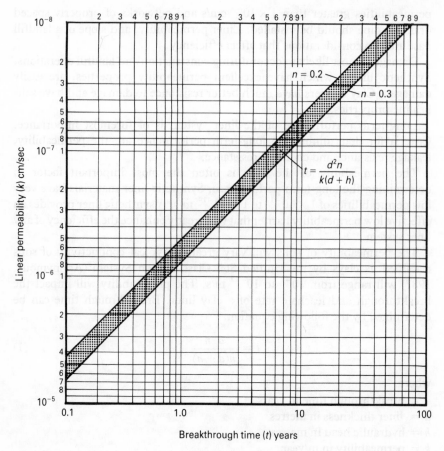

**Figure 1.** Calculated breakthrough time for a 1 m thick clay liner and 0.3 m water head (Anonymous, 1987).

sanitary and hazardous landfills (Haxo *et al.*, 1982; Daniel and Liljestrand, 1984). The results of these studies indicate that synthetic and soil materials show only minor changes in the physical properties when exposed to leachate.

The breakthrough time and the leakage rate can be used to evaluate the leakage resistance of a liner system. The leakage resistance is inherent to the design selected and is closely associated with liner permeability, damage resistance and chemical resistance.

A two-dimensional saturated flow model was used to evaluate the breakthrough time of liner system options for a hazardous waste disposal facility. The system consisted of composite liners with leachate collection systems. One system was assumed to have 1.6 m of $1 \times 10^{-8}$ cm/s clay and the

**Table 1.** Summary of the effect of chemical constituents on various synthetic liner materials (Koerner, 1982).

| Chemical | Butyl rubber 100°F | Chlorinated polyethylene (CPE) 100°F | Chloro-sulphonated polyethylene (CSPE) 100°F | Elasticized polyolefin 100°F | Epichlorohydroin rubber 100°F | Ethylene propylene diene monomer (EPCM) 100°F | Poly-chloroprene (neoprene) 100°F | Poly-ethylene 100°F | Polyvinyl chloride (PVC) 100°F |
|---|---|---|---|---|---|---|---|---|---|
| **General:** | | | | | | | | | |
| aliphatic hydrocarbons | × | × | | × | × | × | × | × | |
| aromatic hydrocarbons | × | | | × | × | × | | × | |
| chlorinated solvents | | | | × | × | | × | × | |
| oxygenated solvents | | | | × | × | | × | × | |
| crude petroleum solvents | | × | | × | × | | × | × | |
| alcohols | × | × | | × | × | × | × | × | × |
| **Acids** | | | | | | | | | |
| organic | × | × | × | × | × | × | × | × | × |
| inorganic | × | × | × | × | × | × | × | × | × |
| **Bases** | | | | | | | | | |
| organic | × | × | × | × | × | × | × | × | × |
| inorganic | × | × | × | × | × | × | × | × | × |
| Heavy metals | × | × | × | × | × | × | × | × | × |
| Salts | × | × | × | × | × | × | × | × | × |

× = generally good resistance.

other system was assumed to have 0.9 m of $1 \times 10^{-7}$ cm/s clay. The range of permeabilities for the clays were representative to those available on site. To evaluate leakage resistance, the liners were assumed to have 3 mm circular punctures. Small holes would be the most likely type of damage that would not be discovered. Large holes or tears would be noticed and repaired before the liner is buried with waste.

The results of this analysis are presented in Fig. 2. If leachate levels are minimized to 1.0 m, breakthrough times for both designs should be in excess of 800 y. As the hydraulic head on the liner increases, the breakthrough time is substantially reduced.

The leakage rates for these liner designs after breakthrough were also determined. In comparison, a single synthetic will not only have immediate breakthrough, but also have leakage rates of about 700 000 l/y. The composite liners with clay layers 1.6 and 0.9 m thick were calculated to have leakage rates of 0.2 and 0.4 l/y, respectively.

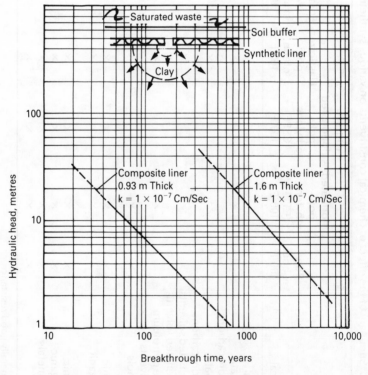

**Figure 2.** Examples of calculation of the breakthrough times of two composite liners.

The availability of materials will impact the liner design and the planned performance. Soil betonite mixtures, asphalt and soil cement are a few examples of fabricated materials that have been used for low permeability barriers.

## Commonly Used Designs

Figure 3 shows several liner systems commonly used by the waste disposal industry in the USA. Figure 3a presents a single clay liner. This liner design is normally used where a substantial thickness of natural low permeability soil

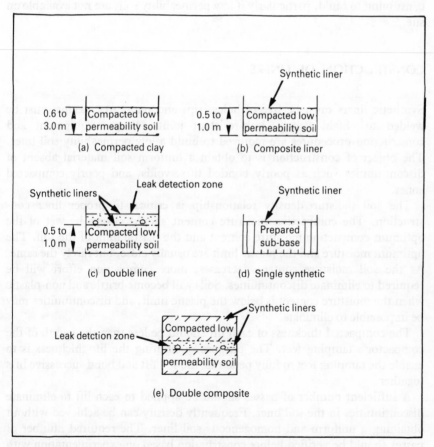

**Figure 3.** Examples of different liner systems used in the USA.

is located beneath the landfill. The purpose of the liner is to provide a homogeneous engineered barrier in addition to the natural soil. This liner system is resistant to damage without total failure of the liner system.

The examples presented in Fig. 3c, d rely principally on the synthetic liners as a hydraulic barrier. These designs are commonly used where low permeability soils are not available. However, synthetic liners are easily damaged and subject to significant leakage. A double liner with a leak detection/collection system (Fig. 3c) is normally preferred by the industry because of the ability to collect leachate that may penetrate the primary liner.

Composite and double composite designs (Fig. 3b, e) are suggested when vulnerable site conditions exist. The composite liner system provides long breakthrough time, low leakage rates, and can withstand substantial damage without total failure of the liner system. These liners are costly and time consuming to build, particularly if low permeability soils are not available on site.

## CONSTRUCTION OF LINERS

Synthetic liners must be free of holes, rips and punctures; seams must be welded to obtain strong, leak-resistant joints. Proper equipment and construction procedures are required to build a low permeability soil liner. The object of construction is to obtain a uniform soil material absent of discontinuities such as poorly bonded lifts, voids, and poorly compacted zones.

The soil moisture/density relationship is critical to proper liner construction. The compaction moisture content should always be wet of the optimum compaction moisture content and the plastic limit of the soil. The optimum moisture and the plastic limit are usually close, but rarely the same. As the soil moisture content decreases, more compaction effort will be required to eliminate discontinuities. Soils will become brittle and non-plastic when the moisture content is below the plastic limit, and discontinuities may be impossible to eliminate.

The compacted thickness of each lift should be less than the length of the compactor's tamping feet. The purpose of limiting the lift thickness is to enable the tamping feet to fully penetrate the soil lift and bond successive lifts together.

A sufficient number of passes should be applied to each lift to eliminate discontinuities in the soil liner. Frequently density can be achieved without obtaining a uniform and homogeneous soil liner. The required number of passes should be verified before construction based on experimentation with the equipment and soil conditions.

## CONCLUSIONS

The long-term performance of a liner system is critical to the protection of the environment. Both proper design and construction are required to achieve the desired long-term performance. A rational approach to liner design is the performance standard method. This method requires a detailed site investigation and a defined set of performance criteria. The purpose of the site investigation is to develop a thorough understanding of the environment and determine the vulnerability of the natural ground-water resources. Performance criteria should include liner efficiency, damage resistance, long-term performance, and material availability.

A low permeability soil liner is an important component of a liner system. Low permeability soil liners can be constructed. Proper construction equipment and procedures must be used to minimize the discontinuities in the soil that can cause seepage.

## REFERENCES

Anonymous (1987). 'American Society for Testing and Materials', Annual Book of ASTM Standards, Section 4, Vol. 04.08, Soil and Rock; Building Stones.

Daniel, D.E. and Liljestrand, H.M. (1984). 'Effects of Landfill Leachate on Natural Liner Systems', Geotechnical Engineering Report GR83-6, Geotechnical Engineering Department, the University of Texas, Austin, Texas.

Haxo, H.E., White, R.M., Haxo, P.D. and Fong, M.A. (1982). 'Liner Materials Exposed to Municipal Solid Waste Leachate', EPA-600/2-82/097 (NTIS No. PB 83-147-801). US Environmental Protection Agency, Cincinnati, Ohio.

Koerner, R.M. (1982). 'Designing with Geosynthetics'. Prentice Hall, N.J.

# 5.2 Surface Capping with Natural Liner Materials

J. HOEKS and A.H. RYHINER

The Winand Staring Centre for Integrated Land, Soil and Water Research,
PO Box 125, 6700 AC Wageningen, the Netherlands

## INTRODUCTION

When the operation of a solid waste disposal site is terminated it is usually covered with a layer of soil. This cover has a number of different functions: it should hide the waste from sight and prevent the blowing away of plastic materials and dust. The soil cover is also needed to keep birds, insects and rats off the waste for reasons of public health (spread of disease). Furthermore, the soil cover has to provide good growing conditions for the vegetation, i.e. plants should have a sufficiently thick rooting zone where adequate water is available.

A further important aspect of the final soil cover is its effect on the infiltration of rainwater. Infiltration of water into a landfill is controlled by rainfall, evapotranspiration and surface and subsurface run-off. Infiltration can only be effectively reduced in the presence of layer of low permeability beneath the soil cover (Hoeks, 1981).

Many waste disposal sites are not sealed at the bottom. Therefore the population is greatly concerned about ground-water contamination in the vicinity of these waste disposal sites, especially since it has become obvious that chemical wastes have been dumped in many sites. For this reason, the construction of an impermeable surface cap could contribute towards preventing further groundwater contamination.

In the Netherlands the guidelines for controlled landfilling prescribe a bottom liner for new waste disposal sites. However, even with a bottom liner the construction of an impermeable surface cap may be attractive for financial reasons. The costs of leachate collection and treatment can be high because special expensive treatment techniques may have to be applied. In such cases it

311

may prove more economical to prevent the production of leachate by surface capping.

This paper describes the application of natural liner materials for surface capping and the effects of surface caps on the infiltration of rainwater. The results are based both on laboratory studies and long-term field experiments.

## SELECTION OF LINER MATERIALS

In principle, various natural and synthetic liner materials can be used for the construction of liners at waste disposal sites. The selection of a certain liner material for a given application involves defining the site requirements, the period of storage required and the waste to be contained. In addition the method of construction as well as the nature, long-term integrity and costs of liner materials should be considered.

The construction in the field should be simple, fast and reliable, even on steep slopes and under less favourable weather conditions (temperature, rainfall, wind). When sheets of synthetic material are seamed together in the field, the seaming method should be relatively simple and insensitive to weather conditions. Seams have to be controlled in the field. In the case of natural liner materials, layer thickness, homogeneity, compaction and, if possible, permeability should be checked during construction.

The permeability of the liner material should be below the limits prescribed. Thickness of the liner is important in relation to mechanical damage and possible leakage. Furthermore, the roughness of the liner surface is of importance because at steep slopes the soil cover may slide down over a smooth liner.

The resistance of a liner material to chemical and physical attack are vital considerations in liner selection. Synthetic liners may loose softeners in contact with water, leachate or organic solvents. Therefore chemical resistance tests should prove that the liner is appropriate for a given application.

This is especially important for bottom liners underneath a waste disposal site. Physical resistance, both at high and low temperatures, is important in order to prevent mechanical damage through punctures and tears due to settlements. For surface caps the physical resistance of the liner is most important because of large and irregular settlements of underlying waste. In this respect the elasticity of synthetic liners should be tested at simultaneous elongation in two directions. It will be obvious that the seams should also be resistant to chemical and physical attack. Synthetic liners will certainly meet the requirements with respect of water impermeability but the long-term durability may be a problem. For that reason natural liner materials may be preferred, provided they can satisfy the permeability requirements.

In the Netherlands there is a lot of competition between liner industries and this results in only small differences in costs of the various natural and synthetic liners. The construction of an adequate liner costs about $10 \$/m^2$ in the Netherlands, excluding soil covers and drainage systems above the liner.

## PERMEABILITY OF CLAY LINERS

### Non-Darcian Flow

In clay liners with a low permeability (often less than $10^{-8}$ m s$^{-1}$) the pore water is bound by the clay particles. In swelling clay soils, for instance, negatively charged clay particles are surrounded by a diffuse double layer where positively charged cations are accumulated. The thickness of this double layer depends on the type of cation present, the ionic strength of the soil solution and the amount of water present. Since water molecules act as dipoles, they are electrically bound in these diffuse double layers.

In soils with a large amount of swelling clay minerals the water molecules are bound so strongly that the hydraulic gradient needs to be quite high before all water molecules move freely and the flow obeys Darcy's law (Gödecke, 1980; Hoeks *et al.*, 1987). At low gradients there may be no flow at all because the hydraulic gradient cannot overcome the binding forces.

For low hydraulic gradients the effect of the binding forces results in a non-linear relation between the filter velocity ($v$) and the hydraulic gradient ($i$). This is illustrated in Fig. 1. According to Gödecke (1980) flow in swelling clay soils can be described as (see also Fig. 1):

$$v = a_m\, i^m \quad \text{for} \quad i \leqslant i_G \tag{1}$$

$$v = k(i - i_0) \quad \text{for} \quad i_G \leqslant i \leqslant i_D \tag{2}$$

where:

$v$ = filter velocity (m s$^{-1}$)
$i$ = hydraulic gradient ($-$)
$a_m$ = filter velocity at $i = 1$ (m s$^{-1}$)
$m$ = exponent ($-$)
$k$ = hydraulic conductivity in the linear range (m s$^{-1}$)
$i_0$ = start gradient for the linear relation ($-$)
$i_G$ = lower transition gradient above which flow is linearly related to the hydraulic gradient ($-$)
$i_D$ = upper transition gradient above which flow does not obey Darcy's law due to turbulence effects ($-$).

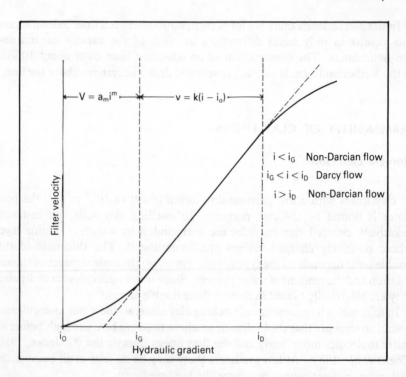

**Figure 1.** Relationship between filter velocity (*v*) and hydraulic head gradient (*i*) in clay liners:

$i < i_G$ non-linear flow due to binding forces
$i_G < i < i_D$ linear flow according to Darcy's law
$i > i_D$ non-linear flow due to turbulence effects.

This means that the traditional hydraulic conductivity, according to Darcy equal to $v/i$, is not a constant value for a clay liner. For the curve in Fig. 1 the conductivity, calculated as $v/i$, increases with increasing gradient. Therefore *k*-values measured in the laboratory cannot be used to predict leakage through clay liners in the field. This is only possible when the relation between *v* and *i* has been established and the range of *i* in the field is know.

It is therefore recommended to express the efficacy of clay liners as the amount of leakage under field conditions, e.g. leakage will be lower than $10 \text{ mm y}^{-1}$ if the gradient will be less than 5. In that case the non-linear effect has to be established and extrapolated to field conditions, taking into account the water level above the liner and the thickness of the liner.

**Permeability Tests**

Before a certain clay material can be used for the construction of a liner, a number of laboratory tests are necessary. Initially, Proctor tests have to be carried out to establish the relation between dry soil density and moisture content. From the curve obtained the optimum moisture content for compaction in the field is determined. It was found that the lowest permeability is reached when the clay is compacted on the wet side of the Proctor curve, indicating that the moisture content should be somewhat higher than the optimum moisture content for compaction (Reuter, 1985).

The next step is the establishment of the relation between filter velocity ($v$) and hydraulic gradient ($i$). Because of the low permeability laboratory tests are often performed at gradients of 20–50. Permeability tests at low gradients, as occurring in the field, would give directly the amount of leakage under field conditions, but measurements at such low gradients are often inaccurate and time-consuming. Reuter (1985) and Hoeks *et al.* (1987) describe permeameters that can be used to determine the relation between filter velocity and hydraulic gradient.

Another aspect of permeability tests is the effect of time on the permeability. Hoeks *et al.* (1987) showed that the permeability of bentonite–sand mixtures decreased during 2–3 months. One explanation may be that the swelling of the bentonite occurred over a certain length of time. A further explanation for this phenomenon may also be that of mechanical clogging by internal erosion of bentonite particles. For practical applications it is recommended to perform the permeability tests during a period of 3–4 weeks.

During the construction of a clay liner field measurements have to be done to check the homogeneity of the clay and degree of compaction, assuming that there is a relation between permeability and compaction. Further research is required on the development of quick and reliable field measurements.

For practical application of clay liners guidelines with respect to type of clay, compaction, permeability tests during construction, have to be formulated. In the Netherlands this is required especially when clays will be used as a bottom liner, because leakage through bottom liners should be practically zero.

## PRACTICAL EXPERIENCE WITH CLAY LINERS

**Silt and Clay**

Earlier research (Hoeks and Agelink, 1982) showed that most clay and silt materials are not appropriate for the construction of surface caps. The hydraulic conductivity of these materials is often too high, in the range of $10^{-8}$ to $10^{-9}$ m s$^{-1}$. In situ certain clay layers may have a lower conductivity but for surface capping the clay has to be excavated and compacted again at the waste disposal site. With such disturbed soil it is practically impossible to re-establish the original degree of compaction and low conductivity.

Under the Dutch climatic conditions, leakage of water through silt or clay layers amounts to 100–250 mm y$^{-1}$. Table 1 shows that the annual leakage rate through a clay liner depends on the hydraulic conductivity and gradient over the layer. In the Dutch climate, leakage through a surface cap can only take place in the winter period when rainfall exceeds evapotranspiration. This leakage period is prolonged slightly at decreasing permeability of the clay liner. Even at low hydraulic conductivities the leakage can be fairly high because the mean hydraulic gradient increases as conductivity decreases. In a field experiment Hoeks and Agelink (1982) measured the leakage through a silty clay liner with a mean hydraulic conductivity of $2.3 \times 10^{-8}$ m s$^{-1}$. The leakage was found to be in the order of 1 to 2.5 mm day$^{-1}$ and continued throughout the entire winter period, which means that almost the entire rainwater surplus was percolating through the liner.

For practical applications it was decided that construction of surface caps would only be of interest if the infiltration of rainwater can be reduced to less than 50 mm y$^{-1}$. Model calculations showed that leakage through clay liners is less than 50 mm y$^{-1}$ when the hydraulic conductivity under field conditions is lower than $5 \times 10^{-10}$ m s$^{-1}$ ($i = 5$, leakage period 200 days y$^{-1}$). Such low values can only be realized with certain special clay materials containing swelling clay minerals (e.g. montmorillonite). Figure 2 presents the ideal build-up of a surface cap with soil cover based on this research.

**Table 1.** Annual leakage through a 25 cm clay liner covered with 50 cm of sand, as a function of hydraulic conductivity and gradient. Under Dutch climatic conditions leakage takes place in the winter period, when rainfall exceeds evapotranspiration.

| Hydraulic conductivity | | Mean hydraulic gradient | Leakage period (d y$^{-1}$) | Annual leakage (mm y$^{-1}$) |
|---|---|---|---|---|
| (mm d$^{-1}$) | (m s$^{-1}$) | | | |
| 2.0 | $2.3 \times 10^{-8}$ | 1–1.2 | 150 | 300 |
| 1.0 | $1.2 \times 10^{-8}$ | 1.5 | 180 | 250 |
| 0.5 | $5.8 \times 10^{-9}$ | 2 | 200 | 200 |
| 0.2 | $2.3 \times 10^{-9}$ | 3 | 200 | 120 |
| 0.1 | $1.2 \times 10^{-9}$ | 4 | 200 | 80 |
| 0.05 | $5.8 \times 10^{-10}$ | 5 | 200 | 50 |
| 0.02 | $2.3 \times 10^{-10}$ | 6 | 200 | 25 |

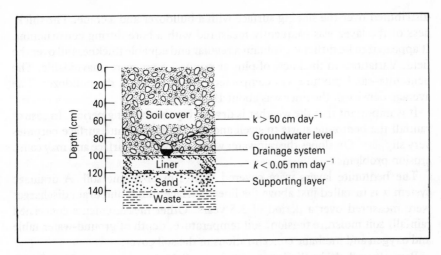

**Figure 2.** Schematic presentation of the final cover with surface cap at a waste disposal site reducing the infiltration of rainwater to less than 50 mm $y^{-1}$.

## Bentonite–Soil Mixtures

Laboratory studies have indicated that hydraulic conductivity values below $5 \times 10^{-10}$ m $s^{-1}$ can be realized quite well with bentonite, a pure clay consisting of the clay mineral montmorillonite. Bentonite is available in powder form and can be mixed with soil. The permeability of such mixtures depends on the bentonite content, the grain size distribution of the soil and the degree of compaction. Furthermore, the quality of the infiltrating water may affect the permeability of a bentonite liner (Hoeks *et al.*, 1987). For practical applications the appropriate mixing ratio has to be established experimentally in the laboratory.

In September 1982 a large-scale field experiment was started (waste disposal site in Wijster, NL) in which bentonite was used as a liner material for surface capping of the landfill. Test fields of 1800 m² each were situated on a slope of 30%. At two fields the liners consisted of sand–bentonite mixtures; one field with a mixing ratio of 100 : 10 (layer thickness 10 cm) and another field with a mixing ratio of 100 : 7.5 (layer thickness 20 cm). In laboratory experiments the hydraulic conductivity of these mixtures proved to be rather below the limit of $5 \times 10^{-10}$ m $s^{-1}$. A third field with an asphalt liner was constructed as a reference field.

The bentonite was mixed with sand in an asphalt mixing installation and from there it was transported to the experimental fields. The mixture was distributed over the sloping surface with a bulldozer and a crane. The thickness of the layer was frequently measured with a bore during construction. It appeared to be difficult to obtain a regular and equable thickness all over the field. Variations in thickness of plus or minus 2.5 cm were unavoidable. The bentonite–sand mixture was compacted in the field with the bulldozer. The average density of the liner was about 1800 kg m$^{-3}$.

It is important that the weather is dry during the working period. In case of rainfall the bentonite starts to swell and the sand–bentonite surface becomes very slippery. On slopes the rainwater will flow over the surface and may cause erosion problems.

The bentonite layers were covered with 1 m of humic soil. A drainage system was installed just above the liner. Surface runoff and drain discharges were measured over a period of 3.5 years. Other measurements concerned rainfall, soil moisture tension, soil temperature, depth of ground-water table and oxygen and methane concentrations in the soil cover.

Run-off and drain discharges from the fields were also calculated with REDRAM, a model developed by Rijtema et al. (1986) for calculation of water balances and discharges at waste disposal sites. In these calculations it was assumed that the liners are impermeable to water. The data measured were compared with the data calculated to establish the leakage through the liner. Comparison with the data of the reference field with the asphalt liner did not supply adequate information about leakage through the bentonite liners because the water balance studies showed that after two years leaks were present in the asphalt liner. The yearly leakage amounted to 50 mm y$^{-1}$. The results of these field experiments are summarized in Table 2. During the first winter the fields were not yet covered with vegetation. On the field with the 10 cm liner a lot of surface runoff occurred causing serious erosion. The topsoil on this field was more sensitive to slaking and erosion than the topsoil on the field with the 20 cm liner. In the model calculations this process could not be simulated.

The data measured for the period April 1984–March 1985 are very inaccurate because of failures in the measuring devices and clogging of the drain discharge system, causing uncontrolled leakage of drain-water and surface run-off.

When the data concerning the period April 1984–March 1985 are left out of consideration, then the total discharge measured for the two fields amounts to 649 and 745 mm respectively, while the discharge calculated for a completely impermeable liner amounts to 734 mm. This means that the 10 cm bentonite liner intercepted 88% of the rainwater surplus, while the 20 cm bentonite liner intercepted 101% of the rainwater surplus. Leakage through the 10 cm liner started after about two years. Probably this liner was too thin to resist mechanical damage by irregular settlement (total settlement was about 1.50 m at the upper part of the slope).

**Table 2.** Measured and calculated discharges (surface runoff + drain discharge) from experimental fields with surface caps of bentonite at the waste disposal site in Wijster (NL). Field 1, surface cap = 10 cm, 9.1% (w/w) bentonite; Field 2, surface cap = 20 cm, 7.0% (w/w) bentonite.

| | Discharge (mm) | | | | Leakage (mm) | |
| --- | --- | --- | --- | --- | --- | --- |
| | Field 1 | | Field 2 | | Field 1 | Field 2 |
| | Measured | Calculated | Measured | Calculated | | |
| Nov. 82–March 83 | 175[a] | 144 | 174 | 165 | −31 | −9 |
| Apr. 83–March 84 | 292 | 302 | 286 | 281 | 10 | −5[b] |
| Apr. 84–March 85 | (97) | (226) | (138) | (227) | [b] | [b] |
| Apr. 85–March 86 | 176 | 274 | 269 | 273 | 98 | 4 |
| Apr. 86–May 86 | 6 | 14 | 16 | 15 | 8 | −1 |
| Nov. 82–May 86[c] | 649 | 734 | 745 | 734 | 85 | −11 |

[a] High because of much surface run-off (no vegetation during first winter).
[b] Not calculated, measured data inaccurate.
[c] Excluding the period April 1984–March 1985.

## Other Swelling Clay Soils

Apart from bentonite, other clay soils may also contain swelling clay minerals. A special Tertiary clay found in the Netherlands and known as 'Reuverse Klei' is used at the waste disposal site in Linne (NL) as a bottom liner. This clay consists of 38.7% clay particles (<2 $\mu$m) and 61.0% silt particles (2–50 $\mu$m). The clay is compacted under optimum moisture conditions in two 25 cm thick layers. Laboratory studies showed that the non-Darcian behaviour at low hydraulic gradients is of utmost importance here (Fig. 3). In this particular case the permeability tests were performed with leachate from a landfill. From

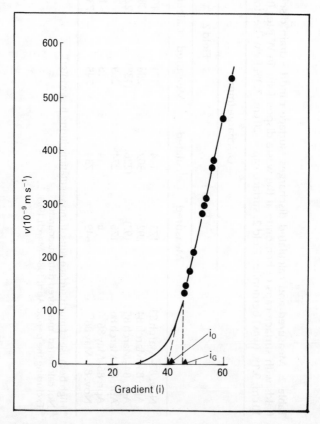

**Figure 3.** Relationship between filter velocity ($v$) and hydraulic gradient ($i$) for Reuver clay, percolated with leachate from a waste disposal site and measured according to the so-called 'falling-head method' ($i_O$ = 39.8, $i_G$ = 45).

the curve the characteristic $i$-values (see also Fig. 1) can be derived, resulting in $i_O = 39.8$ and $i_G \approx 45$. According to Gabener (1983) the following relation for the exponent $m$ in Equation 1 can be derived:

$$m = \frac{i_G}{i_G - i_O} \tag{3}$$

Putting the above measured values for $i_O$ and $i_G$ into Equation 3 the exponent $m$ becomes $m = 9$. The filter velocity can now be described as a function of the hydraulic head gradient:

$$i \geqslant 45 \quad v = 2.43 \times 10^{-8} (i - 39.8) \tag{4}$$

$$i \leqslant 45 \quad v = 1.67 \times 10^{-22} \, i^9 \tag{5}$$

Thanks to the high $i_O$ value this clay liner is practically impermeable under field conditions. For a maximum gradient of 5 the annual leakage is calculated as $1 \times 10^{-5}$ mm y$^{-1}$. This means that no leakage of leachate will occur through this clay liner. Further research is required using various types of clay to indicate which of these can be used for the construction of impermeable bottom liners.

## CONCLUSIONS

Natural clay liners can be used to isolate waste disposal sites from the environment. Adequate permeability tests must be performed to prove that leakage through the clay liner is below the acceptable leakage rate prescribed by governmental guidelines. In the Netherlands the maximum allowed leakage rate through a surface cap is not prescribed but it is recommended that it should be lower than 50 mm y$^{-1}$. For new waste disposal sites bottom liners are now obligatory in the Netherlands and leakage should be zero in this case. In view of the non-linear relation between filter velocity and hydraulic gradient under field conditions it is recommended to express the leakage through a clay liner in mm y$^{-1}$ at the maximum gradient occurring under field conditions. Usually the permeability of a clay liner is expressed with the hydraulic conductivity, traditionally calculated in m s$^{-1}$ according to Darcy's law and with no mention of the gradient. This has little meaning because conductivity is not constant here, but decreases with decreasing gradient.

Normal illitic clay soils and silty soils are not suitable for the construction of liners. Leakage through such liners is in the order of 100–250 mm y$^{-1}$ under Dutch climatic conditions.

Bentonite liners used for surface capping proved to be completely impermeable over a 3.5 year period in spite of large settlements. For surface capping Hoeks *et al.* (1987) recommended a bentonite content of at least 5 weight % and a layer thickness of at least 15–20 cm. For bottom liners higher bentonite contents are recommended as bentonite liners are more permeable for contaminated leachate from a waste disposal site than for clean water.

Research with special Tertiary clays showed that clay liners can be completely impermeable under field conditions. Apparently the pore water in the clay liner is bound so strongly that up to gradients of 5–10 the water in the pores does not flow at all. This means that certain special clays could also be used as a bottom liner below waste disposal sites.

## REFERENCES

Gabener, H.G. (1983). 'Untersuchungen über die Anfangsgradienten und Filtergesetze bei bindiger Böden', Mitt. Fachgebiet Grundbau und Bodentechnik, Univ. GHS, Essen (D), Heft 6.

Gödecke, H.J. (1980). 'Fliessgesetz für die Porenwasserdurchströmung feinkörniger Böden', *Die Bautechnik* 6, 184–193.

Hoeks, J. (1981). 'Measures to control groundwater pollution near waste disposal sites', Proceedings ISWA Symposium, Munich (FRG), June 1981). Also: Miscell. Reports, 262, ICW, Wageningen (NL).

Hoeks, J. and Agelink, G.J. (1982). 'Hydrological aspects of sealing waste tips with liners and soil covers', Proceedings Exeter Symposium, IAHS Publ. no. 139, Techn. Bull. 14, Institute for Land and Water Management Research (ICW), Wageningen (NL).

Hoeks, J., Glas, H., Hofkamp, J. and Ryhiner, A.H. (1987). 'Bentonite liners for isolation of waste disposal sites', *Waste Management and Research* 5, 93–105.

Reuter, E. (1985). Entwurf, Prüfung und Eigenschaften mineralischer Basisabdichtungen. *In* (H. Meseck, ed.), 'Abdichten von Deponien, Altlasten und kontaminierten Standorten', Heft Nr. 20, Inst. Grundbau und Bodenmechanik, TU Braunschweig (D).

Rijtema, P.E., Roest, C.W.J. and Pankow, J. (1986). 'Onderzoek naar de waterbalans van vuilstortplaatsen', Rapporten n.s. 19, Institute for Land and Water Management Research (ICW) Wageningen (NL).

# 5.3 Containment of Landfill Leachate with Clay Liners

DAVID E. DANIEL* and CHARLES D. SHACKELFORD**

*The University of Texas, Austin, Texas, USA
**Colorado State University, Fort Collins, Colorado, USA

## INTRODUCTION

The purpose of this paper is to outline the current state of knowledge with respect to transport of landfill leachate through clay liners. The issues of primary concern are advective transport, diffusive transport, effective porosity of clay liners, and degradation of clay liners exposed to landfill leachate for long periods of time. Also, the relative effectiveness of clay liners versus geomembranes (flexible membrane liners or FMLs) will be evaluated.

## TYPES OF LINERS

Earthen liners may be manufactured or naturally occurring. Manufactured liners consist of a horizontal liner at the bottom of a disposal pit and an inclined liner along the side slopes (Fig. 1); such liner systems are often composed of hydraulic barriers as well as drains (Fig. 1). It is not uncommon to employ a mix of materials, including clayey soils and geomembranes for barriers, and sand, gravel, or geosynthetics for drains. Natural earthen liners are formed by aquitards or aquicludes. Wastes may be buried wholly within a natural earthen liner (Fig. 2A), partially within a natural liner (Fig. 2B), or above but not within a natural liner (Fig. 2C).

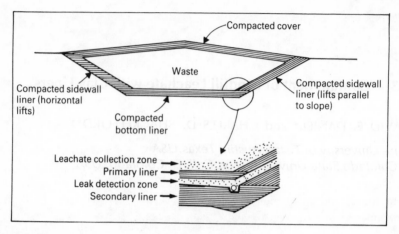

**Figure 1.** Types of compacted liners.

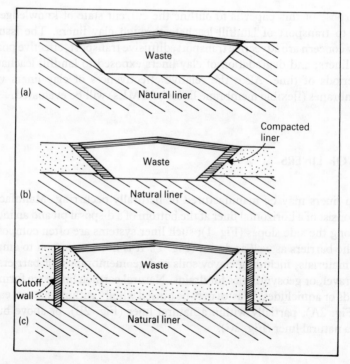

**Figure 2.** Examples of natural liners.

## TRANSPORT OF CONTAMINANTS THROUGH CLAY LINERS

### Molecular Diffusion

Diffusion is a process in which solutes in a solution flow in response to a gradient of concentration. The solution itself need not flow for diffusive transport to occur. An example of solute transported purely by molecular diffusion is given in Fig. 3. In this case, there is no advective flow because the hydraulic gradient is zero. However, since there is a gradient in solute concentration, there will be solute transport via molecular diffusion.

The process of diffusion is usually assumed to occur in accord with Fick's first and second laws. For free solutions (i.e., no porous matrix), Fick's first law states that one-dimensional diffusion occurs as follows:

$$\mathcal{J} = -D_0 \frac{\partial c}{\partial x} \tag{1}$$

where $\mathcal{J}$ is the mass flux (mass transported per unit area perpendicular to the direction of transport, per unit time, or $M\,L^{-2}\,T^{-1}$, $D_0$ is the free-solution diffusion coefficient ($L^2\,T^{-1}$), $c$ is the concentration of a solute ($M\,L^{-3}$), and $x$ is the direction in which the diffusion is occuring ($L$). A more fundamental basis for this law may be described as follows.

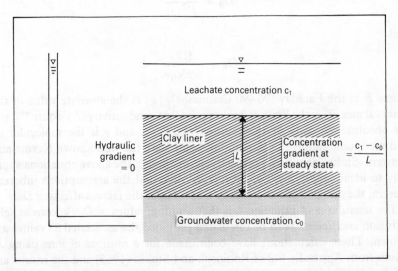

**Figure 3.** Example of a clay liner system in which molecular diffusion is the transport mechanism.

When considering individual ions or molecules, the driving force for diffusion may be taken as the gradient in chemical potential. Also, the limiting velocity of a particle may be defined as the product of the absolute mobility of the particle and its driving force (Robinson and Stokes, 1965). When these relations are combined with the definition of mass flux, the following expression results:

$$J = - \frac{vRT}{N} \frac{\partial c}{\partial x} \tag{2}$$

and, by comparison with Equation 1, the expression for the free-solution coefficient at infinite dilution becomes:

$$D_0 = \frac{vRT}{N} \tag{3}$$

where $R$ is the Universal Gas Constant ($8.134 \, J \, mol^{-1} \, K^{-1}$), $T$ is the absolute temperature (K), $N$ is Avogadro's number ($6.022 \times 10^{23} \, mol^{-1}$) and $v$ is the absolute mobility of a particle ($LT^{-1} F^{-1}$). This expression is known as the Nernst–Einstein equation (Jost, 1960). By combining Equation 3 with expressions relating the absolute mobility to the limiting ionic equivalent conductivity (Robinson and Stokes, 1965) and to the viscous resistance to the solvent molecules, i.e. Stokes Law (Bird *et al.*, 1960), two additional expressions for $D_0$ result:

$$D_0 = \frac{RT\lambda^0}{F^2|z|} \tag{4}$$

and

$$D_0 = \frac{RT}{6\pi N \eta r} \tag{5}$$

where $F$ is the Faraday (96 490 Coulombs), $|z|$ is the absolute value of the ionic valence, $\lambda^0$ is the limiting ionic equivalent conductivity ($LT \, equiv^{-1}$), $\eta$ is the absolute viscosity of the solution ($FTL^{-2}$), and $r$ is the molecular or hydrated ionic radius ($L$). Equations 4 and 5 are the well-known Nernst and Einstein–Stokes equations, respectively. Although the above equations apply only to ideal conditions, i.e. infinite dilution and the assumptions inherent therein, they do provide an indication of some of the factors affecting $D_0$.

The usefulness of Equation 4 is that limiting values of $D_0$ (known as self-diffusion coefficients) can be calculated provided the associated $\lambda^0$ values are known. These calculations have been made for a number of ions using $\lambda^0$ values from Appendix 6.2 of Robinson and Stokes (1965) and the results are indicated in Table 1. A similar table has been provided by Quigley *et al.* (1987).

The values for $D_0$ reported in Table 1 should be considered to be maximum values attainable under ideal conditions (i.e. microscopic scale, infinite dilution, etc.). Under non-ideal conditions (e.g. macroscopic scale, concentrated solutions), a number of these effects, negligible for ideal conditions, become important. A good example of one of these effects is given by Robinson and Stokes (1965). When two oppositely charged ions are diffusing in solution, a microscopic charge separation or electrical potential gradient is set up between the ions due to their different mobilities. The effect of this charge separation is to speed up the slower moving ion and to slow down the faster moving ion. Since, on a macroscopic scale, electroneutrality must be satisfied, the resultant speeds of both ions must be equal. This electrical potential effect is responsible, in part, for the differences between the simple electrolyte diffusion values, shown in Table 2, and their respective component self-diffusion coefficients given in Table 1. Other effects responsible for the difference in $D_0$ values under non-ideal conditions include solute–solute and solute–solvent interactions. In summary, the factors affecting the magnitude of $D_0$, include:

1. temperature of the solution (Equations 3, 4, 5),
2. viscosity of the solution (Equation 5),
3. ionic radius (Equation 5)
4. ionic valence (Equation 4),
5. equilibrium chemistry (dissociated vs. undissociated species),

**Table 1.** Calculated self-diffusion coefficient for representative ions at infinite dilution in water at 25°C.

| Anions | $D_0 \times 10^{10}\,m^2/s$ | Cations | $D_0 \times 10^{10}\,m^2/s$ |
|---|---|---|---|
| $OH^-$ | 52.76 | $H^+$ | 93.07 |
| $F^-$ | 14.74 | $Li^+$ | 10.27 |
| $Cl^-$ | 30.31 | $Na^+$ | 13.33 |
| $Br^-$ | 20.79 | $K^+$ | 19.56 |
| $I^-$ | 20.43 | $Rb^+$ | 20.70 |
| $HCO_3$ | 11.84 | $Cs^+$ | 20.54 |
| Acetate | 10.88 | $Be^{2+}$ | 5.98 |
| $SO_4^{2-}$ | 10.64 | $Mg^{2+}$ | 7.05 |
| $CO_3^{2-}$ | 9.22 | $Ca^{2+}$ | 7.92 |
| | | $Sr^{2+}$ | 7.90 |
| | | $Ba^{2+}$ | 8.46 |
| | | $Pb^{2+}$ | 9.245 |
| | | $Cu^{2+}$ | 7.13 |
| | | $Zn^{2+}$ | 7.02 |

$D_0$ values calculated from the Nernst equation.

**Table 2.** Representative values of free-solution diffusion coefficients (CRC Handbook of Chemistry and Physics, 66th edn, 1985–86).

| Solute | Concentration | $D_0 \times 10^9 \, m^2/s$ |
|--------|---------------|---------------------------|
| HCl | 0.1 M | 3.1 |
| LiCl | 0.1 M | 1.3 |
| NaCl | 0.1 M | 1.5 |
| CaCl$_2$ | 0.1 M | 1.1 |
| Glucose | 0.4% | 0.7 |
| Citric acid | 0.1 M | 0.7 |

6. dielectric constant of the solution, and
7. concentration of the solute (dilute vs. concentrated).

Solutes will not diffuse as quickly in soil as they will in free solutions. For soil, Equation 1 becomes:

$$J = -(\tau D_0) \theta \frac{\partial c}{\partial x}$$ (6)

or

$$J = -D^\star \theta \frac{\partial c}{\partial x}$$ (7)

where $\tau$ is a tortuosity factor (dimensionless), $\theta$ is the volumetric moisture content of the soil (dimensionless), and $D^\star$ is the effective diffusion coefficient $(L^2 \, T^{-1})$. It should be noted that some researchers include $\theta$ in the tortuosity factor. The volumetric moisture content is equal to the product of the porosity, $n$, and the degree of saturation, $S_r$, or

$$\theta = nS_r$$ (8)

and accounts for the reduced cross-sectional area of flow for solutes in soil. All other factors being the same, the above expression for $\theta$ indicates that the greatest solute flux will occur when the soil is saturated, i.e. $\theta = n$.

In general, the tortuosity factor is considered to account for the increased distance of flow and the more tortuous pathways experienced by solutes diffusing through soil. It is expressed as:

$$\tau = (x/x_e)^2$$ (9)

where $x$ $(L)$ is the straight-line distance between two points defining the flow path, and $x_e$ $(L)$ is the actual, effective distance of flow through the soil between the same two points. Since $x_e > x$, $\tau < 1$. Some typical values of $\tau$ reported in the literature are provided in Table 3.

In reality there may be other effects associated with the tortuosity factor. For example, Olsen *et al.* (1965) considered two other effects, the variation in the viscosity of the solvent within the pore space, represented by the factor $\alpha$ and the negative adsorption (i.e., exclusion) of ions, represented by the factor $\gamma$. They combined all of these effects (including the volumetric moisture content) into one factor and termed it the 'transmission factor', or

$$\tau = \alpha\gamma(x/x_e)^2 \theta \qquad (10)$$

While these and other effects may be present in many situations, it is neither possible nor feasible in most cases to separate them. In short, the tortuosity factors reported in the literature may account for more than just the pore geometry of the porous matrix.

**Table 3.** Representative tortuosity factors taken from the literature.

| $\tau$ Values | Soil | Tracer | Reference |
|---|---|---|---|
| 0.08–0.12 | 50% sand–bentonite | $^{36}$Cl | Gillham and Sharma (1985) |
| 0.04–0.49 | 0, 10, 25, 50, 75% bentonite–sand | $^{36}$Cl | Johnston *et al.* (1984) |
| 0.01–0.22 | 0, 10, 25, 50, 75% bentonite–sand | $^{3}$H | Johnston *et al.* (1984) |
| 0.59–0.84 | 0, 5, 10, 15, 25, 50, 100% bentonite–sand | $^{36}$Cl | Gillham *et al.* (1984) |
| 0.33–0.70 | 0, 5, 10, 15, 25, 50, 100% bentonite–sand | $^{36}$Cl | Gillham *et al.* (1984) |
| 0.20–0.33 | Silty clay loam, clay | Br | Barraclough and Tinker (1982) |
| 0.08–0.31 | Silty clay loam, sandy loam | Br | Barraclough and Tinker (1981) |
| 0.022–0.539 | 6 natural soils | $^{65}$Zn | Warncke and Barber (1972a) |
| 0.01–0.58 | 5 natural soils | $^{36}$Cl | Warncke and Barber (1972b) |
| 0.0032–0.023 | Loam | $^{86}$Rb | Patil *et al.* (1968) |
| 0.027–0.31[a] | Loam, silty clay loam, clay | Cl | Porter *et al.* (1960) |
| 0.04–0.45 | 75 $\mu$m and 200 $\mu$m glass beads | $^{86}$Rb | Klute and Letey (1958) |

[a] Transmission factors.

Fick's first law describes steady-state flux of solutes. For unsteady (transient) transport. Fick's second law applies:

$$\frac{\partial c}{\partial t} = D^\star \frac{\partial^2 c}{\partial x^2} \tag{11}$$

Equation 11 is integrated for appropriate initial and boundary conditions to obtain a description of the solute concentration changes with respect to time and space. Complex error functions facilitate the integration.

An example of the application on Fick's second law is as follows. A 1 m thick clay liner retains leachate containing a particular solute at a concentration of 10 000 mg/l. The underlying ground water is completely free of this solute. The effective diffusion coefficient is $10^{-10}$ m$^2$/s. The resulting distribution of solute in and at the bottom of the liner as a function of time solely due to diffusional transport is illustrated in Fig. 4. The solute distribution was determined from the following solution to Equation 11 (Crank, 1975) which assumes the leachate concentration is constant with time:

$$\frac{c}{c_0} = \mathrm{erfc} \left( \frac{x}{2\sqrt{(D^\star t)}} \right) \tag{12}$$

where erfc is the complementary error function. The concentrations of solute reaching the base of the liner at 10, 20, 40 and 80 years are approximately 50, 460, 1600, and 3200 mg/l, respectively. Since, in some cases, these concentrations exceed allowable values, it is apparent that diffusion of chemicals through fine-grained materials can be an important transport mechanism even over relatively short (20–30 years) periods.

## Advective Transport

Advective transport of contaminants through clay liners is generally assumed to occur in accordance with Darcy's law in which

$$v = ki = -k \frac{\partial h}{\partial x} \tag{13}$$

where $v$ is the Darcian velocity ($L/T$), $k$ is the hydraulic conductivity ($L/T$), $i$ is the dimensionless hydraulic gradient, $h$ is the total hydraulic head ($L$), and $x$ is the distance along the path of flow ($L$). The seepage velocity ($v_s$) is the average linear velocity along the path of flow, and for non-sorbed or conservative solutes, is given by:

$$v_s = \frac{ki}{n_e} = -\frac{k}{n_e} \frac{\partial h}{\partial x} \tag{14}$$

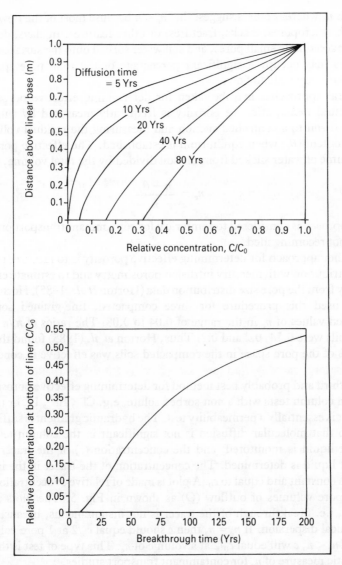

**Figure 4.** Concentration profiles for 1 m thick clay liner and $D^* = 2 \times 10^{-10}$ m$^2$/s.

where $n_e$ is the effective porosity, i.e. the volume of pore fluid that is effective in conducting flow divided by the total volume.

Considerable controversy exists over typical values of $n_e$ for compacted clays. Some have suggested that $n_e$ is more-or-less equal to the total porosity of the soil, $n$, whereas others suggest that $n_e < n$ because most of the flow occurs through macropores, cracks, fractures, or other features. In clays, dead-end pores, exceedingly small pores, and soil water sorbed onto the surfaces of clay particles may occupy a significant percentage of the total pore space but contribute very little to flow.

Several approaches may be taken for determining effective porosity. In agricultural fields, effective porosity is frequently measured by subjecting saturated soil to a controlled suction and measuring the residual volumetric water content ($\theta_r$) when equilibrium is established. The effective porosity is the volume of water sucked from the soil divided by the total volume, or:

$$n_e = \frac{n - \theta_r}{n} \tag{15}$$

This approach is considered too empirical for contaminant transport problems and is not recommended.

Another approach for determining effective porosity is to measure the pore size distribution with mercury intrusion porosimetry and to estimate effective porosity from the pore size distribution data (Horton *et al.*, 1985). Horton *et al.* (1985) used this procedure for three compacted, fine-grained soils and estimated values of $n_e$ in the range of 0.04 to 0.08. The ratios of $n_e/n$ for the three soils were 0.14, 0.2 and 0.2. Thus, Horton *et al.* (1985) found that only 14–20% of the pore space in the compacted soils was effective in conducting flow.

The third and probably best method for determining effective porosity is to perform column tests with a non-sorbed solute, e.g. $Cl^-$, $Br^-$, $I^-$, or tritium. The test is essentially a permeability test. The hydraulic gradient is sufficiently large so that molecular diffusion is not significant in the column test. The effluent liquid is monitored, and the concentration ($c$) of the tracer in the effluent liquid is determined. The concentration of the tracer in the influent liquid is constant and equal to $c_0$. A plot is made of relative concentration, $c/c_0$, versus pore volumes of outflow ($Q$) as shown in Fig. 5. The plots are not vertical, i.e. breakthrough of the tracer is not instantaneous, because of the mechanical dispersion. If $n_e = n$, then $c$ should equal $c_0/2$ at 1 pore volume of flow. If $n_e < n$, $c$ will equal $c_0/2$ at a volume of $n_e$. This type of test is the most diagnostic measure of $n_e$ for contaminant transport studies.

## Attenuation

Contaminants (solutes) may migrate slower than the transporting solution (solvent) for a variety of reasons. Some of the causes of contaminant

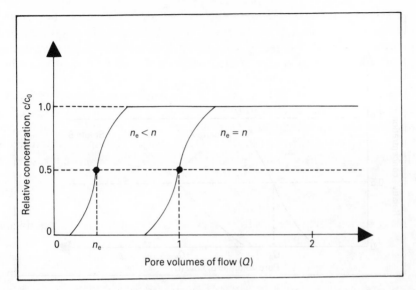

**Figure 5.** Breakthrough of solutes in column tests.

attenuation in soil include ion exchange, precipitation, biological reactions, and radioactive decay.

Suppose a solution contains two solutes, A and B. Solute A is conservative, and therefore not attenuated by soil; solute B is attenuated. A column test is performed and effluent analyses yield the results shown in Fig. 6. We define the 'retardation coefficient', $R$, as follows:

$$R = \frac{Q_B}{Q_A} \qquad (16)$$

where $Q_A$ and $Q_B$ are the pore volumes of flow for solutes A and B, respectively, at a relative concentration $c/c_0$, of 0.5. Since $Q_B > Q_A$, $R > 1.0$ for a retarded solute.

The retardation coefficient may also be calculated independently by the following relationship:

$$R = 1 + \frac{\rho_d}{n} K_p \qquad (17)$$

where $\rho_d$ is the dry bulk density of the soil $(ML^{-3})$, $n$ is the soil porosity, and $K_p$ is the partition coefficient $(L^3 M^{-1})$. The partition coefficient relates mass of solute sorbed per mass of soil, $q$, to the concentration of solute in solution, $c$, at equilibrium. Batch adsorption tests are often used to determine the

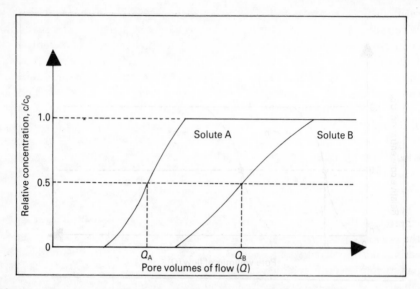

**Figure 6.** Breakthrough for retarded (B) and unretarded (A) solutes.

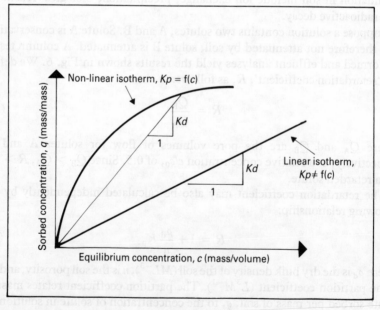

**Figure 7.** Typical adsorption relationships for reactive solutes.

relationship between $q$ and $c$ over an appropriate range of concentrations. When the $q$ vs. $c$ relationship is constant (i.e. linear isotherm), the partition coefficient is given a specific name, i.e., the distribution coefficient, $K_d$. Unfortunately, $K_p$ is often dependent upon the equilibrium concentration of the solution (Fig. 7). None the less, it is usually convenient to assume a linear relationship for the purpose of calculating retardation coefficients, as indicated in Fig. 7. The procedures for performing batch isotherm tests and the factors affecting those procedures are discussed extensively in US EPA (1985).

The sorption of hydrophobic organic pollutants also can be described adequately by Equation 17 provided the distribution coefficient is defined as follows:

$$K_d = f_{oc} \cdot K_{oc} \qquad (18)$$

where $f_{oc}$ is the fraction of organic-carbon in the soil (dimensionless) and $K_{oc}$ is the organic carbon partition coefficient (Karickoff *et al.*, 1979). The organic-carbon partition coefficient has been empirically correlated with a number of parameters, especially the octanol-water partition coefficient, $K_{ow}$. These empirical correlations are covered in detail by Griffin and Roy (1984).

## ATTACK OF CLAY LINERS BY LEACHATE

Waste liquids may attack and effectively destroy earthen liners. It is convenient to consider acids and bases, neutral inorganic liquids, neutral organic liquids and leachates separately.

### Acids and Bases

Strong acids and bases can dissolve solid material in the soil, form channels and increase $k$. Some acids, e.g. hydroflouric and phosphoric acid, are particularly aggressive and dissolve soil readily. Concentration of acid, duration of reaction, liquid–solid ratio, type of clay, and temperature are also important variables.

When concentrated acid is passed through clayey soil, the results depicted in Fig. 8 are commonly observed. Initially, there is a drop in hydraulic conductivity ($k$) that is caused by precipitation of solid matter from the permeating liquid as the acid is neutralized by the dissolved soil. The precipitates plug the pores a short distance into the specimen. With continued permeation, fresh acid enters the soil, redissolves the precipitates and eventually causes an increase in $k$. Soils have a high capacity to buffer acid;

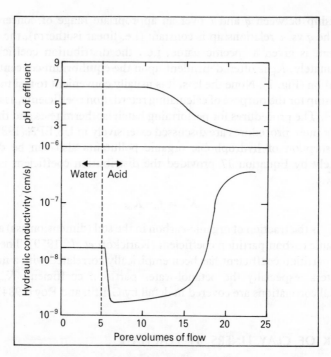

**Figure 8.** Typical variation in hydraulic conductivity for a sample of compacted clay permeated with concentrated acid.

many pore volumes of flow are usually needed before the full effect of the acid is observed. Soils that are composed primarily of sand, with a small amount of betonite, are particularly susceptible to attack by acids because the small mass of bentonite is readily dissolved. Strong bases may also dissolve soil, but data are lacking.

**Neutral, Inorganic Liquids**

The effects of neutral, inorganic liquids may be evaluated with the Gouy-Chapman theory (Mitchell, 1976), which states that the thickness ($t$) of the diffuse double layer varies with the dielectric constant of the pore fluid ($\varepsilon$), the electrolyte concentration ($n_0$), and the cation valence ($z$) as follows:

$$t = \alpha[\varepsilon/(n_0 z^2)]^{1/2} \tag{19}$$

For solutions containing mainly water, the dielectric constant of the liquid is relatively constant, and thus the main parameters are $n_0$ and $z$. As the diffuse double layer of adsorbed water and cations expands, $k$ decreases because flow channels become constricted. Attempts to validate quantitatively the effect of $\varepsilon$, $n_0$, and $z$ on $k$ have generally failed. Qualitatively, however, the Gouy-Chapman theory explains the observed patterns. Aqueous solutions with few electrolytes, e.g. distilled water, tend to expand the double layer and to produce low $k$. Solutions with monovalent cations, e.g. $Na^+$, tend to produce lower $k$ than solutions with polyvalent catons, e.g. $Ca^{++}$. A strong (high $n_0$) solution containing polyvalent cations tends to produce the largest $k$.

## Neutral Organic Liquids

Most organic chemicals have lower dielectric constants than water. Low $\varepsilon$ tends to cause low $t$ (Equation 19) and thus high $k$. In addition, low-dielectric-constant liquids cause clay particles to flocculate and to cause the soil to shrink and to crack. Numerous studies have shown that organic chemicals can cause large increases in $k$ (Anderson, 1982; Foreman and Daniel, 1986; and others). Clays with high negative charge and high activity are more susceptible to attack by organic chemicals than clays with low negative charge (Acar and Ghosn, 1986). Sodium bentonite is particularly susceptible to attack.

Dilute organic liquids do not tend to alter $k$ significantly. If a small amount of low-dielectric-constant liquid, e.g. trichlorethylene ($\varepsilon = 3$), is mixed with water, the dielectric constant of the mixture is only slighly less than that of water (80). Tests have indicated that the dielectric constant must be less than 30 to 50 for $k$ to increase. Experience indicates that $k$ of clay soils is not likely to be adversely affected by an organic liquid if: (1) the solution consists of at least 50% water; and (2) there is no separation of phases, i.e. all of the organic liquid is dissolved in the water and none exists as a separate phase.

## Actual Landfill Leachates

Griffin and Shimp (1978) report data for municipal solid waste while Daniel and Liljestrand (1984) tested leachate from hazardous industrial waste. Both investigators found steady or decreasing $k$ with time. Dilute liquids appear to be incapable of increasing $k$ of most soils.

## COMPARISON OF CLAYS AND GEOMEMBRANES

An interesting question is how the performance of geomembrane liners compare to clay liners. Some calculations may help to answer this question. For example, consider two liners, a geomembrane and a compacted-clay liner, each with the properties shown in Table 4.

These values are assumed, and some discussions of their magnitudes is warranted. The hydraulic conductivity and thickness of the clay are the maximum and minimum values, respectively, currently being allowed by the US EPA for double-liner systems. The clay porosity is a representative, average value, and the diffusion coefficient for the clay is representative of values currently being measured at the University of Texas. The thickness of the geomembrane is typical of current US practice. The diffusion coefficient for the geomembrane is about the same as a value reported by Lord and Koerner (1984) based on water absorption tests. This diffusion coefficient is from 2 to 4 orders of magnitude less than other reported values which were measured using radioactive tracers (Hughes and Monteleone, 1987; Lord and Koerner, 1984). Also, the hydraulic conductivity for the geomembrane is slightly less than the typical value reported by Koerner (1986). Geomembranes are non-porous and the porosity of 0.1 was selected for computational convenience. Finally, the hydraulic gradient in each case represents 0.152 m of ponded leachate assuming atmospheric pressure at the base of the liner and no suction within the liner.

If the values stated above are used in conjunction with the well-known Ogata (1970) solution to the differential equation describing solute transport, breakthrough curves may be calculated and compared. This calculation has been performed and the results are presented in Fig. 9. Note that the breakthrough time (at $c/c_0 = 0.5$) for the geomembrane is exceedingly fast, about 1.7 years, especially when compared with that of the clay liner, or about 11.5 years. This difference results even though the geomembrane has a

**Table 4.** Comparison of geomembrane and clay liners.

| Property | Geomembrane | Clay |
|---|---|---|
| Porosity | 0.10 | 0.50 |
| Hydraulic conductivity, $k$ (m/s) | $1.0 \times 10^{-14}$ | $1.0 \times 10^{-9}$ |
| Diffusion coefficient, $D^*$ (m$^2$/s) | $3.0 \times 10^{-14}$ | $2.0 \times 10^{-10}$ |
| Thickness (m) | 0.00152 (60 ml) | 0.91436 (3 ft) |
| Hydraulic gradient, $i$ | 100 | 1.16 |

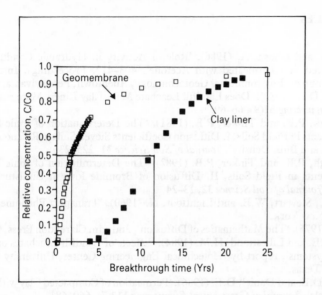

**Figure 9.** Solute breakthrough curves for geomembrane and clay liners.

hydraulic conductivity which is 5 orders of magnitude less than the clay and an effective diffusion coefficient which is about 4 orders of magnitude less. Obviously, the reason for this difference is the relative thickness of the two liners, and not the relative transport properties. No material is impervious, and the question of which liner is more effective, like most questions, is ultimately related to one of economics and the realities of construction practices.

## CONCLUSIONS

Clay liners can slow the movement of pollutants out of land disposal facilities in several ways. If the hydraulic conductivity is low, release will be primarily by molecular diffusion. Attenuation processes also work to slow the transport of many contaminants. Although much attention has been focused on attack of clay liners by leachate, only concentrated chemicals are of serious concern. In most instances, landfill leachate will not degrade clay liners. Based on transit time calculations, clay liners can be very effective barriers and minimize release of pollutants from landfills.

## REFERENCES

Acar, Y.B. and Ghosn, A. (1986). 'Role of Activity in Hydraulic Conductivity of Compacted Soil Permeated with Acetone'. 403–412. Proceedings, International Symposium on Environmental Geotechnology, Allentown, Pennsylvania.

Anderson, D.C. (1982). 'Does Landfill Leachate Make Clay Liners More Permeable?' *Civil Engineering* 52(9), 66–69.

Barraclough, P.B. and Tinker, P.B. (1981). 'The Determination of Ionic Diffusion Coefficients in Field Soils, I. Diffusion Coefficients Sieved Soils in Relation to Water Content and Bulk Density', *Journal of Soil Science* 31, 225–236.

Barraclough, P.B. and Tinker, P.B. (1982). 'The Determination of Ionic Diffusion Coefficients in Field Soils, II. Diffusion of Bromide Ions in Undisturbed Soils Cores', *Journal of Soil Science* 32, 13–24.

Bird, R.B., Stewart, W.E. and Lightfoot, N. (1960). 'Transport Phenomena'. John Wiley, New York.

Crank, J. (1975). 'The Mathematics of Diffusion', 2nd edn. Clarendon Press, Oxford.

Daniel, D.E. and Liljestrand, H.M. (1984). 'Effects of Landfill Leachates on Natural Liner Systems', report by Geotechnical Engineering Center, University of Texas, Austin, Texas.

Foreman, D.E. and Daniel, D.E. (1986). 'Permeation of Compacted Clay with Organic Chemicals', *Journal of Geotechnical Engineering* 112(7), 669–681.

Gillham, R.W., Robin, M.J.L., Dytynyshym, D.J. and Johnston, H.M. (1984). 'Diffusion of nonreactive solutes through fine-grained barrier materials', *Canadian Geotechnical Journal* 21(3), 541–550.

Gillham, R.W. and Sharma, H.D. (1985). 'The Effect of Temperature and Groundwater Composition on the Diffusion of Sr in a 50% Mixture of Silica Sand and Avonlea Bentonite'. Institute for Groundwater Research, University of Waterloo, Waterloo, Ontario, Canada, N2L 3G1.

Griffin, R.A. and Shimp, N.F. (1978). 'Attenuation of Pollutants in Municipal Landfill Leachate by Clay Materials'. US Environmental Protection Agency, Washington, DC, Report No. EPA-600/2-78-157.

Griffin, R.A. and Roy, W.R. (1984). 'Interaction of Organic Solvents with Saturated Soil–Water Systems'. Environmental Institute for Waste Management Studies, University of Alabama, Tuscaloosa, Alabama.

Horton, R., Thompson, M.L. and McBride, J.F. (1985). 'Estimating transit times of noninteracting pollutants through compacted soil materials', 275–282. Proceedings, Eleventh Annual Research Symposium on Land Disposal of Hazardous Waste, US Environmental Protection Agency, Cincinnati, Ohio, EPA/600/9-85/013.

Hughes, J.W. and Monteleone, M.J. (1987). *In* (D.J.A. van Zyl *et al.* eds) 'Geomembrane/Synthesized Leachate Compatibility Testing', Geotechnical and Geohydrological Aspects of Waste Management, 35–50, Lewis Publishers, Chelsea, Michigan.

Johnston, H.M., Gillham, F.W., Robin, M.J.L., Sharma, H.D. and Wilmot, D.J. (1984). 'Evaluation of Diffusion Coefficients for Strontium in Buffer Materials'. Report No. 84-298-K, Ontario Hydro Research Division, Ontario, Canada.

Jost, W. (1960). 'Diffusion in Solids, Liquids and Gases'. Academic Press, New York.

Karickoff, S.W., Brown, D.S. and Scott, T.A. (1979). 'Sorption of Hydrophobic Pollutants on Natural Sediments'. *Water Research* 13(3) 241–248.

Klute, A. and Letey, J. (1958). 'The Dependence of Ionic Diffusion on the Moisture

Content of Nonadsorbing Porous Media'. *Proceedings, Soil Science Society of America* **22**, 213–215.

Koerner, R.M. (1986). 'Designing with Geosynthetics'. Prentice-Hall, Englewood Cliffs, NJ.

Lord, A.E. Jr and Koerner, R.M. (1984). 'Fundamental Aspects of Chemical Degradation of Geomembranes', 293–298. Proceedings, International Conference Geomembranes, Denver, Colorado, 20–24 June.

Mitchell, J.K. (1976). 'Fundamental of Soil Behaviour'. John Wiley, New York.

Ogata, A. (1970). 'Theory of Dispersion in a Granular Medium'. US Geological Survey Prof. Paper.

Olsen, S.R., Kemper, W.D. and Van Schaik, J.C. (1965). 'Self-Diffusion Coefficients of Phosphorus in Soil Measured by Transient and Steady-State Methods', *Proceedings, Soil Science Society of America* **29**, 154–158.

Patil, A.S., King, K.M. and Miller, H.M. (1968). 'Self-diffusion of rubidium as influenced by soil moisture tension'. *Canadian Journal of Soil Science*, **43**, 44–51.

Porter, L.K., Kemper, W.D., Jackson, R.D. and Stewart, B.A. (1960). 'Chloride Diffusion in Soils as Influenced by Moisture Content', *Proceedings, Soil Science Society of America* **24**(6) 460–463.

Quigley, R.M., Yanful, E.K. and Fernandez, F. (1987). Ion Transfer by Diffusion Through Clayey Barriers (R.D. Woods, ed.) *In* 'Geotechnical Practice for Waste Disposal', pp. 137–158. ASCE.

Robinson, R.A. and Stokes, R.H. (1965). 'Electrolyte Solutions', Revised 2nd edn, Butterworth, London.

US EPA (1985). 'Batch-Type Adsorption Procedures for Estimating Soil Attenuation of Chemicals'. Draft Technical Resource Document (TRD), EPA/530-SW-85, Office of Solid Waste and Emergency Response, Washington, DC.

Warncke, D.D. and Barber, S.A. (1972a). 'Diffusion of zinc in soil: I. The influence of soil moisture', *Proceedings, Soil Science Society of America* **36**, 39–42.

Warncke, D.D. and Barber, S.A. (1972b). 'Diffusion of zinc in soil. II. The influence of soil bulk density and its interaction with soil moisture', *Proceedings, Soil Science Society of Ameria* **36**, 42–46.

# 5.4 Leachate Collection Systems

HANS GÜNTER RAMKE

*Leichtweiß Institut, Technical University of Braunschweig, Beethovenstrasse 51a, 3300 Braunschweig, W. Germany*

## INTRODUCTION

Much investigation has been carried out to improve both the design and efficacy of bottom liners but little research has taken place in the field of leachate collection and transportation systems installed above the bottom liner. Experience shows that drainpipes often clog after a certain period of time and therefore no longer function satisfactorily; these clogging effects can also be observed in the gravel surrounding the drainpipes.

The dewatering system is equally important to the liner, as it is necessary to avoid emission of leachate into the groundwater; the leachate which forms above the bottom liner must be collected and transported out of the landfill with a certain celerity in order to avoid the build up of a water-head, thereby resulting in a lesser load to be supported by the liner. For this reason the bottom leachate collection system requires an improved design and maintenance program. In order to meet the latter aims, the clogging mechanism must be investigated.

### Need for a Leachate Collection System

The leachate collection system (LCS) has the following tasks of collecting leachate and discharging it at defined sites outside the landfill; avoiding leachate build up at landfill bottom.

The second point is of particular importance as the build up of leachate could

343

provoke the following problems:

1. High water tables in the landfill would result in a more intensive leaching and consequently higher concentrations of pollutants in the leachate.
2. The hydrostatic pressure head above the bottom liner will be increased thus leading to higher leachate emission into the groundwater–soil system with all sealing materials.

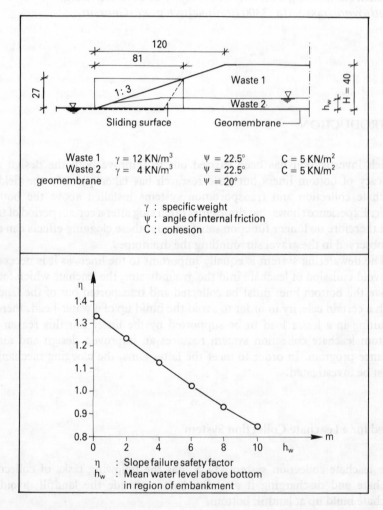

**Figure 1.** Influence of water build up on the stability of a landfill.

3. The stability of the landfill may be affected.
4. Leachate may migrate out of the landfill onto the slopes.

The consequences which concern landfill stability in case of leachate build up are illustrated in Fig. 1. A landfill with a height of 40 m and a slope inclination 1 : 3 was used as an example. The slope failure safety factors were calculated for different water levels according to DIN 4084. The critical sloping sliding surface (precalculated, see Fig. 1) and the landfill bottom (plastic liner) were taken as sliding surfaces. The slope failure safety factors calculated are dependant on the build up of leachate head. In the case of no leachate build-up at the bottom, the required safety factor 1.4 is scarcely reached. If the leachate builds up increasingly the safety factor will decrease below 1 at a water table of 8 m. This rough calculation shows the potential risk with regard to landfill stability. This is of course only true if the landfill is above ground.

This example shows how important a long-term functioning of the leachate collection system is. Considering landfill stability, the leachate collection system has to be effective for as long as leachate is produced, independent of leachate quality.

At large operating landfills it is only possible to restore parts of the leachate collection systems and even this causes enormous technical and financial efforts. In many cases this means the excavation and redumping of part of the waste.

## COMMON PRACTICE IN THE DESIGN OF LEACHATE COLLECTION SYSTEMS IN W. GERMANY

Figure 2 demonstrates four different examples of design of leachate collection systems suggested by the 'LAGA' (Anonymous, 1979).

A leachate collection system basically consists of a drainage layer of inert material with a high permeability and of the drain pipes which have to collect the leachate and to discharge it outside the landfill mound.

Types 1 and 2 of the examples shown in Fig. 2 show the drainage layer implaced with roof-like slopes provided with drain pipes at the deep points. Type 1 would be applied in the case of impermeable native subsoil or an artificial clay liner. The drainage layer is installed directly on the clay liner. In the case of there being no filter stability of the drainage layer material against the clay, both layers must be separated by a geotextile. The Type 2 design is similar to that of Type 1 but a protective layer of fine-grained material for the plastic liner is installed. A further layer of fine material is placed above the drainage layer (fine-grained waste) to protect the plastic liner from perforation by bulky wastes.

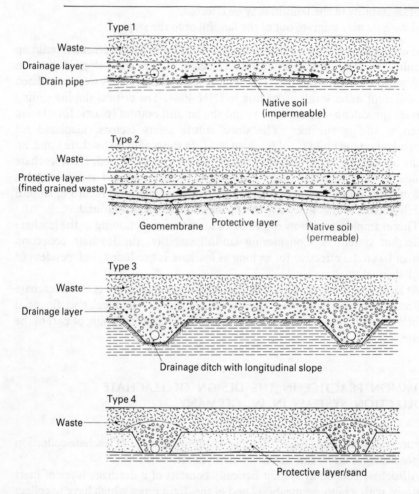

**Figure 2.** Examples of different leachate collection systems (Anonymous, 1979).

The third type of leachate collection system shows drain pipes situated in ditches. Considering also the requirements of minimum liner thickness, this type should only be applied under most specific conditions.

Type 4 combines the protection layer with the drainage layer. Quite simply, the drain pipes are surrounded by a highly permeable material. A less permeable material is used as protection and drainage layer.

## Cross Slopes and Drain Spacing

In flat areas the cross slope between the drain pipes and drain spacing have to be properly designed. Figure 3 illustrates the water level above the bottom in relation to drain spacing with respect to different permeabilities of drainage

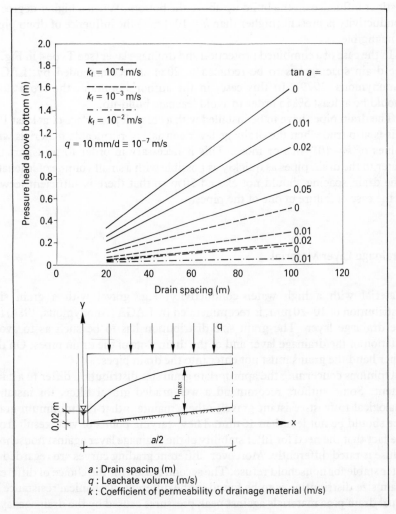

**Figure 3.** Relation between drain spacing, slopes and pressure head.

material and different cross slopes. The calculation is based on a leachate volume of $10 \, \text{mm/d} \approx 10^{-7} \, \text{m/s} = 1 \, \text{l/s ha}$. The value is ten times higher than the average value to be expected for uncovered dumping areas. The model used is based on the simple drain spacing formula (drain pipe on impermeable bottom) which has been extended by one term to respect the cross slope. The maximum leachate build up with respect to drain pipe spacing is presented in Fig. 3. This figure shows that in the case of low hydraulic conductivity ($k = 10^{-4} \, \text{m/s}$ or $k = 10^{-3} \, \text{m/s}$) the cross slope and the drain spacing show a distinct influence on the build up above the bottom. When a higher degree of conductivity is present (higher than $k = 10^{-2} \, \text{m/s}$) the influence of drain pipes is negligible.

In the case of a combined protection and drainage layer (see Type 4 in Fig. 2) the drain spacing has to be reduced to 20 m as recommended by 'LAGA' (Anonymous, 1979). In this case, in the author's opinion, the cross slope should be at least 5% in order to avoid leachate build-up.

The drain pipes have to be installed with a cross slope surface of at least 1%; this is also true when the drainage layer consists of permeable materials with values of $k = 10^{-2} \, \text{m/s}$ or more. This is necessary in order to transport the water to the drain pipes as quickly as possible with a small volume of leachate. The drain spacing should not exceed 50 m so that there is sufficient leeway in the case of failure of one of the pipes.

## Drainage Layer Material

Material with a high water conductivity, e.g. gravel with a grain size distribution of 10–20 mm, is recommended by LAGA (Anonymous, 1984) for the drainage layer. The grain size distribution has to be such as to avoid siltation of the drainage layer and of the drain slots of the drain pipes. On the other hand the grains must not enter into the drain pipes.

Opinions concerning the appropriate grain size distribution differ to a large extent. Some authors recommend a well-graded gravel filter, for instance analogical to frost-resistant gravel; others maintain that the minimum grain size should be not less than 16 mm. These varying points of view result from the fact that the need for filter stability of the drainage layer against household refuse is rated differently. Moreover, different grading curves are regarded as filter stable for household refuse. The second reason for the choice of different grain size distribution is to be found in the different mechanical resistance of usual drain pipe materials against peak pressures caused by the drainage layer material. The problem of suitable grain size distribution will be further discussed.

The drainage layer material must be resistant to landfill leachate and gas. Carbonate rocks as well as calcitic sandstones are totally unsuitable. Natural gravel is generally appropriate. In case of doubt concerning chemical resistance a mineralogist or geological chemist should be contacted. Usually the pressure strength of the rocks proves to be adequate but this should be proven if any doubt exists.

## Drain Pipes

Pipes with holes or slots are used as drain pipes. When choosing the pipes and designing the leachate collection system it is important to bear in mind the possibility of inspection and maintenance of the drain pipes. Consequently, the pipes should be installed in a straight line with respect to ground plan and longitudinal section. Branches within the landfill mound have to be avoided (Anonymous, 1984). The minimum diameter of the pipes should be 200 mm in order to allow inspection by TV cameras and maintenance by high pressure flushing equipment. The distance between two manholes should not exceed 300 m, however, shorter distances are recommended.

Currently, intensive efforts are being made to develop flushing equipment with a higher flushing pressure at the nozzle end and an increased working length. Using modern equipment, a pipe length of up to 300 m can be reached with a flushing water volume of approximately 300 l/min; the pressure at the pump is up to 150 bar. Furthermore, equipment has been developed which is adjustable to the diameter of the pipe for the removal of very hard incrustations.

The minimum slope of the drain pipes should be 1% in order to obtain a sufficiently rapid flow also in the case of a low volume of leachate. For hydraulic reasons alone this slope in combination with pipes of a diameter larger than 200 mm and drain spacings of 50 m does not require the discharging of leachate volume during landfill operation. Hydraulic dimensioning, however, has to consider the fact that during the preparation phase of a new landfill section, the total rainfall may have to be transported out of the landfill.

At a minimum slope of 1% it will seldom be possible to avoid sedimentation of particles washed into the pipe (Ramke, 1986). When designing the leachate collection system, settlement of the subsoil caused by the pressure of the landfill mound has to be considered. The slope must be elevated to accommodate settlements. The drain slots must be as wide as possible in order to avoid an untimely clogging due to precipitation of leachate constituents. Bearing in mind that the drain slots are subject to a flushing effect, it stands to reason that a better flow is obtained through a wide slot than through a smaller channel. Considering this aspect, the relation of width of holes and slots respectively to wall diameter should be a minimum of 1 : 1, 1.5 : 1. The ultimate size of slots or holes furthermore depends on the grading curve of the drainage layer material. The smallest grain from the drainage layer must not pass the drain slots. The filter criteria to be applied are determined by the slope, the uniformity coefficient as well as by the part of tail grains. In general, the diameter of drain slots should not be less than 10 mm.

The drain pipe material has to be resistant against leachate, landfill gas and wastes. Possible attacks by organic solvents have to be considered. The intensive microbial activities in the landfill mound require pipe materials with biological resistance. The expected thermic strain of the pipe material is discussed elsewhere.

The requirements of major importance concerning manholes are stated by LAGA (Anonymous, 1984) as follows:

1. manholes have to be installed both at the beginning and end of a drain pipe;
2. manholes should not be installed inside the landfill under any circumstances;
3. pipes entering the manhole should be provided with a water seal to avoid the access of air.

## FAILURE AND FAILURE MECHANISMS

### Failure of Leachate Collection Systems of Sanitary Landfills

Table 1 lists failures that occurred in leachate collection systems of sanitary landfills. Leachate discharge from the first two sections of landfill V has been rather constant since starting operation in 1976 although leachate volume has remained below the volume estimated for five years. A camera inspection ascertained that there were no incrustations in the pipes. Excavation showed that the filter material around the pipes (mixture of gravel and sand) had become almost concreted and was therefore impermeable. The parts controlled were repaired by replacing the original filter material with a coarser material. The drainpipes had not been equipped with a syphon before excavation, so that air could enter.

On landfill H1 leachate discharge decreased which was deemed to be due to clogging of the drain pipes. As flushing of the drain pipes did not ease the situation, excavation was carried out. This excavation provided further proof that the drain pipes were clogged and that the graded gravel filter surrounding

**Table 1.** Failures of leachate collection systems (sanitary landfills).

| Landfill | Failure | Restoration |
|---|---|---|
| E | Leachate pressure pipe-line partly clogged by incrustations | Flushing |
| H1 | Drain pipes clogged | Flushing (only partly successful) |
| | Drainage layer (well graded gravel filter) concreted and impermeable | Filter material partly replaced |
| | Drain pipes partly destroyed | |
| H2 | Drain pipes partly clogged by incrustations | Flushing (only partly successful) Milling (successful) |
| O | Drain pipes partly clogged by incrustations | Flushing (only partly successful) |
| S | Drain pipes partly clogged by incrustations | Flushing (only partly successful) |
| | Manholes inside the landfill partly broken | Excavation/reconstruction |
| T | Drain pipes partly clogged by incrustations | Flushing (only partly successful) Milling (only partly successful) |
| | Pipe encasement of concrete (uniform grain size material) dissolved | |
| V | Untapped intrusion of water at the base of the landfill | |
| | Drain pipes free of clogging material | |
| | Drainage layer (mixture of gravel and sand) concreted and impermeable | Excavation/filter material partly replaced |

the pipes had been transformed into a concrete-like material. In this case the drain pipes had not been protected against air intrusion either.

The Leichtweiß-Institute for Water Research of the TU Braunschweig participated in comprehensive inspections concerning the state of the leachate collection system of landfill H2. Details are included in a later chapter. Nearly all drain pipes were clogged by incrustation. In the first sections of the landfill, cleansing by means of flushing proved unsatisfactory and the incrustations had to be removed by milling.

Excavations carried out on some of the other landfills showed that drainage materials had not become clogged, although the drain pipes had been filled up with clogging material. On landfill T the pipe encasement made from uniform grain-size cement had dissolved along some parts of the line. Mechanical destruction of parts of the leachate collection system could only be observed in two cases where unsuitable pipes had been used for landfill draining or static dimensioning of the manholes was insufficient. These few examples show that the main problems were caused by clogging of the drain pipes or of the drainage layer.

## Analysis of Clogging Material

The results of analyses carried out on incrustations in leachate collection systems can be observed in Table 2. A preliminary investigation to this regard has been carried out in the Federal Republic of Germany by Essig *et al.* (1981). Only in the case of landfill H2 analyses of the corresponding leachate as well as of other environmental conditions proved possible. On examination of Table 2, therefore, it can be concluded that in most cases the main cationic components are calcium and iron. Carbonate is the most frequent anionic component. The concentrations of manganese and magnesium are significantly lower. The volatile solids vary from approximately 30 to 300 g/kg of dry solid matter. Apart from the varying organic contents, this can most probably be attributed to differences in the hydrate water content. Most of the samples obtained show a calcium and iron concentration of approximately 150 g/kg. Significant deviations can be observed comparing sample 2 (Essig *et al.*, 1981) with the sample obtained from landfill E: an iron content of only about 10 g/kg was observed, whereas the calcium content reached 240 and 300 g/kg respectively.

The sample analysed by the Leichtweiß-Institute for Water Research was taken from a leachate transportation pipe-line under pressure. In this case clogging mechanisms seemed to be of a different nature.

The analyses of the clogging materials taken from the drainage area are similar to those obtained from the pipes. It can be concluded from the results presented in Table 2 that, apart from special cases, the clogging mechanisms involved were of the same nature.

## Physicochemical Reasons for Clogging

To date it has only been possible to hypothesize on the mechanisms of clogging in leachate collection systems. The results discussed in the last chapter however lead to the conclusion that the precipitation of carbonate and iron (caused either biochemically or by air intrusion) is particularly critical.

Physicochemical processes may take place when changes in environmental

**Table 2.** Composition of clogging material found in drain pipes and drainage layers of sanitary landfills.

| | | Literature | | | Leichtweiß-Institute for Water Research | | | | |
| --- | --- | --- | --- | --- | --- | --- | --- | --- | --- |
| | | Bass et al. (1984) | Essig et al. (1981) 1 | Essig et al. (1981) 2 | Landfill H1[b] | Landfill S | Landfill T | Landfill E | Landfill H2[a] |
| Mg | g/kg TS | 29.2 | 4.5 | 21.0 | 8.4 | 2.8 | 4.9 | 8.2 | 18.0 |
| Ca | g/kg TS | 241 | 117 | 240 | 140 | 137 | 153 | 307 | 223 |
| Fe | g/kg TS | 156 | 208 | 8.5 | 150 | 157 | 125 | 9.0 | 48 |
| Mn | g/kg TS | 5.7 | 0.6 | 3.7 | n.b. | 3.3 | 4.7 | 5.5 | 3.0 |
| $CO_3^-$ | g/kg TS | n.b. | 360 | 560 | n.b. | 207 | 294 | 282 | 396 |
| GR | g/kg TS | n.b. | 700 | 900 | n.b. | 844 | 870 | 968 | 853 |
| Drain/Filter (D) (F) | | F | D | D | F | D | D | D | D |

TS = dry solid matter.
GR = ignition residue.
[a] Mean values.
[b] Analysed by Institute f. Siedlungswasserwirtschaft TU Braunschweig.

conditions occur within the leachate collection system:

1. access of air;
2. decrease of partial pressure of carbon dioxide;
3. drop in temperature.

The intrusion of air into the drainage layer will mainly influence the solubility of iron. There are reducing conditions within the landfill. If the iron-containing leachate comes into contact with better oxidizing conditions in the leachate collection system, the soluble $Fe^{2+}$ will be transformed into the insoluble $Fe^{3+}$, thus causing iron precipitations.

Figure 4 illustrates these relationships. Leachate from anaerobic landfills has a pH-range between 6 and 8 and a redox potential $E_h < 0$. Following intrusion of oxygen the redox potential increases and the iron will then be transformed into an insoluble state.

A drop in partial pressure of the carbon dioxide in leachate caused, for example, by the access of air could be the reason for the precipitation of $CaCO_3$ (calcite). Figure 4 demonstrates the soluble calcium concentrations for undiluted water relative to the partial pressure of $CO_2$. At $CO_2$-partial pressure of 0.4 atmospheres the calcium solubility is approximately 10 times higher than under normal atmospheric conditions ($P_{CO_2} = 3 \times 10^{-4}$). A decreasing partial pressure, similar to the difference between landfill gas and the normal atmosphere, will cause a precipitation of calcium as $CaCO_3$.

Temperature influences the process of solubility of different salts. Usually the solubility of salt decreases as temperature lowers. Temperature gradients between landfill and drainage layer may cause precipitation of leachate compounds. These considerations, along with the diagrams presented in Fig. 4, are only valid for unpolluted water. It is far more difficult to compute chemical equilibria for a mixture of multiple components such as leachate.

**Biological Clogging**

Biological mechanisms such as slime and filamentous growth, formation of biomass and ferric incrustations (ochre) may also cause clogging of drain pipes. Furthermore, sulphide- and carbonate-precipitations originate in the bacterial production of $H_2S$ and $CO_2$.

**FIRST RESULTS OF IN SITU INVESTIGATIONS**

As already mentioned, the state of the leachate collection system of landfill H2 was examined in Summer 1986. The landfill receives the waste from

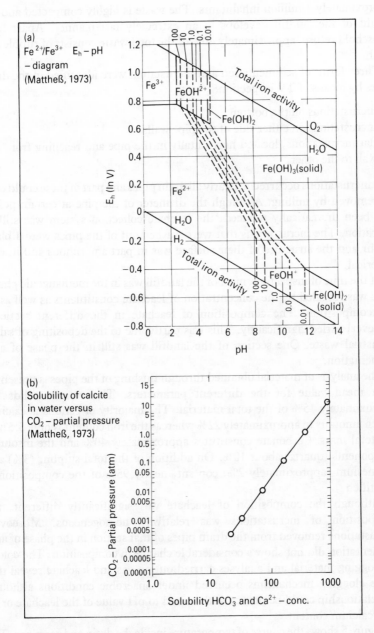

**Figure 4.** Solubility of iron and calcium.

approximately 1 million inhabitants. The waste is highly compacted and the height of the landfill develops in an extremely fast manner. Apart from household refuse, approximately 40% of the total waste consists of rubble and earth.

Three forms of sedimentation and incrustation were identified in the drain pipes by means of TV inspection:

1. incrustations on the bottom;
2. incrustations on either side of the pipe walls;
3. flat incrustations, located horizontally in the pipe and reaching from one wall to the other.

The incrustations occurred in nearly all drain pipes and part of these could only be removed by milling. Although the diameter of the pipe at certain points had been dramatically reduced, the leachate collection system was still in operation. The incrustations that were washed out of the pipes were a black colour and the structure of their surface was in part amorphous and in part spherical.

At the time of inspection, most of the landfill was in the methangenic phase. This was revealed by the concentration of leachate constituents as well as by gas composition. The composition of leachate in the different sections, however, varied significantly. This was partly due to the depositing of saline industrial waste. One section of the landfill was still in the phase of acid fermentation.

The analysis of material obtained through flushing of the pipes is presented as a mean value for the different parameters. The volatile solids are approximately 15% of the total material. The majority of cations is calcium which amounts to approximately 22% whereas the iron content is only 5.5% of the total mass. Carbonate constitutes approximately 40% and the insoluble components (quartz) about 10%. On addition of the total sulphur (3%) and magnesium (approximately 2%) content, nearly 95% of the composition is identified.

Although the composition of leachate was significantly different, the composition of incrustations was relatively homogeneous. Moreover, incrustations removed from the drain pipes of that section in the phase of acid fermentation did not show a considerable change in composition. The colour of clogging material and analyses carried out on gas and leachate reveal that these clogging mechanisms occurred under anaerobic conditions although no relationship could be proven with regard to pH value of the leachate or its iron/calcium content.

Figure 5 shows the course of temperature inside the drain and gas pipes. The temperature measured at the manhole was approximately 20°C and rose

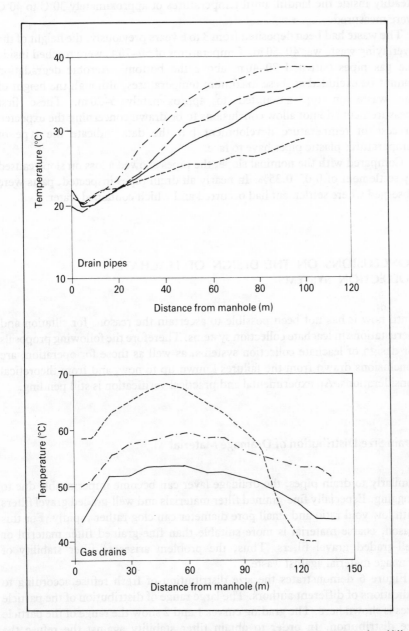

**Figure 5.** Temperature measurements inside drain and gas pipes in a landfill.

steadily inside the landfill until temperatures of approximately 30°C to 40°C were measured.

The waste had been deposited from 3 to 6 years previously, the height of the overlaying waste was 40–50 m. Temperatures of 50–70°C were reached inside the gas pipes (situated 20–40 m above the bottom). Aerobic degradation cannot be excluded at these maximum temperatures, although the height of the waste on top amounted to approximately 5–20 m. These first measurements do not allow conclusions to be drawn concerning the expected process of temperature development but the data indicate the type of temperatures plastic pipes have to face.

Compared with the nominal slope, the pipes showed a loss on slope caused by settlement of 0.05–0.35%. In nearly all drain pipes inspected, parts were observed where settlement had occurred and which contained water.

## CONCLUSIONS ON THE DESIGN OF LEACHATE COLLECTION SYSTEMS

Until now it has not been possible to ascertain the reasons for siltation and incrustations in leachate collection systems. Therefore the following proposals for design of leachate collection systems, as well as those for operation, are conclusions drawn from the failures known up to now, and from theoretical considerations. An experimental and practical verification is still pending.

### Grain Size Distribution of Drainage Material

Similarly to drain pipes, the drainage layer can become impermeable due to clogging. Especially fine-grained filter materials and well-graded gravel filters with low void ratio and small pore diameter can clog rather rapidly. For this reason, coarse material is more suitable than fine-grained filter material or well-graded gravel filters. Thus, the problem arises of filter stability of drainage material against waste.

Figure 6 demonstrates the size distribution of fresh refuse according to indications of different authors. The large range of distribution of the particle sizes is shown here. The grading curves 1 and 2 show the range of the particle size distribution. In order to obtain filter stability against the refuse the grading curve 1 which contains a greater percentage of fine material was used for calculation.

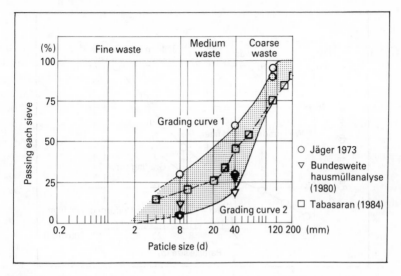

**Figure 6.** Particle size distribution of household refuse (different authors).

The application of filter criteria according to Terzaghi

$$\frac{D_{15}}{d_{85}} \leqslant 4 \quad \frac{D_{15}}{d_{15}} \geqslant 4 \tag{1}$$

where:

$d$ = diameter of grain of soil to be filtered
$D$ = diameter of grain of filter material

results in a maximum grain size diameter $D_{15}$ of more than 200 mm. The minimum diameter can amount to approximately 10 mm whereas the use of such high values is not recommended for the maximum diameter considering the embedding of the pipe material.

In accordance with these latest sieve analyses of waste, the grading curve of a well-graded gravel filter, used here for a question of comparison, is extremely unsuitable. It has been shown in Fig. 7 that the high percentage of fine solids of this filter material is not necessary for filter stability. The permeability of such a gravel filter can be a thousand times lower than that of a uniform gravel filter. The possibility of incrustations in a drainage layer constructed from this type of material has been explained previously.

Drainage materials with a $D_{15}$ diameter greater than 10 mm are also suitable with regard to filter stability. Gravel of a grain size distribution 10–13 or

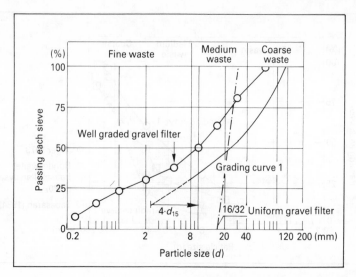

**Figure 7.** Comparison of particle size distribution of household refuse, well graded gravel filter and uniform gravel filter with steep grading curve.

16–32, which the author prefers due to their higher pore diameter, are appropriate.

The danger of fine material from the biological decomposed waste being washed into the uniform graded gravel filter is considered to be unimportant. Leachate volume is fairly small and the resulting flow velocity is low. Furthermore the aggregate stability of the waste is higher than that of soil. The presence of a certain amount of fine constituents when using this coarse filter material is acceptable.

**Design of the Leachate Collection System**

The exact design of the drainage layer depends on the topography of the landfill site and on the design of the bottom liner. As already mentioned, with regard to life expectancy of the drainage layer, the best results are obtained with a drainage layer of coarse material (grain size 16–32 and coarser). In the case of installation of a combined drainage and protection layer, the utilization of a finer material may be necessary in order to protect the plastic liner. Furthermore, it may be difficult to obtain the required quantities of drainage material of coarse grain size distribution. The use of better graded material might then become necessary too. The life span of the leachate collection system can be improved in these cases by the installation of split gravel trenches transverse to the drains or in the slope line.

Figures 8 and 9 present the layout and cross section of a leachate collection system. The landfill bottom is roof-shaped with a minimum cross slope of 1%. The drain pipes are placed at the deep points. The longitudinal slope should be 2% or more to guarantee a deposit-free transportation inside the drainpipes. The drains should be spaced at 50 m intervals and the maximum distance between two manholes must be less than 300 m. The split gravel trenches suitable for a drainage layer of fine material should be placed at a distance of

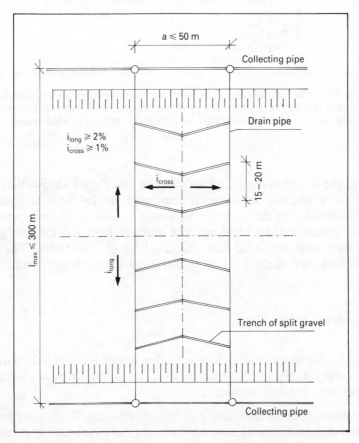

**Figure 8.** Proposal for the design of leachate collection systems.

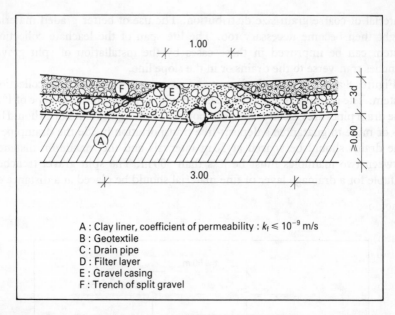

**Figure 9.** Proposal for the design of leachate collection systems (detail, cross-section): (A) clay liner coefficient of permeability, $k_f \leq 10^{-9}$ m/s; (B) geotextile; (C) drain pipe; (D) filter layer; (E) gravel casing; (F) trench of split gravel.

15–20 m and should have a minimum width of 2 m. Figure 9 shows the cross-section of a leachate collection system when a clay liner is installed. Underneath the pipe the bottom is excavated in a mould-shaped fashion in order to reduce leachate build-up. The drainage layer and the split gravel trench have to be separated from the clay liner by a geotextile. The filter stability between drainage layer and trench of split gravel has to be maintained.

**Consequences for the Pipes**

If a lengthy life span of the leachate collection system is meant to be obtained using coarse material for the drainage layer, the following consequences for the pipes must be taken into account:

1. The diameter of the drain slots may become enlarged. When using washed gravel of grain size 16–32 mm, the diameters may be enlarged by as much as 15 mm or more depending on the shape of the material used.

2. The strength of the pipes has to be adapted to the peak loads resulting from use of the coarse gravel.

The influence of the large drain slots on the bearing capacity has yet to be proved.

## Operation and Maintenance

The drain pipes have to be controlled by TV cameras immediately after implacement of the first waste lift. Mechanical failure caused by compacting machines can be repaired without any problems at this stage. Investigation into the failure of leachate collection systems showed that concreted clogging material could only be removed in part by flushing.

It has been observed in agricultural draining that ochre deposits can easily be removed in the initial phase by means of flushing equipment. After concretion of clogging material, the incrustations cannot be dissolved from the pipe. It is important therefore to commence regular flushing of the leachate collection system as soon as possible. The frequency of flushing necessary may be ascertained by TV inspection.

The leachate discharge from the various pipes must be kept under strict control. A significant decrease in leachate discharge may be caused by clogging of the leachate collection system.

The main inorganic components of the incrustations are calcium and iron compounds. Landfills with leachates of increased pH-value contain less iron and calcium compounds than relative to their solubility. The pH-value of leachate increases as soon as the landfill reaches the methane phase; for this reason landfill operation should be optimized in order to reach this phase in the shortest time possible.

The development of landfill height should take place gradually in thin layers in order to keep the acid phase as short as possible (Ehrig, 1985).

## CONCLUSIONS

The leachate collection system is an essential part of a sanitary landfill. It has to be designed with particular regard to maintenance and life span. Failure of a leachate collection system may reduce the efficacy of bottom liners and also cause stability problems. The current state of the art of leachate collection system design has been presented in this contribution.

Over recent years an increasing number of leachate collection systems have failed either partially or totally: the main problem has invariably proved to be the clogging of pipes and drainage layer. Techniques of flushing and restoration have only met with partial success. Analyses carried out on clogging material have shown that iron and calcium compounds are the main constituents. The mechanisms of clogging are still unknown but biochemical and physicochemical precipitations would seem to play an important role.

Recently, investigations have been started by the authors to study these phenomena and special equipment for on site measurement has been developed (e.g. devices for sampling up to a distance of 150 m, measurement of temperatures at the same distance, surveying the bottom of pipes, etc.).

Because of the possibility of clogging, only coarse grain is suitable for the drainage layer. The statics of the pipes and the diameter of the drain slots must be adapted to the coarse material. Inspection and maintenance of the pipes is necessary on a routine basis; it is of specific importance during the phase of commencing landfill operation.

## REFERENCES

Anonymous (1984) 'Informationsschrift Sickerwasser aus Hausmüll- und Schlackendeponien', Länderarbeitsgemeinschaft Abfall (LAGA), Stand Oktober 1984.

Anonymous (1979). Merkblatt 'Die geordnete Ablagerung von Abfällen', Länderarbeitsgemeinschaft Abfall (LAGA), Stand September 1979.

Bass, J.M., Ehrenfeld, J.R. and Valentine, J.N. (1984). 'Potential Clogging of Landfill Drainage Systems'. Project Summary EPA-600/s2-83-109.

Ehrig, H.J. (1985). 'Auswirkungen des Deponiebetriebes auf Sickerwasserbelastungen — Messungen an Deponien und Lysimetern' Sickerwasser aus Hausmülldeponien — Einflüsse und Behandlung, Veröffentlichungen des Institutes für Stadtbauwesen der TU Braunschweig, Heft 39.

Essig, Köhloff, Laber, Limbach and Schickel (1981). 'Untersuchung von Wirkung und Langzeitverhalten von Basisabdichtungen aus tonigem Erdmaterial und von Sickerwasserdränagen zur Sohlentwässerung bei Hausmülldeponien', Teilbericht 2: Sickerwasserdränagen Umweltforschungsplan des BMI, Forschungsbericht 1 03 02 210.

Mattheß, G. (1973). 'Die Beschaffenheit des Grundwassers'. Gebrüder Bornträger, Berlin, Stuttgart.

Ramke, H.G. (1986). 'Überlegungen zur Gestaltung und Unterhaltung von Entwässerungssystemen bei Hausmülldeponien', Fortschritte der Deponietechnik, Abfallwirtschaft in Forschung und Praxis, Band 16. Erich Schmidt Verlag, Berlin.

# 6.  ENVIRONMENTAL  IMPACTS

6.5 ENVIRONMENTAL IMPACTS

# 6.1 A Method for the Assessment of Environmental Impact of Sanitary Landfill

G. ANDREOTTOLA*, R. COSSU** and R. SERRA**

\* Institute of Sanitary Engineering, Polytechnic of Milan,
Via F.lli Gorlini 1, I-20151 Milano, Italy
\*\* Institute of Hydraulics, University of Cagliari, Piazza d'Armi,
I-09100 Cagliari, Italy

## THE PROCEDURE OF ENVIRONMENTAL IMPACT ASSESSMENT

The term 'environmental impact' is used to define the alteration of the environment caused by the activities involved in the implementation of a programme. In this context, the concept of environment includes the complex of physical, social, cultural and aesthetic factors regarding the individual and the community which determine both the form, the character, the relation and the development. The environmental impact, in its above-defined most vast meaning, caused by the presence of an installation may evoke both positive and negative effects. The procedure of Environmental Impact Assessment (EIA) aims at evaluating (in the sense of weighing-up and marking) the relationship which exists between the installation proposed and the environment in which it is to be implemented. This is carried out by considering the largest amount of information available which involves technical, legal, economic, social and environmental aspects in order to form a judgement of feasibility and acceptability. For the assessment of environmental impact numerous methods have been proposed, many of which emerged during the first half of the 1970's.

The methods proposed and generally in use can be schematically summarized in two groups:

1. formal methods, standardized as a guide and work–tool used to arrange the environmental information deriving from the impact study;
2. *ad hoc* methods, to be developed according to the situation without any preconstituted scheme.

367

With regard to the formal methods, the most significant examples may be grouped into 4 classes:

—the 'overlapping map' method
—lists of questions and controls
—correlation matrices
—networks

The first method consists in the overlapping of transparent maps relating to all elements of impact considered onto maps of various themes (socio–economical, morphological, etc.) in order to show up the areas of minimum and/or maximum impact. This method has proved to be valid and has mainly been applied to studies concerning the siting of infrastructures, roads, motorways, oilducts, etc.

The list of questions is a series of queries concerning the various environmental problems which may be encountered. These lists are necessarily compiled by involving as many experts, corporations and populations as possible and focus on the aspects which will become the object of the study. The control lists represent an evolution of the list of questions and permit the individualization of those activities and elements of impact which may influence the environment as well as the environmental categories present.

The matrices may be considered as two-dimensional control lists in which on one dimension the individual characteristics of a project are shown (proposed activities, elements of impact, etc.) whereas on the other dimension the environmental categories which may be affected by the project are listed. The effects or potential impacts are therefore individualized by a confrontation of the two control lists. The differences between the various proposed types of matrix are to be observed mainly in the variety, number and specificity of control lists, as well as in the system of evaluation of the individualized impact. With regard to assessment, this ranges from the mere individualizing of impact (marked with some sort of sign, a cross, dash, asterisk, etc.) to a qualitative evaluation (good, moderate, sufficient, reasonable) or to a numerical evaluation which may be either absolute or relative: generally an assessment is effected regarding the result of impact (positive or negative). The numerical evaluation is often subject to criticism as it would seem to introduce a criteria of objective judgement that, in reality, is impossible to achieve. Among the better-known examples of matrix is Leopold's matrix (Leopold, 1971). This is made up of two control lists including respectively 100 possible actions linked to the proposed project and 88 environmental components which are susceptible to impact. The impacts to be analysed therefore are 8800. This method requires that the intersection between each action and each environmental characteristic is diagonally barred. This having been carried out, a number (from 1 to 10) is inserted in the upper part of the barred square to indicate the size of impact. In the lower part another number (from 1 to 10) indicates the importance. Other examples are afforded by Moore's matrix (Moore *et al.*, 1973) which requires an assessment on a four–level ordinal scale (negligible, low, moderate, high) and Clark's matrix (Clark *et al.*, 1976) which gives not a numerical assessment but a qualitative evaluation based on a five–polarity nominal scale:

—beneficial/adverse
—short term/long term
—reversible/irreversible
—direct/indirect
—local/strategic

A further tool in use is that of networks which introduce a falling sequence of cause/effect conditions: in this way permitting the identification of primary, secondary and tertiary impact thereby enabling assessment of cumulated impact as long as the probability of the event happening (the occurrence of impact), its degree of importance and its dimension are taken into account. On closer examination, the latter two factors represent the analogue of the two numbers examined according to Leopold's matrix. Application of this method, known also by the name of 'impact tree' was undertaken by Sorensen (Sorensen, 1971) with regard to impact caused by the construction of a new road.

One of the major problems evidentiated in literature is that of obtaining a quantitative assessment of impact in such a way as to allow confrontation of the effects provoked by various proposed projects by means of numerical values. This tendency is clearly shown in the application carried out by Sorensen and in the procedure which was first set up by the Battelle Institute (Dee Norbert *et al.*, 1972). The latter method is aimed at assessing the effects on the environment in commensurable units. The method consists in the subdivision of the environment into four main categories: ecology, pollution, aesthetic factors, human environment. The impact on the various environmental categories is described by means of parameters, each of which is given a certain value, with regard to importance, so that the overall sum is equal to 1000. Subsequently, an index of 'quality of the environment' is introduced for each parameter with values ranging between 0 and 1 which are plotted on graphs. Therefore, by mutliplying the index of quality by the relative weight and adding the values obtained, a numerical assessment of the impact to which each environmental category is subject can be obtained.

It should be underlined that, in our opinion, it is somewhat difficult to individualize indicators of environmental quality which may be effectively measured thereby rendering the above-mentioned approach impractical or at least problematical.

## METHOD OF IMPACT ASSESSMENT USING CHROMATIC MATRICES

This chapter describes a method for the assessment of environmental impact which uses chromatic tonalities in order to facilitate the understanding of the final results of the study. This represents the perfecting of a method already used for assessment of the impacts caused by a wastewater treatment plant (Cossu *et al.*, 1986). The following sections will consider the more significant steps of the assessment item by item.

### General Description

As a reference point the EEC normative has been applied (EEC, 1985). This underlines that information is to be provided by means of environmental impact assessment, but it does not supply indications concerning the way in which the impact is to be evaluated nor regarding the choice of criteria to be used to limit eventual negative effects. Therefore a new '*ad hoc*' method has been developed, which above all met the fundamental requirement of flexibility of use.

In fact, EIA may be applied with various objectives:

1. a single project or a single site;
2. more than one project but relative to only one site;
3. one project at different sites;
4. more than one project at more than one site;
5. a project that has already been carried out.

The chromatic matrix scheme proposed here may be used for each of the five EIA objectives. This is made possible by the presence of five matrix schemes which evidentiate, each for their own sector, interaction between the causes, elements of impact and environmental categories. In order to carry out a more detailed analysis of the causes of impact, several phases of the evolution of the project are characterized and linked to different types of impact with regard to duration, entity and probability of the event taking place. The phases taken into consideration are as follows:

(a) temporary phase (pre-project period and construction of the installation)
(b) phase of ordinary exercise
(c) phase of extraordinary exercise

For each of these phases a group of five matrices may be prepared, as explained later.

Considering the difficulty often encountered in quantifying the entity of interaction between the various control lists present in each matrix, we used a chromatic representation to describe them in a qualitative form. Two different chromatic scales were used to which positive or negative influences corresponded and which include four assessment levels (expressed by different tonalities). The four chromatic tonalities correspond to negligible, low, moderate or high qualitative levels.

The chromatic representation of impact consents an immediate and synthetic individualization of the critical elements of impact where action may eventually be taken.

## Matrix of the Causes and Elements of Impact (Matrix A)

The first matrix of the series evidentiates the activities of the plant which cause the elements of impact. As the method refers to three different phases of the project, the three relative matrices must focus on the various elements of impact for each phase. The latter were identified as those factors able to modify the state of environmental categories. The importance of the causes with regard to determination of a specific element of impact is assessed by means of different chromatic tonalities.

This first matrix (or better, this first group of matrices) enables us to individualize the environmental impact of a project or of a plant and thereby identify the need for improvements. In the case of more alternative projects to be assessed, the matrix represents an efficient tool in establishing a sound background for decision-making.

## Matrix of Indicators and Environmental Categories (Matrix A1)

The environmental categories may be defined as those components of the environment which suffer from the effects generated by the elements of impact. These include not only the physical components of the environment (air, water, flora, fauna, etc.) but also those more strictly related to human activity (public health, economic activities, social relations, cultural values, etc.). Contrary to the elements of impact which are characteristics peculiar to the single operational phases of the plant, the environmental categories are clearly invariable.

In order to describe the state of the single environmental categories of the site (or sites) in question, it is necessary to define some relevant indicators. In fact, the definition of an informative background of the existing environmental categories constitutes one of the main phases of the procedure of impact assessment and is specifically demanded in some countries (e.g. USA and France).

The assessment of environmental indicators may be either of a qualitative or a quantitative nature depending on the categories being dealt with and the mathematical or direct measurement instruments available. The chromatic tonality resulting from the intersection between an indicator and an environmental category allows the assessment both of the qualitative state of the environment and the effect of certain causes in rendering the environment more or less compatible to receiving the installation.

By means of this matrix (unique as to its independence from the activities linked to the plant) it will be possible to express a judgement of suitability of one or more sites with regard to the type of plant which is to be installed, which makes the method useful in solving the problem of the best location.

## Matrix of Potential Impacts (Matrix B)

This matrix presents as control lists the elements of impact and environmental categories already defined in matrices A and A1 respectively.

From the intersection of these two lists, the potential impact manifested by the installation with regard to the environment can be singled out and therefore the B matrix is capable of globally evidentiating all problems under consideration.

In fact, if it should prove necessary to assess the impact of a particular installation at various sites, the same number of matrices will be obtained and on cross-examination a judgement of acceptability may be passed. Evidently in this case the diversity of the various B matrices is put down to the variations in the matrix A1 (see last section). This latter variation coincides with the different characteristics of the sites whilst the A matrix remains unchanged on account of its dependence on the characteristics of the installation.

Vice versa, if it should prove necessary to assess the impact of more than one installation on one single site, the resulting various B matrices will be influenced by the variation in the A matrices whilst obviously the A1 will remain unchanged.

In the case of a decision which depends on the results of consideration of more than one project based on more than one site, the number of B matrices taken into account increase considerably as they are equal to the combination of all possible cases. However, the examination of the A1 matrices alone will provide enough information in order to immediately discard those sites which are clearly unsuitable to contain a certain type of installation (e.g. a sanitary landfill), thereby allowing the examination of a smaller number of B matrices.

Finally, in the case of an already existing single installation, the B matrix simply demonstrates the situation of potential impact, indicating those points towards which improvements should be directed. Concerning the other cases previously discussed, although the B matrix is capable of allowing a first judgement, the definitive choice should be made only after careful assessment of the situation in the light of the operations of limitation.

## Matrix of Limiting Criteria (Matrix C)

The fourth matrix of the scheme takes into consideration, on the basis of potential negative impact individualized in matrix B, operations and measures to be adopted in order to limit, or rather to eliminate or reduce to acceptable environmental levels, the negative impacts. Their action must have an influence on the causes which greatly contribute towards the onset of negative elements of impact and therefore their singling out must include a careful evaluation of the A matrix. To this regard, two important aspects must be underlined. Firstly, it is necessary to have a clear picture of all possible interrelations as any one single element of impact may be influenced by more than one introduced improvement. Moreover, it must not be overlooked that these same measures may produce simultaneously positive effects on some elements and negative side effects on others. For example, in the field of sanitary landfilling, with the aim of limiting the impact caused by the presence of leachate using a collection system and an on-site treatment process, together with the evident positive effects, some negative effects may arise from odours coming from various parts of the treatment plant. Therefore the efficacy of the measures will have to be assessed using the two different chromatic scales already defined.

## Matrix of Residual Impacts (Matrix D)

On the basis of the limitation measures utilized and of their efficacy, evaluated using the C matrix, the method provides a fifth matrix for assessment of limited impact, that is of the residual impact once the operations of limitation have already been set up to counteract the various causes of impact.

This matrix is entirely analogous to the B matrix used for assessment of potential impact but, diversely from the latter, the examination of this matrix allows us to express a definitive judgement on the degree of compatibility of an installation with regard to the surrounding environment. Moreover, the cross-examination of B and D matrices will consent a visible appreciation of the

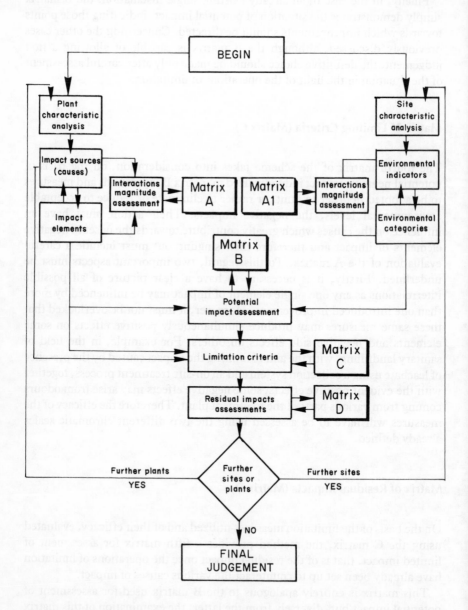

**Figure 1.** Scheme of the environmental impact assessment methodology through chromatic matrices.

efficacy of limitation criteria. This last observation demonstrates how such a tool is undoubtedly suitable in those cases where assessment of environmental impact should require the formulation of proposed restrictive operations in order to obtain less and less residual impact: this aim could be fulfilled by means of an iterative process.

## Comprehensive Scheme of the Method

Figure 1 shows a block scheme of the method described in the previous sections. The iteration evidentiated in the scheme means that the number of evaluations of residual impact will be equal to the product of the number of projects for the number of sites deemed suitable. All these assessments will subsequently be the object of cross-analysis (dotted line in the figure) prior to final judgement. Figure 2 shows a global representation of the five matrices: the arrows represent the logical pattern for their use. Obviously, the whole scheme will be repeated for each of the three phases of the installation, thereby allowing the individualization of residual impact both in the temporary phases and in the phases of ordinary and extraordinary exercise.

Only the person responsible for the final decision may give the correct value to the impact present in each phase; this task is not the responsibility of the person carrying out the impact study. However, once the amount of consideration that must be given to each phase is defined (taking into account politico-strategic reasonings), this method permits the amalgamation into a single series of matrices of all the various phases: obviously this will mean dealing with matrices of a higher dimension.

## APPLICATION OF THE METHOD TO SANITARY LANDFILL

The described methodological tool can easily be applied to various types of installation. Generally the environmental categories taken into account are the same for all types of installation, as underlined also in the EEC directive (EEC, 1985):

1. man, flora and fauna
2. ground, water, air, climate, landscape
3. interaction between the previous factors
4. material goods and cultural patrimony

Vice versa, the environmental indicators will be strictly dependent on the type of installation as their function is to direct the study of the pre-existing

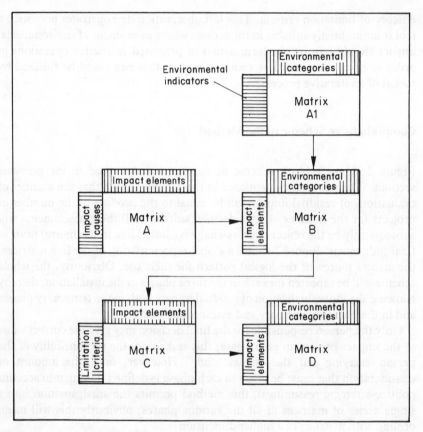

**Figure 2.** General representation of chromatic matrices.

environment in those directions which may better be able to explain interaction with the installation and the environment itself. Similarly, the elements of activities peculiar to the installation, will not be of a general nature but will be specific for each different type. It may prove necessary impact, to consider some specific items of the processes used in any given type of installation.

In the following sections the method is applied to the specific case of a sanitary landfill. It must be borne in mind that the assessment will not only take into account the ordinary working phase but also the preceeding phase and also the phase of extraordinary exercise. Each of these may be examined either as a single phase or together with the others, in which case the results may be summarized into a single group of matrices.

## Matrix of Causes and Elements of Impact

### Temporary Phase

The temporary phase concerns the time from the decision to build the installation to its completion. The necessity of constructing a solid waste disposal plant is linked to the need to solve a situation of environmental degradation caused by the uncontrolled disposal of waste matter together with the requirement of meeting precise legislative dispositions.

Therefore, the environmental degradation and the obeyance of a normative should be considered as causes of the decision to build a sanitary landfill. This decision is itself an element of impact capable of modifying the surrounding environment; it affects social relations with regard to the eventual protests from the population who live near the pre-chosen site and affects the value and order of the territory.

The start of the actual construction, the setting up of the site and the occupation of the area provokes the onset of several elements of impact such as noise, heavy traffic caused by lorries, dust, etc. which, although they do not last very long, must not be ignored.

Table 1 shows the two control lists of causes and elements of impact summarizing the considerations made hitherto.

### Phase of Ordinary Functioning

The singling out of the elements relative to all phases may be conducted by means of examination of all activities connected to the functioning of a sanitary landfill. The elements of impact linked to the phases of landfilling, compacting, covering of waste, etc. are shown in Table 2. The importance of the item 'type of waste' must be underlined as it may be the cause of numerous

**Table 1.** List of causes and elements of impact for a sanitary landfill during the temporary phase.

| Causes | Elements of impact |
|---|---|
| Environmental degradation | Decision to build a plant |
| Obeyance of normatives | Emission of dust and particles |
| Installation and start of construction | Noise |
| | Increase of traffic |
| | Occupation of area and volume |
| | Risk of accidents |

**Table 2.** List of causes and elements of impact for the ordinary functioning phase of a sanitary landfill

| Causes | Elements of impact |
|---|---|
| Cover soil supply | Emission of dust and particles |
| Transport of waste | Noise |
| Type of refuse | Smells |
| Disposal and spreading of waste | Biogas |
| Compacting of waste | Increase of traffic |
| Covering | Dispersion of light matter |
| Height of tip | Leachate |
| Final covering | Stability of ground/waste |
| | Occupation area and volume |
| | Vectors |

elements of impact: the local population is aware of this and makes it a fundamental question in the acceptance of a sanitary landfill.

## Phase of Extraordinary Functioning

Table 3 shows a list of causes and consequent impacts during the extraordinary phase. With this term we mean those situations which may occur as exceptional events, even if already taken into consideration by the containment measures.

It should be specified that above all this phase may be studied independently from the others in order to acquire information which permits the setting-up of maximum limitation measures against catastrophic events and to assess the risk linked to such situations.

**Table 3.** List of causes and elements of impact for the extraordinary functioning phase of a sanitary landfill.

| Causes | Elements of impact |
|---|---|
| Escape of leachate | Leachate |
| Migration of biogas | Biogas |
| Explosions | Stability of ground/waste |
| Clogging of drains for collection of biogas and leachate | Risk of accidents |
| | Odours |
| Breakdown of leachate treatment plant | |
| Breakdown of biogas treatment plant | |

Matrix A

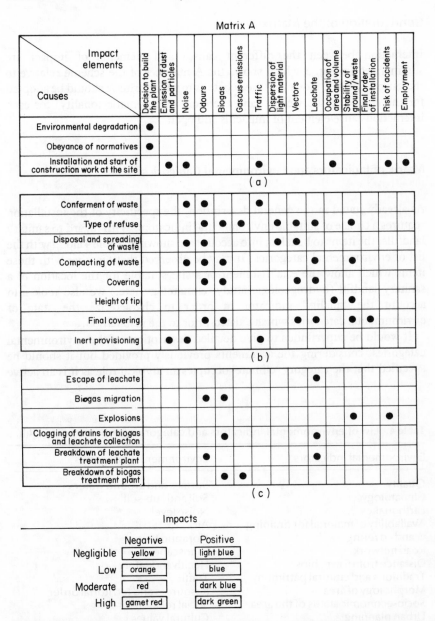

(a)

(b)

(c)

Impacts

| | Negative | Positive |
|---|---|---|
| Negligible | yellow | light blue |
| Low | orange | blue |
| Moderate | red | dark blue |
| High | gamet red | dark green |

**Figure 3.** Representation of A matrices in case of a sanitary landfill: (a) temporary phase; (b) ordinary functioning phase; (c) extra-ordinary functioning phase. Intersection cells between causes and impact elements are pointed out with a stronger frame. The frame will be filled by the colour suggested by the assessment.

## Configuration of the Matrix

Interaction between the different causes and elements of impact are summarized in Fig. 3 which shows the A matrices of the scheme relative to the various phases. However, the debate as to which criteria should be adopted for the choice of the weight (and therefore of chromatic tonality) for each interaction cause/element of impact is still open.

## Matrix of Indicators and Environmental Categories

As already stated in the general description, the specificity of the installation requires a choice of suitable environmental indicators. With regard to sanitary landfills, the items to be taken into account are shown in Table 4 along with the list of environmental categories. It may be observed that together with those items which allow the establishment of a site suitable for the location of a sanitary landfill (hydrogeologic study, climatology, etc.) we also took into account the 'quality' indicator in order to characterize the state of environmental categories which existed prior to the installation.

It would be superfluous to give detailed descriptions of the environmental categories, considering the arguments previously provided but it should be specified that the category 'climate' occupies an atypical role as it is an active

**Table 4.** List of environmental indicators and categories for a sanitary landfill.

| Environmental indicators | Environmental categories |
|---|---|
| Quality | Surface and ground waters |
| Climatology | Soil and sub-soil |
| Earthquakes | Noise level |
| Availability of material for draining and covering | Air |
| | Flora and fauna |
| Road network | Landscape |
| Distance from townships | Health and safety |
| Traditions and cultural patrimony | Traffic |
| Morphology of area | Resources and territorial order |
| Socio-economic status of the area | Social relations |
| Urban planning | Cultural values |
| Sanitary situation | Occupation and economic activities |
| Classification and state of species | Climate |
| Hydrogeologic and geotechnic characteristics | |

element influencing the elements of impact (for example the formation of leachate).

Therefore its presence should have been foreseen in the A matrix among the causes of impact, but considering the close dependence of these items on the type of installation, therefore directly subject to limitation criteria, this position would have also appeared as anomalous. Therefore it is preferable to include the latter in the present matrix as the item 'climate' plays an important role in the judgement of acceptance of a site. In fact, it should be remembered that the global examination of the A1 matrix permits, *a priori*, scrutiny of the more suitable sites.

Matrix A1

| Environmental categories / Environmental indicators | Surface and ground water | Soil and sub-soil | Sound level | Air | Flora and Fauna | Landscape | Health and safety | Viability and traffic | Resources and territorial order | Social relations | Cultural values | Occupation and eco-nomical activities | Climate |
|---|---|---|---|---|---|---|---|---|---|---|---|---|---|
| Quality | ● | ● | ● | ● | | | | | | ● | | | |
| Climatology | | | ● | | | | | | | | | | ● |
| Earthquakes | ● | | | | | | ● | | ● | | | | |
| Availability of material for draining and covering | ● | | | | | | | ● | ● | | | | |
| Road network | | | ● | | | | | ● | | | | | |
| Distance from townships | | | | | | | ● | ● | ● | | | | |
| Traditions and cultural patrimony | | | | | | | | | | | ● | | |
| Morphology of area | | | | | ● | | | ● | | | | | |
| Socio-economic status of the area | | | | | | | | ● | ● | | ● | | |
| Urban planning | | | ● | | | ● | | ● | ● | ● | | | |
| Sanitary situation | | | | | | | ● | | | | | | |
| Hydrogeologic and geotechnic characteristics | ● | ● | | | | | ● | | | ● | | | |
| Classification and state of species | | | | | ● | | | | | | | | |

Impacts

| | Negative | Positive |
|---|---|---|
| Negligible | yellow | light blue |
| Low | orange | blue |
| Moderate | red | dark blue |
| High | gamet red | dark green |

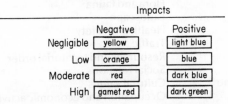

**Figure 4.** Representation of A1 matrix for a sanitary landfill.

Concerning the chromatic tonality to be inserted in the intersection square, the choice will depend on the specificity of the sites and will have to be supported by a series of data not only regarding the quality of air, water, soil, etc., but also on traditions, socio-economic state, urban standards etc. Figure 4 shows the overall matrix in which the intersecting squares are simply individualized by a darker border.

## Matrix of Potential Impacts

From the confrontation between the elements of impact and environmental categories, the situation regarding potential impact of a sanitary landfill is obtained. The chromatic matrix allows a clear and immediate visual assessment of the entity of various impacts and represents the conclusion of the examination both of the project characteristics and of those regarding the site where the installation is to be located. In order to be thorough, Table 5 shows the control lists relating to this matrix, or better a list of elements of impact and environmental categories, whilst Fig. 5 shows the B matrix scheme valid for the phase of ordinary exercise. The compilation of those relating to the other two phases will follow the same procedure: the elements of impact will differ but not the environmental categories that, representing the environment, will remain unchanged. A first analysis of the matrix shows how some categories are affected by numerous elements of impact: this with regard to 'Resources

**Table 5.** List of elements of impact and environmental categories for a sanitary landfill.

| Elements of impact | Environmental categories |
|---|---|
| Decision to build the plant | Surface and ground waters |
| Emission of dust and particles | Soil and sub-soil |
| Noise | Sound level |
| Smells | Air |
| Biogas | Flora and fauna |
| Gaseous emission | Landscape |
| Movement of lorries | Health and safety |
| Dispersion of light matter | Traffic |
| Vectors | Resources and territorial order |
| Leachate | Social relations |
| Occupation of area and volume | Cultural values |
| Stability ground/waste | Occupation and economic activities |
| Definitive order of installation | |

## Matrix B

| Impact elements \ Environmental categories | Superficial waters | Soil and sub-soil | Sound level | Air | Flora and Fauna | Landscape | Health and safety | Viability and traffic | Resources and territorial order | Social relations | Cultural values | Occupation and economical activities |
|---|---|---|---|---|---|---|---|---|---|---|---|---|
| Decision to build the plant | | | | | | | | | ☐ | ☐ | | |
| Emission of dust and particles | | | | ☐ | | | | | | ☐ | | |
| Noise | | | ☐ | | | | ☐ | | ☐ | ☐ | | |
| Odours | | | | ☐ | | | ☐ | | ☐ | ☐ | | |
| Biogas | | ☐ | | ☐ | ☐ | | ☐ | | ☐ | ☐ | | |
| Gasous emissions | | | | ☐ | ☐ | | ☐ | | | ☐ | | |
| Movement of lorries | | | ☐ | | | | | ☐ | ☐ | ☐ | | |
| Dispertion of light material | | | | ☐ | | ☐ | | | | ☐ | | |
| Vectors | | | | | | | ☐ | | ☐ | ☐ | | |
| Leachate | ☐ | ☐ | | | ☐ | | ☐ | | ☐ | ☐ | | |
| Occupation of area and volume | | | | | | ☐ | | | ☐ | ☐ | ☐ | |
| Stability of ground/waste | | ☐ | | | | ☐ | ☐ | | | | ☐ | |
| Final order of installation | | | | | | ☐ | | | ☐ | ☐ | | |
| Risk of accidents | | | | | | | ☐ | | | | | |
| Employment | | | | | | | | | | | | ☐ |

### Impacts

| | Negative | Positive |
|---|---|---|
| Negligible | yellow | light blue |
| Low | orange | blue |
| Moderate | red | dark blue |
| High | garnet red | dark green |

**Figure 5.** Representation of B matrix for a sanitary landfill, in the phase or ordinary exercise.

## Matrix C

## Impacts

| | Negative | Positive |
|---|---|---|
| Negligible | yellow | light blue |
| Low | orange | blue |
| Moderate | red | dark blue |
| High | garnet red | dark green |

**Figure 6.** Representation of C matrix for a sanitary landfill, in the phase of ordinary exercise.

and territorial order' and 'Social relations' which evidentiate the most delicate problem of the location of a sanitary landfill.

## Matrix of Limitation Criteria

Figures 6, 7 and 8 show the matrices of limitation (Type C) valid for the three phases of the landfill. With regard to the temporary phase, emphasis is put on proper information being given to the public since this is the only thing limiting the impact caused by the decision to build a landfill.

Concerning the ordinary functioning phase, the limitation criteria will mainly affect the manner of management of the plant, but can also propose new protective installations such as artificial lining, a plant for the collection and treatment of leachate, a plant for the collection and treatment of biogas, etc. As already pointed out, such installations may produce negative effects and so themselves be the cause of further elements of impact. This same aspect is immediately evidentiated from the matrix when also the tonality of the chromatic scale indicates negative effects.

For the phase of extraordinary functioning, the criteria used are aimed at limiting as far as possible the entity (wells, drainage system under synthetic liner) or the risk of occurrence (measuring of the quality of biogas, flame-breaking gear, etc.) of exceptional events. The limitation criteria which have been identified and proposed in the three C matrices certainly do not exhaust all problems relative to control: new proposals will consent the amplification both of the extent and efficacy of actions of limitation.

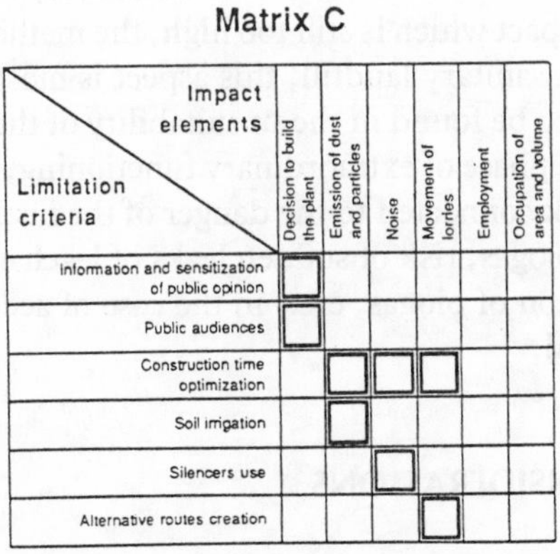

**Figure 7.** Representation of C matrix for a sanitary landfill, during the temporary phase.

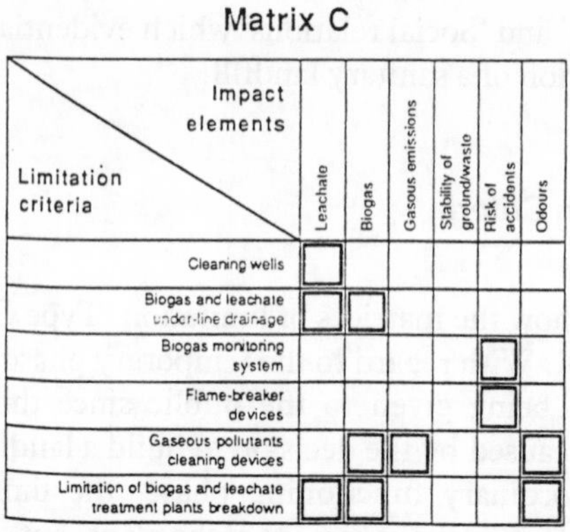

Matrix C

**Figure 8.** Representation of C matrix for a sanitary landfill, in the phase of extraordinary functioning.

## Matrix of Residual Impact

As previously explained, the aim of the D matrix is that of showing the entity of the residual impact still left after having applied the measures of limitation. Therefore this is the matrix which reassumes all information concerning environmental impact and directs the final decision.

The control lists are analogous to those of the B matrix (Fig. 9) and their cross-examination permits the evidentiation of the efficacy of limitation measures adopted without obliging the user of the study to examine the C matrix.

If, after having made use of limitation measures, the D matrix should present a residual impact which is still too high, the method consents arrival at the conclusion (for a sanitary landfill, this aspect is most important) that the principal causes are to be found in the unsuitability of the site.

With regard to the phase of extraordinary functioning, the matrix facilitates assessment of the risk connected to the danger of the installation in itself (risk of explosion due to biogas, risk of sudden leaks of leachate, risk of landslides, uncontrolled migration of biogas, etc.) in the case of accidents which are not completely controlled.

## CONCLUSIVE CONSIDERATIONS

The impact assessment scheme using chromatic matrices, similarly to the other matrix systems from which it is derived (Leopold's matrix, Moore's

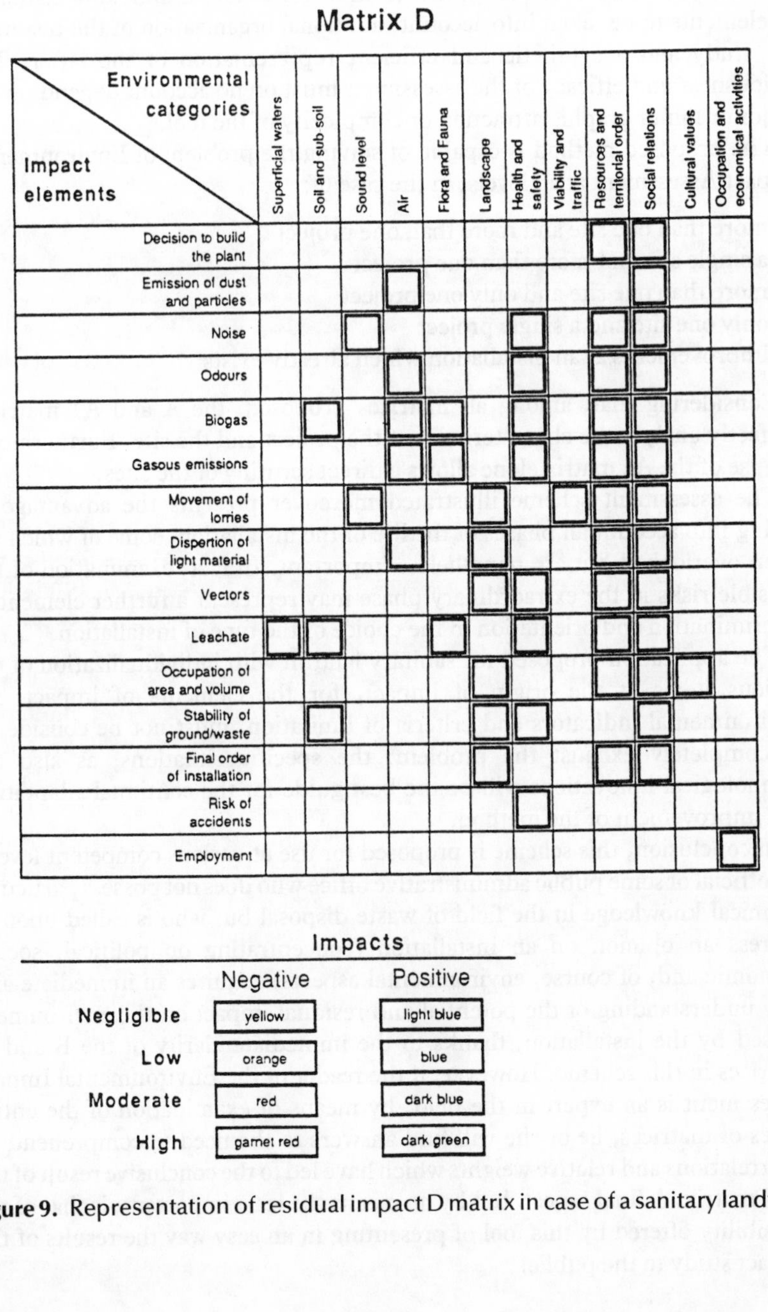

## Matrix D

| Impact elements \ Environmental categories | Superficial waters | Soil and sub soil | Sound level | Air | Flora and Fauna | Landscape | Health and safety | Viability and traffic | Resources and territorial order | Social relations | Cultural values | Occupation and economical activities |
|---|---|---|---|---|---|---|---|---|---|---|---|---|
| Decision to build the plant | | | | | | | | | ■ | ■ | | |
| Emission of dust and particles | | | | ■ | | | | | | ■ | | |
| Noise | | | ■ | | | | ■ | | ■ | ■ | | |
| Odours | | | | ■ | | | ■ | | ■ | ■ | | |
| Biogas | ■ | | | ■ | ■ | | ■ | | ■ | | | |
| Gasous emissions | | | | ■ | ■ | | ■ | | ■ | | | |
| Movement of lorries | | | ■ | | | | | ■ | ■ | | | |
| Dispertion of light material | | | | ■ | | ■ | ■ | | ■ | | | |
| Vectors | | | | | | ■ | ■ | | ■ | | | |
| Leachate | ■ | ■ | | | ■ | | ■ | | ■ | | | |
| Occupation of area and volume | | | | | | | ■ | | ■ | ■ | ■ | |
| Stability of ground/waste | | ■ | | | | ■ | ■ | | ■ | | | |
| Final order of installation | | | | | | | ■ | | ■ | ■ | | |
| Risk of accidents | | | | | | | ■ | | | | | |
| Employment | | | | | | | | | | | | ■ |

## Impacts

| | Negative | Positive |
|---|---|---|
| Negligible | yellow | light blue |
| Low | orange | blue |
| Moderate | red | dark blue |
| High | garnet red | dark green |

**Figure 9.** Representation of residual impact D matrix in case of a sanitary landfill.

matrix) must be considered exclusively as a tool for environmental impact assessment. This method is applied to allow for a careful individualization of all elements to be taken into account, a rational organization of the results of the study and a synthetic and efficient representation of the latter. The seriousness and efficacy of the assessment must on no account depend on the major or minor graphic attraction or complexity of the tool.

The proposed method is capable of solving the problem of Environmental Impact Assessment at all levels, in the case of:

1. more than one site and more than one project
2. a single site and more than one project
3. more than one site and only one project
4. only one site and a single project
5. improvements at an installation which already exists

Considering that, among all matrices proposed, the A and A1 matrices perfectly analyse the characteristics of the project and the site. Furthermore, the use of the A1 matrix alone allows a direct scrutiny of the sites.

The assessment scheme illustrated moreover presents the advantage of taking into account all phases of the life of the installation, some of which are often overlooked but are nonetheless important. In fact, examination of the possible risks in the extraordinary phase may represent a further element of discrimination and orientation in the choice of the type of installation.

The application proposed for sanitary landfill with individualization of the various items at the origin of impact, for the elements of impact, for environmental indicators and criteria of limitation, must not be considered to completely exhaust the problem: the specific situations, as also the technological innovations will be the best guide for the continual adaptation and improvement of the method.

In conclusion, this scheme is proposed for use at various competent levels. An official of some public administrative office who does not possess particular technical knowledge in the field of waste disposal but who is called upon to express an opinion on an installation (concentrating on political, social, economic and, of course, environmental aspects) acquires an immediate and easy understanding of the potential and residual impact on the environment caused by the installation, thanks to the immediate clarity of the B and D matrices in this scheme. However, if the reader of the Environmental Impact Assessment is an expert in the field, by means of examination of the entire series of matrices, he or she will find answers to the need to comprehend all interrelations and relative weights which have led to the conclusive result of the final matrix. A final aspect, but by no means the least important, is that of the possibility offered by this tool of presenting in an easy way the results of the impact study to the public.

## REFERENCES

Clark, B.B. *et al*. (1976). 'Assessment of major industrial applications: a manual' – Research Report No. 13. Department of Environment, London.

Cossu *et al*. (1986). 'Studio di Impatto Ambientale dell'impianto di depurazione di Pero', Ctip, Roma, Cogefar, Milano.

Dee Norbert *et al*. (1972). 'Environmental evaluating system for water resource planning'. Battelle – Columbus, USA.

EEC (1985). 'Council Directive No. 85/377', 27.06.1985.

Leopold, L. (1971). 'A procedure for evaluating environmental impacts'. US Geological Survey Circular 645/1971. Washington.

Moore, J.L. *et al*. (1973). 'A methodology for evaluating manufacturing environmental impact statements for Delaware's coastal zone'.

Sorensen, J.C. (1971). 'A framework for identification and control of resource degradation and conflict on the multiple use in dicoastal zone'. University of California, Berkeley.

# 6.2 Water and Element Fluxes from Sanitary Landfills

HASAN BELEVI and PETER BACCINI

*Swiss Federal Institute for Water Resources and Water Pollution Control, CH-8600 Dübendorf, Switzerland*

## INTRODUCTION

A municipal solid waste landfill can be defined as a 'chemical and biological fixed bed reactor'. The first educt is the solid waste fixed in the landfill and the mobile second educt is the water which enters the reactor continuously due to the precipitation.

The products of the complex reactions in the landfill are transported by percolating water and by produced gas. During these transports physical processes, such as adsorption and diffusion, take place beside chemical and biological reactions. Consequently, educts, which have not reacted, and products leave the landfill by the output fluxes 'leachate' and 'gas'. Residual solid mixture consists of inorganic and organic matter, which becomes chemically inert and less soluble in water.

To prevent a pollution of the environment, the output fluxes from a landfill must be treated. This treatment will be continued until output fluxes are directly compatible with the environment. During this period the municipal solid waste landfill is called a 'reactor landfill'. After this period the landfill has a 'final storage quality'.

The technique of controlled sanitary landfills was first applied at the beginning of the 1970s. Even the oldest landfills have not reached the final storage quality yet; therefore, the time required to reach this quality has to be extrapolated from landfills which are still in their 'reactor phase'. This study describes the assessment of the time to reach such a final storage quality for municipal solid waste landfills.

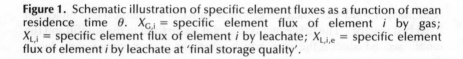

**Figure 1.** Schematic illustration of specific element fluxes as a function of mean residence time $\theta$. $X_{G,i}$ = specific element flux of element $i$ by gas; $X_{L,i}$ = specific element flux of element $i$ by leachate; $X_{L,i,e}$ = specific element flux of element $i$ by leachate at 'final storage quality'.

## METHODS

The method chosen in the present work is based on field investigations involving four comparable landfills of different ages followed during a short period (1 year). The inputs and the operational conditions are practically equal for all investigated landfills. It can therefore be postulated that they represent different steps of development of the same landfill. A more detailed description is given elsewhere (Baccini *et al.*, 1987).

In Fig. 1 the time-dependent behaviour of specific element fluxes by gas and by leachate is shown schematically. The specific element flux is defined as the annual flux per mass unit of municipal solid waste; $\theta$ is a mass-weighted mean residence time as defined by Baccini *et al.* (1987).

Gas production lasts about one to two decades (Ehrig, 1986; Stegmann, 1978/79). In this period elements are exported by gas as well as by leachate. Afterwards the export continues practically only by leachate. The oldest of the four studied landfills had a mean residence time of about 10 y; therefore the time dependent behaviour of the landfill could be elucidated only in the first ten years of its evolution.

The non-metals, carbon, nitrogen, fluorine, phosphorous, sulfur and chlorine and the metals, iron, copper, zinc, cadmium, mercury and lead were chosen to study the characteristics of the element transfer and transport in the landfills.

## RESULTS AND DISCUSSION

Precipitation data and leachate measurements, which were carried out since the beginning of the filling of the investigated landfills, permit the elucidation of the water household of a municipal solid waste landfill (Baccini *et al.*, 1987). Annual leachate per kg municipal solid waste decreases with increasing amount of municipal solid waste per area. After the filling is stopped it reaches a constant value. In a moderate climate and the chosen specific area, the ratio leachate/precipitation reaches a value of about 0.4 (Baccini *et al.*, 1987).

**Table 1.** Water content and element concentrations of Swiss municipal solid wastes (MSW).

| Element | $c_{MSW}$ (g/kg) | $p_{VAG}$ (%) | $p_{PCT}$ (%) |
|---------|------------------|---------------|---------------|
| $H_2O$  | 250   | 60    | 26    |
| $C_{org}$ | 290 | 25    | 45    |
| N       | 4[a]  | 40[a] | —     |
| F       | 0.2   | —     | —     |
| P       | 1[a]  | 40[a] | —     |
| S       | 2     | 15    | 30    |
| Cl      | 7.2   | 6     | 8     |
| Fe      | 50    | 0     | 0.2   |
| Cu      | 0.4   | 5     | 5     |
| Zn      | 1.2   | 4     | 8     |
| Cd      | 0.011 | 3     | 10    |
| Hg      | 0.004 | 2     | 2     |
| Pb      | 0.4   | 4     | 6     |

$c_{MSW}$ = measured mean concentrations in Swiss MSW (Brunner and Moench, 1986; Brunner and Ernst, 1986).
$p_{VAG}$ = contribution from VAG (vegetable and animal household garbage), calculated values (Diener, 1987).
$p_{PCT}$ = contribution from paper/cardboard/textiles, calculated values (Diener, 1987).
[a] Calculated from MSW compost data (Obrist, 1985).

The concentration ranges of the indicator elements in Swiss municipal solid wastes are shown in Table 1. They were determined in several mass balance studies of municipal solid waste incinerators (Brunner and Moench, 1986; Brunner and Ernst, 1986). With respect to chlorine and the metals the contributions from vegetable and animal household garbage (VAG) and paper/cardboard/textiles (PCT) are relatively small.

Table 2 shows transfer coefficients for different elements into gas and leachate. After a mean residence time of about ten years ($\theta$ = 9.4) 22% of the carbon is exported in the gas, 5% of chlorine and 6% of nitrogen are transferred to the leachate, and 5% of fluorine is exported in the gas. In the range from 0.1 to 1% carbon, phosphorus and sulphur are exported in the leachate. In the same range chlorine and sulphur are exported in the gas. About 0.09% of fluorine is transferred to the leachate. More than 99.9% of metals are still in the residual solid. All investigated non-metals show higher transfer coefficients than metals.

The metals are mainly transferred in the leachate. The corresponding values for the gas are two or three orders of magnitude lower. The only exception is

**Table 2.** Transfer coefficients of twelve elements for gas ($k_G$) and for leachate ($k_L$) in a municipal solid waste landfill at a mean residence time of about 10 years ($\theta$ = 9.4 y).

| Element | $k_G$ | $k_L$ |
|---------|-------|-------|
| C | 0.22 | 0.003 |
| N | — | 0.06 |
| F | 0.05 | 0.0009 |
| P | — | 0.002 |
| S | 0.003 | 0.001 |
| Cl | 0.006 | 0.05 |
| Fe | $2 \times 10^{-7}$ | $1 \times 10^{-4}$ |
| Cu | $2 \times 10^{-7}$ | $2 \times 10^{-4}$ |
| Zn | $4 \times 10^{-7}$ | $2 \times 10^{-4}$ |
| Cd | $4 \times 10^{-6}$ | $6 \times 10^{-5}$ |
| Hg | $4 \times 10^{-5}$ | $6 \times 10^{-5}$ |
| Pb | $1 \times 10^{-7}$ | $7 \times 10^{-5}$ |

Transfer coefficients $k$ are defined as $M_{product}/M_{educt}$ where:
$M_{product}$ = specific total output until mean residence time $\theta$ by leachate or by gas respectively;
$M_{educt}$ = specific total input by municipal solid waste.

mercury. Mercury is transferred by gas in the same order of magnitude as by leachate.

The output fluxes have to be treated until they are compatible with the environment. The compatible emissions depend on the environment of the landfill site. In a first approach the quality standard for leachates compatible with the environment is chosen on the basis of quality standards for running waters (SOWW, 1975). Thus, the 'final storage quality' concentrations must have values in the same order of magnitude as quality criteria concentrations for Swiss running waters, i.e. not higher than ten times the standard values. Phosphorus, sulphur, fluorine and chlorine are exceptions. For phosphorus the upper concentration of phosphate for oligotrophic lakes was chosen. For sulphur, fluorine and chlorine, the quality criteria of running waters for sulphate, fluoride and chloride were chosen. These concentration ranges are shown in column 3 of Table 3. The measured flux-weighted mean concentrations of twelve elements at $\theta = 9.4\,y$ are show in column 2 of Table 3 (Baccini *et al.*, 1987).

After about ten years of mean residence time, the mean concentrations of

**Table 3.** Estimated mean residence time $\theta_E$ of municipal solid waste in landfills required to reach 'final storage quality' with respect to leachate.

| Element | $c_{mean}$ (mg l$^{-1}$) | QS (mg l$^{-1}$) | $\theta_E$ (y) |
|---|---|---|---|
| $C_{org}$ | 1000 | 2–20 | 400–4000 |
| N | 1200 | 0.5–5[a] | 20–100 |
| F | 0.65 | 1 | ≤10 |
| P | 6.8 | 0.04–0.4[b] | 200–2000 |
| S | 2.7 | 30[c] | ≤10 |
| Cl | 1300 | 100 | 45–80 |
| Fe | 8.0 | 1–10 | ≤10 |
| Cu | 0.1 | 0.01–0.1 | ≤10 |
| Zn | 0.6 | 0.2–2 | ≤10 |
| Cd | 0.002 | 0.005–0.05 | ≤10 |
| Hg | 0.0005 | 0.001–0.01 | ≤10 |
| Pb | 0.00008 | 0.05–0.5 | ≤10 |

$c_{mean}$ = mean concentration at $\theta = 9.4\,y$ in leachate (Baccini *et al.*, 1987).
QS = quality standard for running waters (SOWW, 1975).
$\theta_E$ = time required to reach final storage quality.
[a] Sum of $NH_3 + NH_4^+$.
[b] Upper concentration range for oligotrophic lakes.
[c] Quality criteria for sulphate.

the metals Cd, Hg and Pb in the leachate are less than quality standards for running waters. The concentrations of Fe, Cu, Zn, F and S are in the same order of magnitude as the chosen quality standards. The concentrations of C, N, P and Cl are one to three order of magnitude higher than the chosen quality standards.

It is reasonable to concentrate on the assessment of the time required to reach the quality standard concentrations for C, N, P and Cl. For this purpose, however, the fractions of these elements, which can be transferred to the leachate, should be known. Such data are not yet available; therefore, a series of assumptions are to be made which are still very speculative.

About 40–70% of C, N and P exists in a chemical form which can be degraded within the reactor period (data in Table 1). However, only a part of this amount can be mobilized. The mobilization of the other part occurs at such a slow rate that produced concentrations are much lower than quality standards. Therefore it is assumed that the fraction of C, N and S, which can be still mobilized after $\theta = 10$ y, is in the range from 1 to 10% of the residual content of the municipal solid waste in the landfill. For carbon it is $2–20 \, \text{g kg}^{-1}$, for nitrogen $40–400 \, \text{mg kg}^{-1}$ and for phosphorus $10–100 \, \text{mg kg}^{-1}$.

Chlorine is an exception. About 85% of the chlorine content in municipal solid waste comes from polyvinylchloride. PVC is considered to be practically inert or degradated at so slow a rate that only chloride concentrations at quality standards are produced. Since about 5% of the initial content has been transferred until $\theta = 10$ y, it is assumed that 5–10% can be still mobilized after this time.

The landfill is hydrologically stabilized after $\theta = 10$ y (Baccini *et al.*, 1987). Afterwards the specific water flux remains constant. Therefore, as a first approximation, the degradation process can be simulated by a first order reaction, i.e. the reaction rate only depends on the specific amount of the element. Consequently the time $\theta_E$ required to reach the final storage quality can be determined for all twelve elements. A more precise estimation procedure supported by laboratory experiments will be given elsewhere (Belevi and Baccini, 1989).

In column 4 of Table 3 the time $\theta_E$ is given for all 12 elements. According to this assessment the organic carbon would demand the longest period to reach the chosen quality standard. Therefore, the organic species are the most important substances to control until the final storage quality is reached.

## CONCLUSION

The accomplished evolution of compound fluxes from landfills with gas and leachate, based on data from four Swiss landfills up to ten years old, indicates that organic compounds may be the most critical. This first assessment indicates that it may take several centuries before the organic compounds in the leachate reach the level of surface waters.

## REFERENCES

Baccini, P., Henseler, G., Figi, R. and Belevi, H. (1987). 'Water and Element Balances of Municipal Solid Waste Landfills', *Waste Management and Research* 5, 483–499.

Belevi, H. and Baccini, P. (1989). 'Long-term Behavior of Municipal Solid Waste Landfills', *Waste Management and Research* 7, 43–56.

Brunner, H.P. and Ernst, W. (1986). 'Alternative Methods for the Analysis of Municipal Solid Waste', *Waste Management and Research* 4, 147–160.

Brunner, H.P. and Moench, H. (1986). 'The Flux of Metals Through Municipal Solid Waste Incinerators', *Waste Management and Research* 4, 105–119.

Diener, H.P. (1987). Personal communication.

Ehrig, H.-J. (1986). 'Study of Gas Production from Municipal Solid Waste' ('Untersuchungen zur Gasproduktion aus Hausmüll'), *Müll und Abfall* 5, 179–183.

Obrist, W. (1985). Personal communication.

Stegmann, R. (1978/79). 'Gases from Controlled Landfills' ('Gase aus geordneten Deponien'), *ISWA-Journal* 26/27, 11–24.

SOWW (1975). 'Swiss Ordinance on Waste Water'. BUS, CH-3000 Berne.

## 6.3 Landfill Gas Migration, Effects and Control

DAVID CAMPBELL

*Harwell Laboratory, Environmental Safety Centre, Building 146.3, Didcot, Oxfordshire, OX11 0RA, UK*

### INTRODUCTION

The adoption of high standards for landfill management and site practices involves proper control of generated landfill gas (LFG) in order to limit migration and detrimental effects of hazardous and/or obnoxious gases.

The generation and potential migration of landfill gas (LFG) would appear to be a relatively new phenomenon of increasingly acute concern within landfill management, as judged by a review of references or bibliographies contained within published literature on the subject. There are many factors which will have contributed to this erroneous impression, although modern landfill practices and current concerns about our environment will undoubtedly have exacerbated a problem which nevertheless always existed. Before the 1960s there was surprisingly little published evidence (Jones and Owen, 1932; Eliassen *et al.*, 1957) to indicate that LFG was either a major product of waste degradation processes or, more importantly, that it might then migrate beyond site boundaries giving rise to external environmental problems. There are, it should be acknowledged, many differences between past landfilling practices, and the need or ability to monitor the effects of so doing, when compared with those appropriate to today's technology. These changes include various factors:

1. Landfill sites were in the past often small in area and generally shallow, with little attempt made to achieve adequate compaction of deposited wastes. Thus aerobic degradation of organic matter predominated, whereas in modern large, and often deep, landfills anaerobic processes of waste decomposition occur.

399

2. Landfilled waste composition has substantially altered over the last few decades with much higher proportions of potentially degradable organically based materials being deposited.
3. The composition and quantities of LFG have altered due, at least in part, to the above factors. Whereas carbon dioxide would have been the major product (still a potential hazard) in the past, methane and carbon dioxide are now the major components of LFG, with many other minor components also present, some of which are evolved as a direct consequence of and contained within the deposited wastes themselves.
4. Little or no monitoring was carried out either within or around landfill sites to determine whether gases were being evolved or not.
5. Suitable monitoring equipment of sufficient sensitivity and reliability was less readily available than is currently the case.
6. There was much less pressure or need rapidly to redevelop sites and their immediate surrounds than there is today and, as a consequence, there was little perceived need to examine the ground to determine its suitability for re-use.

## LITERATURE SUMMARY

As a background for the current state of technology for providing effective means of LFG monitoring and control a brief survey of the literature is presented.

By the late 1960s literature references related to gas generation and migration both within and around landfill sites began to increase rapidly. Various field investigations were undertaken (Fungaroli, 1970; Callahan and Gurske, 1971) and, together with other studies examining factors influencing gas production from a variety of organically based waste materials, knowledge of gas production and migration potential increased markedly. The 'current' interpretations of waste degradation processes and LFG production developed during the 1970s (Farquhar and Rovers, 1973; Rees, 1980) are still valid today.

During the late 1970s and early 1980s there was an increasing number of published articles, papers and reports referring to case histories of gas migration problems, many of which provided actual or conceptual design criteria for controlling LFG movement beyond site boundaries (Hill, 1986; Anon., 1983).

Several mathematical models were developed (Metcalfe and Farquhar, 1987; Moore, 1979; Anderson and Callinan, 1970) to assess or predict the likely effects and extent of gas migration within strata around landfill sites. In some instances these models included assessments of the impact of gas production rates and yields on the gas pressures generated within wastes or strata around sites, and on the potential life time over which gas migration might occur. It is significant, however, that in order to verify the validity of predicted effects, considerable practical field data are often required. It is probable that regulatory and control authorities will continue to require actual data as evidence of efficient gas migration control and they will not accept the results of predictive models alone.

There are currently many published papers, conference proceedings (GRCDA, 1978; Anon., 1986a), etc. which contain reference to gas migration problems and control, often coupled with descriptions of gas recovery projects to provide energy for industrial applications. LFG migration potential is becoming increasingly important to regulatory authorities in determining revised guidance or regulations for the proper management and disposal of wastes in landfill sites. The UK government, through the Department of the Environment, has recently published a waste management paper (Anon., 1986b) which has specific sections devoted to LFG environmental issues. Among the points covered are the potential causes of gas migration, methods for their abatement and safety issues related to suitable plant and equipment for monitoring and control of LFG. A very useful flow diagram is provided which, when a potential problem or complaint arises, allows the investigators to follow a whole series of Yes/No questions and answers to allow them to take appropriate actions to identify the cause and overcome the problem.

In concluding this brief review of the published literature which has led to the current understanding and state of the technology, it is apparent, unfortunately, that there are still far too many site operators who fail to monitor within, and in the vicinity of, their landfill sites for the presence of LFG. Many sites of course will not give rise to LFG migration problems but it must nevertheless be proven that this is indeed the case. There are several examples of unforeseen and serious problems which have occurred as a consequence of gas migration (Fricker, 1986; Stearns and Beizer, 1985). Such examples invariably could have been avoided with proper monitoring being undertaken to provide early warnings of developing problems. The financial penalties in having to provide remedial measures and/or compensation may be severe and, at least in the UK, ignorance or inadequate prevention will no longer be accepted as a reasonable defence.

Current technology has been developed largely on an ad hoc basis at specific sites, as an emergency response in order quickly to overcome identified problems of LFG migration. Often the recommended LFG controls introduced have been constructed without a full appreciation of the cause and source of the problem or without an assurance that the measure will fully overcome the problem or will not introduce other problems. Thus at what seem to be two similar types of site, a shallow passive venting trench may be used in one case, while a series of positively pumped wells may be used at the other site. The possibility that LFG migration may, initially, be occurring vertically beneath one and moving laterally at the other may not have been fully investigated. An appreciation of the long-term effectiveness of proposed or installed LFG control measures related to changing circumstances both within the wastes and in the surrounding strata may be lacking. The location of, and sampling frequencies from, installed bore holes may be insufficiently accurate to provide data on the potential pathways and extent of any LFG migration plume.

## POTENTIAL PROBLEMS CAUSED BY GAS MIGRATION

The main problems associated with migrating LFG can be summarized as:

1. Air depletion in soils causing vegetation die back.
2. The build-up of flammable and, within confined spaces, explosive concentrations of methane-rich gas within property, services and other enclosed spaces.
3. The build-up of gas accumulations (not necessarily containing high methane concentrations) which may cause asphyxiation.
4. The creation of an odour and/or possible health risk.

Damage to crops, trees or other vegetation is a common early indicator that significant gas migration is occurring, and regrettably it is often the first and only indication in the absence of any provision for proper facilities for LFG monitoring. The damage to vegetation is usually caused by depletion of oxygen in the root zone. The odour problem is primarily a nuisance, since the health effects are negligible at normal dilution rates (Young and Parker, 1983; Zimmerman et al., 1985). Even more serious are the increasing number of instances of LFG migration causing hazard, and sometimes damage, to property. There are many examples where prospective property developers are not being made aware, or are simply ignorant, of the potential hazards. The redevelopment of completed landfill sites for building purposes, which are still actively producing LFG, is becoming increasingly common. While advice in the UK (Anon., 1986b) suggests that building should only be considered on shallow landfill sites following a period of some 20 years after site restoration, it is further accepted that with proper precautions in building design more recently completed sites may also be acceptable.

In assessments of potential problems caused by LFG migration and emissions it is important to recognize that gases other than methane may be the principal components present. It is therefore essential that, in addition to methane monitoring, carbon dioxide concentrations are measured. In some circumstances, especially when entering confined spaces suspected to contain LFG, it may be more relevant to measure any oxygen deficiency in the atmosphere before assessing the extent of LFG contamination.

## FACTORS AFFECTING GAS MIGRATION POTENTIAL

There will always be a site-specific course of action required to prevent LFG migration, nuisance or hazard, but there are some general principles which should always be applied. These rely on already having, or being able to obtain, basic facts about the site and an understanding of the factors which can have an impact on LFG migration potential. There are two basic aspects to consider, namely influences on and within the landfilled wastes, and influences on and within the surrounding strata.

### The Landfill Site

There are of course a wide diversity of landfill site types and designs each of which will produce localized differences within them. Among the more important variables which will affect gas production and migration potential are the following:

1. Sites may be small, large, shallow or deep; they may be predominantly 'above-ground landfills' or be contained within quarries, pits, canyons, etc.
2. Sites will receive and contain a variety of waste types and different input rates, not only between sites but often within a site, over a period perhaps extending to several decades.
3. Site operating procedures will vary. These will include using area, terrace or cellular methods of infilling, various compaction methods, and use of a variety of types of 'soils' for daily, intermediate or final cover and/or for bund formation.
4. Sites will have differing water/leachate management policies, and climatic conditions (particularly rainfall) will be different in various localities.

These and perhaps many other site-specific variables will affect conditions within the site. These conditions will, in turn, have a major impact on waste degradation processes, gas production rates and yields, and gas pressures generated, each of which affect the potential for gas migration.

The subsequent discussion will not be referring to any unique conditions which might occur within a specific site, but will concentrate on generalities likely to apply to most if not all sites. It is further presumed that deposited wastes are predominantly of municipal solid waste origin which are likely to produce the 'worst case' conditions in terms of maximizing gas produced and therefore potential for gas migration.

It is the generation within sites of gas pressures above those observed around or above sites which will cause gas to migrate. The rate or extent of migration will not necessarily be a function of gas production rates, but will be related to various site conditions and those pertaining within the surrounding environment.

Within a site such conditions as high waste density and moisture content will reduce porosity and increase gas pressures within deposited waste materials. Excess pressures of at least 8000 $N/m^2$ above atmospheric pressure have been recorded in some large deep sites. The presence of impermeable cover soils on layers of waste, or use of similar materials for local bund wall formation (e.g. in creating operating cells for waste deposits) will further impede uniform gas permeabilities within the landfill and create a wide diversity of gas pressures within different parts or depths of a site. Conversely, the use of very permeable materials, temporary road constructions, drainage or venting/monitoring systems within sites, might create localized zones of high gas permeability where gas pressures are low.

Once generated, LFG will tend to migrate within sites to areas of lower pressure, setting up a range of pressure gradients both laterally and vertically within them. In general the probability of lateral migration within sites will be substantially higher than will that of vertical migration. This is likely for two reasons. First, waste is commonly placed in a layered formation which will tend to produce horizontal zones within a site, where higher compaction and hence lower porosity will be achieved primarily at the surface of each waste layer. This will be particularly noticeable where each layer is several metres in depth, rather than where thin-layering techniques are used. The second reason is perhaps more important. The use of daily, intermediate and final covering materials will additionally create horizontal layers of lower permeability compared with those found within wastes, even where permeable soils are used.

The creation of this horizontal layering effect, as summarized diagramatically in Fig. 1, will further control vertical gas migration rates as rainfall infiltration percolates through the waste. It is a well-established fact that localized zones of perched water (leachate) become established at different horizons, usually at or on one or more of these 'daily' layers of waste. This will further reduce vertical gas migration rates and encourage gas movement laterally towards the boundaries of the site.

In discussing the impact of infiltrating water to affect migration pathways, we should recognize that site-operating procedures also control rates of water

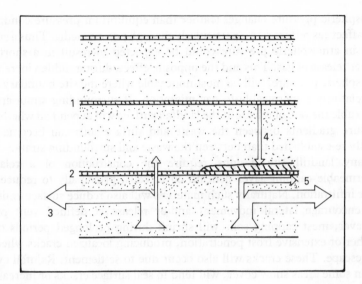

**Figure 1.** Lateral gas migration within deposited waste: (1) permeable daily/intermediate cover; (2) upper portion of waste layers exhibit greater compaction; therefore (3) majority of gas generated within waste layer migrates laterally; (4) impermeable perched leachate table caused by rainfall infiltration; (5) gas that would have migrated vertically is now forced to migrate laterally.

infiltration which in turn affect rates of microbial degradation and hence gas production. The size and rate of infilling and restoration of a phase or limited area of a site will directly control waste moisture contents, and the extent of any aerobic degradation giving rise to elevated waste temperatures. Both temperature and moisture content are important parameters in controlling microbial activity and gas production.

Settlement of wastes will also have an impact on migration pathways within sites. This settlement of wastes could conceivably increase waste porosity locally by creating voids and, similarly, loss of waste materials as leachate or gas could also increase waste porosity and gas movement. Generally, however, settlement will reduce waste porosity, and make both lateral and vertical gas migration pathways more uniform with time.

It must be recognized that processes of waste decomposition, settlement, etc. are dynamic and related to changing conditions such as moisture, temperature and microbial activity; thus gas production and potential rates of migration will change over time, possibly extending over several decades.

In addition to rainfall impacts within deposited wastes, climatic conditions have major impacts on gas emissions via landfill site surfaces. Rapid atmospheric pressure changes (rather than equilibrium pressure conditions) may affect gas release rates, particularly through porous media. Thus a sudden drop in atmospheric pressure could be expected to result in a short-term greater release of gas. This and the opposite effect, during sudden increases in atmospheric pressure, will be most noticeable where the site boundary strata are relatively impermeable to gas migration. If surrounding strata are also permeable the rate of release at the landfall surface will depend on whether the pressure gradients between the waste and surface soils can become more rapidly re-established than between the waste and surrounding strata.

Many landfill site designs require the construction of a relatively impermeable capping layer beneath cover soils, principally to reduce rain-water infiltration. Naturally such materials will also reduce surface emissions and encourage lateral movement and/or release at defined vent points. However, most capping soils will shrink during prolonged periods of dry weather or extensive frost penetration, producing localized cracks where gas may escape. These cracks will also occur due to settlement. Rainfall events, and in some areas snow cover, will tend to seal surface cracks or increase soil moisture thus reducing soil permeabilities and gas migration through them (Fig. 2). Often both rapid atmospheric pressure changes and high rainfall

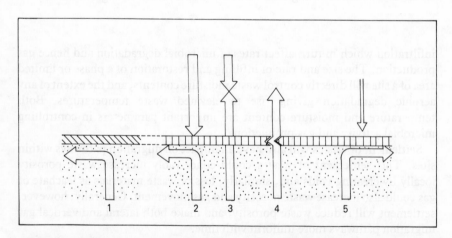

**Figure 2.** Relationship between landfill cap and atmospheric conditions on gas migration routes: (1) impermeable cap. Landfill gas produced migrates laterally; (2) permeable cap, high atmospheric pressure. Landfill gas produced may migrate laterally depending on pressure differential; (3) permeable cap, low atmospheric pressure; landfill gas encouraged to migrate vertically; (4) impermeable or permeable cap; crack in cap encourages vertical migration; (5) sealing of the cap by rainfall forces landfill gas to migrate laterally.

events occur simultaneously and the exact cause of rapid or reduced emissions is less clear. Measurements of gas pressure changes within wastes beneath capping soils can assist in assessing the impact of these climatic changes on the landfill surface.

The discussion has highlighted some of the more important factors or influences which affect gas production and pressure changes within landfills, and which will therefore ultimately have an impact on gas migration potential. In order fully to ascertain and appreciate the potential significance and impact of LFG on the surrounding environment it is essential that monitoring is carried out within the deposited wastes to determine some of these parameters. It is not sufficient simply to measure gas quality and perhaps pressure and flow rates. Monitoring and measurements of gas/waste temperatures, leachate levels and quality, settlement rates, climatic data, and obtaining records of waste disposal practices are all valuable aids to understanding the potential for generated LFG to migrate. This monitoring should be carried out routinely at defined time intervals both during site development and after site restoration.

In conjunction with monitoring carried out beyond the site boundaries, as discussed below, the accumulated data will allow for correct interpretation of results, either to permit adoption of appropriate controls, or to assess the effectiveness of those already designed and in place.

## The Surrounding Environment

There are two basic aspects which are important with respect to gas migration potential beyond site boundaries. The first is the nature of the strata beneath and around the landfill within which, depending on rock permeabilities, gas may migrate. The second aspect relates to potential dilution of gas in the air above and around sites to maintain atmospheres at acceptable component concentrations. In this context, odour nuisance, effects on vegetation, flammability, toxicity and asphyxiant potential are important.

In evaluating the gas migration potential, in particular when applying mathematical models, the interface between the waste and the site boundary strata is considered separately. In any new landfill design, where gas could migrate beyond the site boundary, it is presumed that such an interface would incorporate some form of barrier or venting system (see later discussion), and will therefore be an integral part of the site design. For purposes of the subsequent discussion the importance of the interface will be assumed to be minimal (as perhaps might exist at many completed sites with permeable strata beyond). It will be assumed that LFG at the site boundary will be at some positive pressure in relation to gas (air) pressure within the surrounding strata and that the site base and surrounds are not sealed with impermeable material.

In the same way that a host of variable conditions will exist within the landfill, so will they within the strata surrounding landfill sites. In the most simple hypothetical situation where a completely homogeneous stratum exists and is subjected to a given gas pressure, the potential for gas to migrate within it will be directly related to the gas permeability of the stratum. The distance of migration from the site boundary will also, under such circumstances, be related in particular to its ability to dissipate through the surface soils. Thus a clay soil overlying pure sand strata is likely to result in increased distances compared with those where seepage from a sandy soil cover may readily occur.

This simplistic situation is unfortunately extremely unlikely to exist at most sites. If it does exist the whole process of determining suitable gas monitoring locations, assessment of data, and installation of controls is made extremely simple. Among the factors which will modify gas migration rates and pathways (Fig. 3) are the following:

1. The strata are extremely unlikely to be completely homogeneous. Rock may be fissured, and will almost certainly contain strata horizons of differing gas permeabilities.
2. There may be local shafts, adits, faults, or culverts in the strata, either human-made or naturally occurring. For example in some strata human-made fissures may exist as a consequence of previous rock blasting operations during mineral excavation.
3. The slope planes of any horizons in the geological formations may not be horizontal, leading perhaps to certain more permeable horizons intersecting the land surface at discrete distances from the site boundary.
4. Gas generated at the landfill may only migrate from it at discrete points around site boundaries. For example it is not uncommon for localized sand lenses to exist in what is otherwise predominantly an impermeable stratum such as clay.
5. Localized perched water, or ground water tables in the strata may vary with time, affecting both gas migration pathways and pressure changes within the strata.
6. Climatic conditions may, as occurs within the landfill, alter gas pressures and migration rates and distances within the strata, or affect surface soil porosities or emission rates.
7. Changes in the use of land adjacent to landfill sites may alter surface emission rates and hence gas migration distances (e.g. changes in land use from animal pasture to cultivation for arable crops, or redevelopment for building).

Where predominantly above-grade landfilling is practised (i.e. the formation of refuse mounds or hills), any gas migration into strata which occurs will initially involve some downward penetration of LFG beneath the site. Quarry

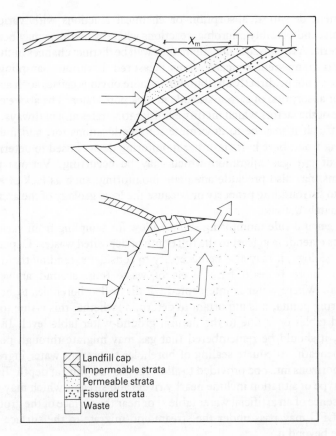

**Figure 3.** The migration of landfill gas through strata and the effect of impermeable subsoils.

infills may cause both downward and direct lateral migration. In both cases it will be assumed that no naturally occurring impermeable layer exists or has been artificially created either beneath or around the site boundary. Under such circumstances LFG migration may be further complicated by migration of leachate. Note only will leachate be saturated in carbon dioxide which may subsequently be released into the strata, but also organic matter (carboxylic acid) in the leachate may degrade anaerobically during leachate migration producing additional sources of methane-rich landfill gas.

Finally, it is possible that stratification and/or separation of migrating LFG components may occur. This will be due to such factors as differences in gas component densities, absorption or chemical reactions with various rock materials, and possibly microbial reactions whereby methane may be oxidized by bacteria. As a consequence, there may well be distinct changes in the ratio of LFG gas component concentrations measured at various sampling points, which may not be directly attributable to more obvious causes such as dilution by air or absorption of carbon dioxide in rock pore water. The above examples of some of the factors which modify gas migration rates and pathways, indicate how difficult it may be to determine suitable locations for, and methods of sampling from, bore holes or other monitoring devices used to determine the extent of any gas migration which may be occurring. Various practical problems may also preclude adequate monitoring, such as lack of adequate access to surrounding property or because the local geology of the area may be inadequately defined.

As a general rule monitoring should allow for sampling from various rock horizons extending at least to the full depth of deposited wastes. Depending on known geology, it may also be necessary to consider extending them to some 10 m or more beneath base levels of waste (e.g. around above-ground landfills). Where water is intercepted within drilled boreholes to be used as monitoring points, it is important to determine whether this is due to shallow perched tables or is due to the natural ground-water table level. If it is the former, it should be remembered that gas may migrate through permeable zones beneath. Adequate sealing of boreholes to prevent water ingress from upper horizons must be provided to allow for gas sampling at depth. Examples of this type of situation include nearby rivers and streams which may result in the presence of an artificial water table at or near the surface of the ground, but where LFG may pass under the stream and migrate to the surface at some distance beyond it.

Finally, gas emissions to atmosphere from landfill surfaces or from surrounding land may either be uniformly distributed or, more probably, be from discrete point or area sources. These will include cracks in soil profiles, installed gas vent points, or areas of high permeability soils. Gas is commonly released at the landfill/rock interface, where waste settlement results in deformation of surface soils.

## LFG MONITORING EQUIPMENT AND PROCEDURES

This section discusses the methods for monitoring both within and around landfill sites either to ascertain the extent of migration, or to provide assurance that migration is not occurring and/or has been controlled.

## Monitoring within Landfill Sites

The principal methods for monitoring LFG within sites is via installed boreholes or gas wells either constructed during site operations or soon after progressive restoration of small areas of a site. Previous discussion has referred to the essential need to obtain data which indicate conditions within the site. In large sites (especially those covering large areas) several such boreholes may be required. In most cases boreholes will consist of lengths of perforated plastic tubing extending to or near to the base of the site (Fig. 4a). It should be remembered that resultant gas data will be a composite of conditions and LFG component concentrations within all depths of wastes. Gas pressures recorded will similarly be average values representative of all depths in adjacent wastes.

More exact data can be obtained by using packed boreholes where samples or readings can be obtained from discrete waste horizons (Fig. 4b). Even with such systems data will probably only be meaningful if taken at 5–10 m intervals due to potential short-circuiting through the waste around the borehole pipe between each sampling depth. In assessing gas migration potential, data will be more relevant if sampling boreholes within sites are located close to site

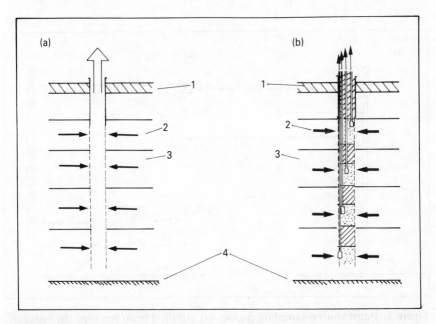

**Figure 4.** Monitoring boreholes within landfilled waste or surrounding strata: (a) integrated samples; (b) individual layers. 1: Cap, 2: Gas pathway or strata, 3: Waste layers, 4: Landfill base.

boundaries rather than within the centre of a site. Pressure gradients within a site will generally radiate from the centre towards the site boundaries. Resultant data obtained will, ultimately, be used for comparison with equivalent data from within surrounding strata, and should therefore closely resemble conditions within wastes near to the site boundaries.

More limited, but locally valuable, data may be obtained by installing probes through the surface of landfilled wastes or by burying probes in situ during waste disposal (Figs 5a and 5b). The advantages of such probes are that they are less expensive to install and they will give rapid results of very localized conditions. The disadvantages are that they are more liable to be 'lost' as waste disposal proceeds, and they will not give an overall picture more relevant to gas migration on a large scale. Probes installed through the surface are also unlikely to penetrate more than a few metres into the wastes.

### Monitoring around Landfill Sites

The number and location of monitoring boreholes required around sites will, as already described, depend on the nature of the rock involved. Despite or

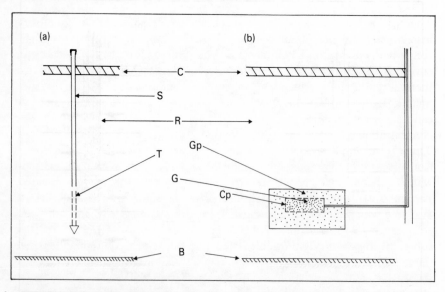

**Figure 5.** Point source sampling probes (a) installed from the top; (b) installed during landfilling. C = Cap; S = Steel probe; R = Refuse; Gp = Gravel pack; T = Perforated steel tube; G = Gravel; Cp = Plastic or galvanized steel perforated cylinder; B = Base of landfill.

perhaps because of this, regulatory conditions are already being applied at many sites in the UK that spacings of about 20 m may be most appropriate and that they should not exceed 50 m. It is often (rightly) considered necessary to prove that gas migration is not occurring. Unless other measures are taken, such as installation of a completely impermeable barrier around the site, it is unlikely that less frequent spacings will suffice. The boreholes should ideally be spaced not only immediately around the site boundary but at increasing distances from it, particularly at completed sites with no previous monitoring provisions. The locations and borehole frequency required may be more problematical where old sites are being monitored to determine the extent of any suspected gas migration. In most instances at new sites a single line of boreholes will be sufficient as other measures to prevent migration positively will have been taken.

The use of packed boreholes may have even more value for monitoring gas quality and pressure beyond site boundaries, where conditions within specific permeable strata horizons may need to be identified.

## Monitoring Equipment and Sampling

There is a wide variety of commercially available portable monitoring instruments capable of measuring the concentrations of most major gas components present in LFG. Several types of methane detection instruments covering various concentration ranges are available, many of which use thermal conductivity principles of detection. It should be recognized that such meters are essentially flammable gas detectors and while they may be calibrated at specific methane in air or methane in carbon dioxide concentrations, they will give positive responses to most flammable vapours. Thus, for example, hydrogen present in LFG will give a positive response. Conversely other components present in LFG, such as carbon dioxide and water vapour, may result in erroneous low readings being obtained, particularly at low methane concentrations. Portable infra-red detectors are available which are specific to the calibrated gas component, and other less reliable (due to interference effects) detectors, such as chemically reactive tubes, are also available. Portable flammable gas detectors, utilizing sensitive flame ionization principles of detection are available for measuring very low flammable gas concentrations. Typically these would be used for atmospheres above or around sites, or where low gas seepage into confined spaces is suspected. Portable oxygen meters can also be used to measure oxygen depletion in air atmospheres, but detectors will often be rapidly poisoned if measuring oxygen concentrations in concentrated plumes of LFG.

Most portable instrumentation involves extracting samples, or maintaining continuous flows of gas past the detector and care must be taken to ensure that rates of extraction do not exceed rates of generation at the sampling point. This may be particularly important when sampling from confined spaces or from within boreholes located beyond site boundaries. Rates of migration to the sampling point may be below normal instrument extraction rates. Consequently air ingress and dilution may be significant, or LFG may be drawn from more extensive areas not directly equivalent to the local conditions at the sampling point.

In determining gas quality it is, in most circumstances, advisable to obtain samples of LFG, or of the atmosphere being monitored, for subsequent analysis by gas chromatography to confirm data obtained in the field.

Various other instruments are available for field evaluation of gas flow rates, pressures, temperatures, etc. Gas emission rates through surface soils above and around landfill sites can be determined by carrying out flux box measurements whereby the rate may be calculated by monitoring the build up of LFG with time within the flux box. More sophisticated techniques allowing rapid evaluation of surface LFG emissions over large areas are being developed based on using lasers to measure infra-red radiation of specific gaseous components.

## GAS MIGRATION CONTROL AND PREVENTION

There are two basic methods for gas migration control. The first relies on passive venting techniques where gas is released either directly, or indirectly after flaring, to atmosphere. The second method involves positive extraction techniques where gas will generally be flared before release to atmosphere.

### Passive Control

There are several techniques which may be employed to prevent gas migration from above or around both active and completed landfill sites.

These techniques include installation of a simple venting trench backfilled with sized stone material (Fig. 6a). Such systems are generally employed beyond the landfill boundary and will, because of safety limitations during construction, be restricted to use at shallow landfill sites. Construction of such trenches within the landfill will generally be undertaken alongside disposal of wastes and depth restrictions under these circumstances will be less stringent. The use of these simple trenches will only be acceptable where strata porosities

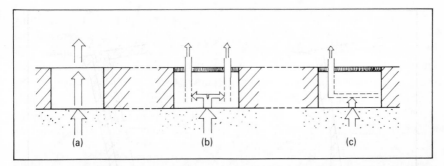

**Figure 6.** Gas venting trenches: (a) passive gas venting trench; (b) passive gas venting trench with impermeable cover and PVC/PE vent pipes; (c) passive or pumped extraction trench with impermeable cover and PVC/PE vent pipes.

are low in relation to the trench backfill materials. Gas may in some circumstances be allowed to vent directly from the trench surface, e.g. at sites remote from population or property where both odour and gas hazards may be diluted readily within the atmosphere. More commonly, however, trenches will be sealed at the surface, with a series of vent stacks to relieve collected gas (Fig. 6b). A further refinement, which allows for better control, would involve the placement of a perforated gas collection pipeline within the trench which could similarly be directed to vent chimneys or flare stacks (Fig. 6c).

Another technique involves installation of low permeability barriers either at the site boundary or just beyond in the strata (Fig. 7a). Naturally occurring materials (e.g. clays, betonites) or artificial membranes may be used to line the site base and/or sides before or during site operation, or they may be installed within excavated trenches. A refinement to this option which may be of value within homogeneous strata is injection at high pressure of cement based slurries within predrilled, closely spaced holes. Where barriers are employed they will usually be backed by one of the above methods of gas collection and venting (Fig. 7b). The appropriateness of these options to particular circumstances will depend on the nature, extent and depth of the landfill, the type and uniformity of strata surrounding the site, and the proximity of property, amenity of agriculture to be protected.

Vertical gas vent pipes may be installed within the waste near to and around the site boundary. This method may be suitable where surrounding strata are relatively impermeable and the objective is to reduce gas pressures within the waste thus discouraging lateral migration.

Low permeability gas barriers may be constructed above landfilled waste surfaces to reduce surface emissions. Such systems will be particularly

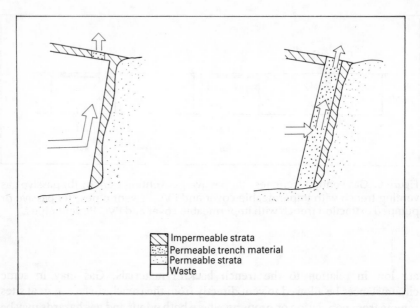

Impermeable strata
Permeable trench material
Permeable strata
Waste

**Figure 7.** Impermeable barrier methods for preventing lateral gas migration.

applicable to proposed development on the site surface (e.g. recreation or building) and will generally require the installation of a gas drainage system beneath the barrier to collect and vent any trapped gas (Fig. 8).

Permeable 'soils' can be used only as cover materials for landfill sites. This option would be restricted to remote sites where leachate generation was not a problem and where odour nuisance or gas hazard by emission through the permeable surface would not be deleterious. Such an option might be suitable, for example, for sites containing largely inert deposits where a limited risk of lateral gas migration potential affecting vegetation was the only concern.

The main advantages of passive control methods include:

1. low cost of construction for most options, although material costs may still be significant;
2. rapid response for providing control of developing gas migration problems due to ease of construction;
3. low operation and maintenance costs. It should be noted that these costs are not zero and will, for example, require expenditure on maintenance of open trenches to prevent blockages by vegetation growth or soil deposition. In

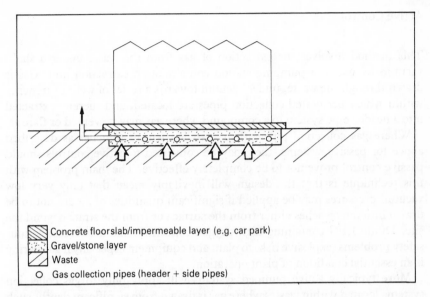

Concrete floorslab/impermeable layer (e.g. car park)
Gravel/stone layer
Waste
○ Gas collection pipes (header + side pipes)

**Figure 8.** Ventilation system to prevent landfill gas build-up under impermeable surfaces.

addition monitoring boreholes installed beyond such barriers must continue to be surveyed at regular intervals.

The main disadvantages of passive control methods are that:

1. the techniques are, on their own, most applicable to shallow landfill sites and will therefore be less acceptable in the future as sites tend to become larger. However, as discussion below indicates, they may be used in conjunction with active pumped gas schemes;
2. it may be difficult to ensure continued operation of gas flares attached to passive controls because both gas rates and quality may be inadequate;
3. where flaring is not feasible odour release or potential gas hazard from localised gas vents may be significant;
4. venting trenches may become blocked with time either with solids or liquids. Synthetic barrier material integrity may, in the longer term, become suspect due to holes, tears and possibly due to chemical attack by minor gas components;
5. costs of low permeability barriers may be high, particularly where natural materials are not available on site or where artificial membranes are used.

## Active Control

This method involves the extraction of gas from the waste under a slight vacuum by use of a pumping station containing an extraction unit. Gas is drawn through the waste, under vacuum towards a series of wells or trenches within which perforated collection pipes are located, and thereby extracted into a header pipe system to a compound where gas may be vented or flared.

Where gas collection pipework is installed within trenches as described above for passive systems, they may be converted to active controls should passive control prove not to be completely effective. The main problem with this technique is that the design will inevitably mean that only very low vacuum pressures may be applied if significant quantities of air are not to be drawn into the trenches either from the surface or from the strata beyond the site. Dilute LFG containing high oxygen concentrations may create serious safety problems (explosive risk) to plant and equipment especially if gas flaring is an essential condition of plant operation.

More typically active pumped systems will apply only to gas collection systems located within wastes where gas is drawn from significant depths such that air ingress may be avoided (Figs 9a, b). A series of gas wells drilled into completed areas of waste, at spacings which will typically be no more than 50 m apart and about 25 m from the boundary, is commonly used.

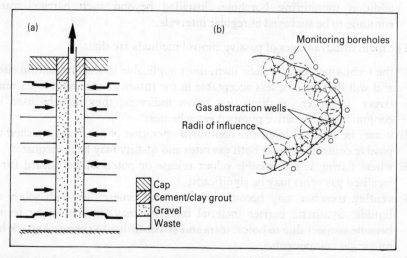

**Figure 9.** Construction of wells to extract gas from all waste depths and achieve complete migration control: (a) gas collection from all waste layers; (b) gas collection from overlapping spheres of influence.

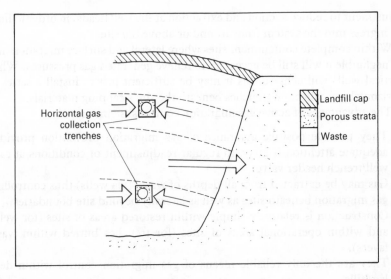

**Figure 10.** The potential problem of using trenches in waste layers to control gas migration.

Alternatively a series of trenches constructed in selected waste layers during waste deposition may be used (Fig. 10). While the former will by design intercept all waste layers, the latter will not and, due to reduced influences on wastes around them, will only be suitable for gas migration control where daily cover materials used are permeable in nature.

If an optimum spacing of gas migration control wells is required it is probable that an initial testing programme will be required, mainly because gas rates and radii of influence around and between individual wells, will not be uniform. More commonly, however, a closer spacing than necessary will be adopted with well head conditions and subsequent operation of wells determined by reference to performance data from the wells, each of which will have its own characteristics of applied suction and rate of extraction related to local conditions within adjacent wastes.

In many cases where positive extraction methods of gas migration control are employed, the potential for gas recovery and utilization as an energy resource will be significant. Thus any actively pumped system should be designed with this future objective in mind even if, ultimately, a separate utilization scheme is adopted. Where this is likely its impact on the gas migration control system should be assessed. For example, when LFG recovery commences, the amount of gas abstracted by the migration system may be substantially reduced. The control settings provided for gas migration control may need adjustment to reduce suction and extraction at the well heads, in order to limit air ingress into the system from around or above the site.

Within complete containment sites where lateral and surface migration may be negligible it will still be necessary to relieve generated gas pressures. While vertical wells will achieve this it may be sufficient just to install a series of horizontal gas collection trenches beneath the site's capping materials.

The advantages of active gas migration control systems are:

1. They provide positive assurance of gas migration prevention provided adequate attention is given to regular readjustment of conditions at each well/trench header valve.
2. Gas may be extracted from all depths of waste (via wells) thus controlling gas migration beneath sites as well as laterally beyond site boundaries.
3. Construction is relatively simple within restored areas of sites (for wells) and within operational areas of sites (for trenches buried within waste layers).
4. They are the only reliable means of gas migration control within deep landfills.
5. They provide a means of gas recovery for subsequent use as an energy resource.

The main disadvantages are:

1. The complete costs of a pumped control scheme will in most examples be higher than for passive control.
2. A higher level of monitoring, control and maintenance of plant will be required for many years and perhaps over several decades.
3. It is often more difficult and costly to install pumped control system within sites while disposal operations are being carried out simultaneously in the same area, due to the great care needed to protect wells, trenches, pipe-work, etc. from damage by site operational plant.

## GAS MIGRATION CONTROL AT FUTURE LANDFILL SITES

It is rapidly becoming standard policy in many countries, in planning and design for future landfill sites, that provisions for gas migration control must form an integral part of the overall project. Furthermore, it is reasonable to assume that this policy will mean in the future that some of the current costs for commercial utilization of LFG will have already been expended in providing environmental controls for LFG. Therefore, most gas migration control schemes will form an integral part of overall gas management and energy recovery for use at future sites.

It is also becoming standard practice in site design to undertake progressive and sequential infilling and restoration of sectors of a landfill site. This policy not only reduces potential leachate generation, but provides evidence to the general public of good management and more effective and rapid afteruse of sites than has been achieved at most sites to date. Such methods will also be beneficial in order to achieve high standards of LFG control.

Monitoring boreholes should be provided during site preparation around all landfill site boundaries extending at least to the base of the infilled waste and, if ground water is not intercepted, to some 10 m below the base of the site within unsatured porous strata.

At many landfill sites it is becoming increasingly necessary to contain all possible sources of potential environmental contamination within the site boundaries. Where such standards are insisted on, complete containment will be required. Thus the base and sides of sites will have to be sealed with very low permeability materials, if they are not already in place.

LFG, inevitably produced where any organic matter is deposited, will require venting or extraction at least to eliminate any significant gas pressure build-up. In large and very deep sites some temporary method for controlled gas release may be required to keep pressures generated to a minimum before wastes can be brought to the surface, the area restored, and long-term gas migration control measures introduced. This may involve construction of venting chambers as waste deposition proceeds but it should be recognized that these vents create a potential safety hazard within operational areas of sites as well as impeding efficient disposal and compaction operations. Horizontal trenches laid in wastes, with pipes led to the site perimeter, may be more appropriate in some locations. Such measures will probably be required in sites where overall site restoration profiles cannot be achieved within a period of about two years.

There are still some uncertainties about the use of trenches within wastes for gas collection purposes. These include adverse impacts on them of long-term waste settlement, the potential blockage of systems due to solids or liquids (perched leachates) and the overall cost-effectiveness and performance related to vertical wells. It is, however, likely that many operators will continue to use trenches for gas recovery and experience may prove these concerns to be unfounded.

On the assumption that within each site landfilled waste volumes continue to increase, the use of gas wells to control landfill gas, or indeed to extract it for use, will continue to be the most commonly adopted method. With progressive restoration and effective monitoring, the early abstraction of LFG will become standard practice and potential adverse impacts of gas on the environment should cease to exist.

Landfill sites will, increasingly, be developed into large hills or mounds above original ground contours. Gas emission potential from the surfaces of sites will assume a greater importance and, as has already happened in parts of the USA (Bogardus, 1986) standards for maximum permissable concentrations for emissions of organic vapours will be set. It is in the interests of the landfill waste management industry in every country to ensure that all current (and many completed) sites have effective gas control methods and/or prove that those sites have no problems, in order that future site operators are not bound by legislative measures and controls which, it could be argued, are a result of justifiable over reaction to today's often inadequate standards.

## CONCLUSIONS

Based on the current experience on controlling LFG migration several conclusions can be made.

The environmental consequences of landfill gas migration are substantial and may range from nuisance to damage to persons, property and vegetation.

Past practice has been very lax with respect to improving understanding and implementating action on matters related to landfill gas and its control.

It is important to understand the processes occurring within deposited and decomposing wastes and the factors which can affect those processes which, ultimately, will have an impact on the potential for generated gas to migrate.

The surface and subsurface impact of gas on the surrounding environment must be fully understood by provision of adequate monitoring facilities within and around sites.

Various methods of control are available involving both passive and actively pumped venting techniques. While passive controls may be acceptable at many shallow and old landfill sites, pumped control methods are likely to be more effective at large deep sites and may well be required as standard practice in the future.

Landfill gas migration prevention and control technology is sufficiently well established by many site operators in several countries that it can no longer be a viable defence to presume that gas migration will not occur, or that one was unaware of methods for its control.

## REFERENCES

Anderson, D.R. and Callinan, J.P. (1970). 'Gas generation and movement in landfills'. Proceedings of National Industrial Solid Waste Management Conference, University of Houston, USA.

Anon. (1983). 'Impact of the Resource Conversion and Recovery Act on landfill gas utilization' (study undertaken by SCS Engineers for US Dept of Energy at Argonne National Laboratory, Illinois, USA). ANL-CNSV-TM-117 (DE 83 016702).

Anon. (1986a). 'Energy from landfill gas'. Proceedings of conference held at Solihull, UK, sponsored by the UK and US Depts of Energy, ISBN 0-7058-1484-X.

Anon. (1986b). 'Landfilling wastes'. Waste Management Paper 26, Dept of the Environment (UK), HMSO. ISBN-0-11-751891-3.

Bogardus, E.R. (1986). 'Landfill gas legislation and control in the United States'. Presented at International Landfill Gas Symposia. GRCDA, Silver Spring, Maryland 20910, USA.

Callahan, G.P. and Gurske, R.H. (1971). 'The design and installation of a gas migration control system for a sanitary landfill'. Presented at the first annual symposium of the Los Angeles Forum on Solid Waste Management, Pasadena, California, USA.

Eliassen, R., O'Hara, F.N. and Monahan, E.C. (1957). 'Sanitary landfill gas control'. *Am City*, 115–117.

Farquhar, G.J. and Rovers, F.A. (1973). 'Gas production during refuse decomposition', *Water Air and Soil Pollution* 30, 161–175.

Fricker, J. (1986) 'The waste tip that blew up a bungalow'. *New Scientist* 110, 24.

Fungaroli, A.A. (1970). 'Instrumentation of two experimental sanitary landfills'. *IEEE Transactions on Geoscience Electronics* 8, 118–125.

GRCDA (1978). Proceedings of International Landfill Gas Symposia held annually since 1978. Organized by GRCDA, Silver Spring, Maryland 20910, USA.

Hill, C.P. (1986). 'Landfill gas migration from operational landfill sites-monitoring and prevention'. *Wastes Management* 76, 169–178.

Jones and Owen (1932). 'Some notes on the scientific aspects of controlled tipping', Manchester, UK.

Metcalfe, D.E. and Farquhar, G.J. (1987) 'Modelling gas migration through unsatured soild from waste disposal sites', *Water, Air and Soil Pollution* 32, 247–259.

Moore, C.A. (1979). 'Landfill gas generation, migration and controls'. *Crit. Rev. Environ. Control* 9, 157–183.

Rees, J.F. (1980). 'The fate of carbon compounds in the landfill disposal of organic matter', *J. Chem. Tech. Biotechnol.* 30, 161–175.

Stearns, R.P. and Beizer, M.B. (1985). 'Landfill gas is a growing concern', *World Wastes* Feb., 34–36.

Young, P.J. and Parker, A. (1983). 'The identification and possible environmental impact of trace gases and vapours in landfill gas', *Waste Management and Research* 1, 213–226.

Zimmerman, R.E., Stetter, J.R. and Altepeter, L.L. (1985). 'Process monitoring for trace sulphur and chlorine compounds in landfill generated methane'. *Anal. Instrum.* 21, 89–101.

# 6.4 Odour Emissions and Controls

F. B. FRECHEN

*HYDRO-Ingenieure, Adlerstr. 34-40, D-4000 Düsseldorf 1, West Germany*

## INTRODUCTION

Landfill sites for domestic and toxic waste show a broad variety of secondary emissions for example noise, dust, leachate and gaseous emissions. Amongst these emissions, odorous gaseous emissions often cause serious annoyance to the vicinity, especially to residential areas, and thus cause public opposition to construction and operation of landfill sites. Thus, construction and operation of landfill sites must consider minimization of nuisance and provide all available measures to reduce the release of odorous emissions.

One of the special problems concerning 'odour' is the difficulty of its measurement. As 'odour' is a human sensation, it depends on the physiological circumstances as well as the psychological influences and emotional structures of each respective person perceiving it. This causes differences in the assessment of a sample's odour when assessed by various persons and leads to the fact that a generally valid relationship between concentrations of odorous substances contained and odour impression caused cannot yet be given. Analytical identification and measurement methods ranging from tube tests to gas-chromotography coupled with multifarious detectors or a mass-spectrometer result in *substance–related* values. However, in the case of annoyance caused by malodours, these substance–related parameters are not of interest, but *effect–related* parameters gained by sensory measurement techniques are required to give a description of the current situation. Substance–related values, usually concentrations of respective air compounds, may of course be of high interest when reaching ranges which are correlated with long-term or even short-term toxic effects. But, since the annoyance caused by malodorous immissions is the point of interest here, odour will be the main subject of discussion in this chapter.

425

## MEASUREMENT TECHNIQUES

In the case of odour, only sensory measurement techniques will be applicable. Nevertheless, most of the measurements carried out at landfill sites are completed with analytical measurements by means of test tubes, flame-ionization detectors or gas chromatography. The analytical methods are well known and thus are not described here.

Sensory methods in the field of odour always involve test panels of different people who may be divided according to the subject of the measurement:

1. *Odorant concentration*: the concept behind this technique, called olfactometry, is to dilute a given air sample until the odour threshold is reached. 'Odour threshold' is defined as the condition at which odour is detected by just 50% of the test panel members. The numerical value of this dilution is the value of the 'odorants' concentration' (OC), and its unit is 'odour units per cubic meter (o.u./m$^3$)' by definition. The German VDI–guideline 3881 describes fundamentals and details of the olfactometric measurement technique.
2. *Odour intensity*: assessment by means of direct scaling, using a given range of predefined and usually numbered categories (category scaling), by means of comparing the odour with a reference odour (reference scaling) or by means of estimating in terms of magnitude (magnitude estimation). The following intensity key scale is used today:

   0—no odour
   1—barely perceivable (odour threshold)
   2—faintly perceivable (recognition threshold)
   3—clearly perceivable
   4—strong
   5—very strong
   6—stronger than 5

   This predefinement of intensity categories and the given coupling with integer numbers is supposed to be prescribed by the German VDI–guideline 3882 also, which is in discussion at the moment. This guideline will give further details concerning odour intensity and hedonic odour tone.
3. *Type of odour*: direct judgement or selection from a given list.
4. *Hedonic odour tone*: assessment of an odour within a rating scale. Hedonic rating scales describe a psychological continuum indicated in the dichotomy 'pleasant' — 'unpleasant'. A series of successive categories of response may be given, or only top and bottom of a 'thermometer' rating scale, which includes an impair quantity of points, are named. It is also

possible to use one or two standard odours as reference odours (hedonic reference scaling). The above mentioned VDI–guideline 3882 will give further details.

It must be kept in mind that each of the methods mentioned can only give one piece in the jig-saw puzzle of the complete description of the odour emission and immission situation. Whatever method is used, the results must be judged considering the whole variety of circumstances which are relevant in the respective case.

Determination of odour intensity, type of odour or hedonic odour tone may be used at the emission source as well as in surveys and questionnaires on the immission side. Olfactometry to determine the odorants concentration usually is only applicable for emission measurements and for testing air treatment devices. Olfactometry currently is the most common technique in West Germany and was used to gain the results presented in the following.

At the moment, olfactometry to determine the odorants concentration of a sample in o.u./m$^3$ according to the German VDI–guideline 3881 is the only standardized instrumental measurement technique and gives quite comparable and reproducible results. Work is carried out within the VDI–Commission on Air Pollution Prevention to standardize further methods, especially hedonic scale rating and describing the odour intensity as a function of the odorants concentration in mixtures of different dilutions, which leads to the $k_w$–factor of the Weber–Fechner law. The respective VDI–guideline 3882 is expected in 1989.

Within the measurement programs presented here, sampling on the landfill surface was done by erecting a cone with a bottom diameter of 0.5 m on the surface, waiting a few moments and then sucking the sample air into sample bags. These were analysed with the olfactometer immediately after sampling. Each of the test panel of four persons had to judge a series of seven differently diluted mixtures three times. Although for determination of the odorants concentration it was sufficient to say 'no, I don't smell anything' or 'yes, I smell something' the panel were asked to judge the respective mixtures presented, according to the intensity key scale listed above.

A calculation was made by evaluating all answers with an intensity key number of 1 and higher as a 'yes' answer, forming the 'characteristic curve' (see VDI–guideline 3881). Then the 50-percentile was calculated by means of transformation, assuming a Gaussian distribution of odour sensitivity over the logarithm of odorants concentration. Although in principal it is possible to do this calculation manually, it is recommended to use available computer software in order to save time.

It is possible to carry out 4 to 6 measurements per day when having four persons and using the described method.

## LEGAL REGULATIONS

At present the legal regulations in West Germany concerning tolerable immissions (AA, 1986a), refer to perceptibility of odour, which represents an odorant concentration of 1 o.u./m$^3$ by definition, or an odour intensity which may be described as 'clearly perceivable' (intensity number 3 in a numbered scale from 0 to 6). Recognizing the lack of knowledge about fundamentals and generally applicable objective standards, no explicit standard values are given, but at the moment it is conventional to recognize an odour immission exceeding the odour threshold of 1 o.u./m$^3$ within more than 3% to 5% of the hours of a year as illegal.

In West Germany, the Gaussian dispersion model is forced by law to do dispersion calculations, and a reference computer program exists made by order of the government.

## ODOUR EMISSIONS AT DOMESTIC WASTE LANDFILL SITES

At landfill site A, one short measurement program was carried out to get an idea of the emissions from the site surface. At landfill site B, a more complex program was executed to examine the emissions from the different parts of the site. The results of the measurements at this landfill site are shown in Table 1.

The odorant concentration is given as an average of all sample bags and as a range of individual bags relative to the average.

Using measurements from both landfill sites, it is possible to characterize typical emission concentrations in terms of odorants concentration (summarized in Table 2).

If no measurements are available, for example when the landfill site is under design, the values listed in Table 2 may be a first attempt to estimate the emission concentrations that may become relevant.

Some remarks are necessary. Measurement on top of the site at a height of about 1–1.5 m represents the odorant concentration in the air above the landfill site, provided that a distance of about 100 m between the site's border and the point of measurement, measured against wind direction, is assured. The range of odorant concentration is quite highly dependent on atmospheric influences, location of measurement point with respect to the extent of the site, respective part of landfill site, actual events during sampling, etc.

Measurements on a slightly covered site surface with the sampling cone set up loose on the bottom represent most parts of the sites examined. The range of the odorant concentration values is not large, and these values may be interpreted as resulting from the gas losses via the landfill site-surface.

**Table 1.** Results of olfactometric and analytic measurement at landfill B.

| Sampling point | Odorants concentration av./rel. range† o.u./m³ | Organic carbon ppm |
|---|---|---|
| Freshly tipped waste (s) | 1478/102% | — |
| Freshly tipped waste (s) | 407/70% | — |
| Freshly tipped waste (a) | 112/88% | — |
| Freshly tipped waste (a) distance 25 m | 60/145% | — |
| Slightly covered waste (s) | 75/60% | 1000 |
| Slightly covered waste (s) | 100/9% | 60 |
| Passive degassing (a) 5 m lee-side | 45/102% | 200 |
| Passive degassing (a) 12 m lee-side | 21/90% | — |
| Slope (s) | 264/79% | 10 000 |
| Slope (s) | 49/114% | 8000 |
| Slope (s) | 72/23% | 6000 |
| Slope (s) | — | 5000 |
| Slope (s) | 41/130% | 50 |
| Slope (a) lee-side of site | 78/283% | |
| Slope (a) lee-side of site | 62/73% | |
| Exhausted gas at blower station (25 ppm $H_2S$) | 38 834/30% | — |
| Gas exhausted from site body, 0.5 m deep (54% $CH_4$, 30% $CO_2$, 400 ppm $H_2S$) | 117 880/8% | ≫20 000 |

(s) = sampling on surface.
(a) = sampling at 1.5 m height.
† Maximum value minus minimum value divided by average.

Measurements from the landfills slopes, which were quite steep in these cases, showed a higher than average and range than those carried out on the surface. The reasons being that covers on steep slopes are more difficult to handle and the probability of fractures is much higher due to settling of the earth body. At fractures high odorant concentration values were detected, which were accompanied by methane concentrations of up to 10 000 ppm. Of course, freshly tipped waste shows quite high emission concentrations due to the biodegrading which has already taken place during collection. In addition,

**Table 2.** Typical odorant concentrations at different parts of domestic waste landfill sites.

| Part of landfill site | Odorant concentration | | Number of samples |
| --- | --- | --- | --- |
| | Average o.u./m$^3$ | Range o.u./m$^3$ | |
| Top of site, 1–1.5 m high | 60 | 20–110 | 5 |
| Site surface, slightly covered | 70 | 55–100 | 6 |
| Freshly tipped waste | 950 | 400–1500 | 2 |
| Site slopes | 90 | 40–265 | 6 |
| Original gas from site body | 118 000 | – | 1 |
| Exhausted gas | 39 000 | – | 1 |

gas is released during compaction of the tipped waste with a much higher specific volume flow rate than found on the other parts of the site. As can be seen from Table 1, a slight daily cover can reduce these emissions significantly.

Two samples of biogas formed inside the domestic waste layers were analysed. One sample was taken out of a domestic waste layer at a depth of 50 cm and may represent the gas quality as produced by biodegradation. It showed an extreme odorant concentration value and a high hydrogen sulphide concentration. The other sample was taken from the landfill's degassing system just before the burner and also showed a very high odorant concentration value as well as a considerable hydrogen sulphide concentration. Nevertheless the somewhat lower values compared to the sample mentioned before indicate different gas qualities in the respective parts of the site as well as dilution effects due to the suction of fresh air by the degassing system.

## ODOROUS EMISSIONS AT A HAZARDOUS WASTE LANDFILL SITE

Due to the problems connected with the operation of hazardous waste landfill sites only a few sites are in operation in West Germany. A measurement program was carried out at a site where mainly salt slags with aluminium content, dusts from aluminium industry and filter sludges are landfilled. It is well recognized that the odour emissions formed are highly dependent upon the chemistry of the materials deposited, circumstances of deposition, humidity, interaction between different waste compounds, etc. Thus, predicting emissions from different types of waste is very difficult and the results presented here should be taken as an example only.

As can be seen from Table 3, there are no critical emissions concerning

**Table 3.** Odorant concentrations at a hazardous waste landfill site.

| Part of landfill site | Odorant concentration | | Number of samples |
| --- | --- | --- | --- |
| | Average o.u./m$^3$ | Range o.u./m$^3$ | |
| Freshly tipped waste (salt slags) | 113 | 54–227 | 6 |
| Site surface, slightly covered | 60 | 59–61 | 2 |
| Site surface, covered, some weeks old | 250 | 50–625 | 13 |
| Same as above, after heavy rainfall (12 800 ppm NH$_3$) | 2340 | – | 1 |
| Site slope, fractured (1580 ppm H$_2$S) | 68 900 | – | 1 |
| Waste gas from leachate stores 2170–3455 ppm NH$_3$, 350–1800 ppm H$_2$S | 63 300 | 25 440– 96 100 | 7 – |

odour from the fresh waste. But after a short time a considerable increase of the odour emissions was noticed. Due to the nature of the waste an increase in humidity produced rapid formation and release of ammonia. With increasing time different compounds are produced and released, as various tube tests and GC–MS–analyses have shown. High concentrations of hydrogen sulphide as well as organic sulphides, toluol, cyclic hydrocarbons and others are present. The gas from the leachate storage, which also includes gas from active degassing from parts of the site, represents this kind of gas quality.

The problem of fracture formation is evident from the data given in Table 3. Problems mainly occur at slopes due to settling and formation of horizontal gas channels according to horizontal extent of the particular layers. In the case mentioned in Table 3, the fractures were clearly visible and around them was an area of about 50 m$^2$ which in contrast to the rest of the slope was not covered with grass.

## WASTE AIR DEODORIZATION TECHNIQUES

In order to deodorize collected gas from landfill sites, different processes are applicable:

1. thermal oxidation
2. wet gas scrubbing
3. activated carbon filtration
4. biofiltration

Usually thermal oxidation is used at domestic waste landfill sites with or without energy recovery. However, this method will be too expensive when involving gas with a very low methane concentration, as was the case at the toxic waste landfill site mentioned above. Thus, at this site the other methods were taken into consideration, and a pilot plant was operated consisting of

1. activated carbon filter 1
2. wet gas scrubber
3. activated carbon filter 2
4. biofilter with garbage/sludge-compost as filter material

This pilot plant erected and operated by the operator of the site was checked within a measurement program with different combinations of the particular devices. As biofiltration is a relatively young technique which has recently been given increasing interest due to low investment and operation costs, deodorization by using only biofiltration was studied. The results of the measurements were presented in detail by Frechen and Kettern (1986). Figure 1 shows the results of the whole plant being in operation and of only biofiltration being in operation.

At a volume flow of $800 \, \text{m}^3$ per hour when operating the whole plant (left part of Fig. 1), the theoretical times were (contact time):

| | |
|---|---:|
| — first activated carbon filter: | 13.5 s |
| — wet gas scrubber: | 1.1 s |
| — second activated carbon filter: | 13.5 s |
| — biofilter: | 223 s |

The biofilter was loaded with $14.6 \, \text{m}^3$ air per $\text{m}^2$ filter surface area and with $16.1 \, \text{m}^3$ air per $\text{m}^3$ filter material, respectively.

For only biofiltration, the right part of Fig. 1, the average contact time was 20.5 minutes, and the load of the biofilter thus was (averages):

| | |
|---|---:|
| — surface area load: | $4.4 \, \text{m}^3/(\text{m}^2/\text{h})$ |
| — specific filter material load: | $2.9 \, \text{m}^3/(\text{m}^3/\text{h})$ |
| — specific ammonia load: | $5.8 \, \text{g}/(\text{m}^3/\text{h})$ |
| — specific hydrogen sulphide load: | $3.1 \, \text{g}/(\text{m}^3/\text{h})$ |
| — specific odorants concentration load: | $172\,500 \, \text{o.u.}/(\text{m}^3/\text{h})$ |

The results are averages over a filter run period of three weeks. After six weeks of operation the biofilter efficiency decreased significantly as indicated by the numbers in Table 4.

Although hydrogen sulphide removal still was very good, it was found that the microbial activity decreased due to an overload of ammonia.

The conclusions derived from the measurement program are that when using specially treated carbon activated carbon filters show high efficiency, but are high in operation costs. The wet gas scrubber was not too effective, but it

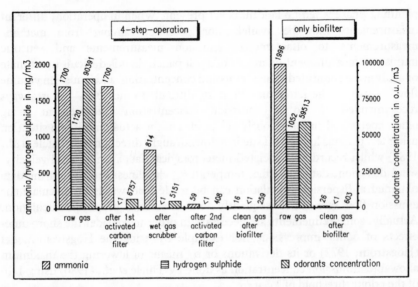

**Figure 1.** Deodorization of exhaust gas from toxic waste landfill.

are high in operation costs. The wet gas scrubber was not too effective, but it may be assumed that it could achieve better efficiency when design, especially size, and operation are optimized. Biofiltration using garbage/sludge-compost was useful as a final step of the plant as well as for sole treatment technique, but must be optimized according to the respective waste air to prevent breakdown of the biological degradation processes caused by particular waste air compounds. Further research concerning design and operation as well as filter materials seems to be very promising.

## PREDICTION OF ODOROUS EMISSIONS AND IMMISSIONS

Prediction of odorous emissions and immissions are needed at various stages of the life of a landfill, often when specific data is scarce, for example at the

**Table 4.** Raw gas and clean gas quality of the biofilters after six weeks of operation.

|  | Raw gas | Clean gas |
|---|---|---|
| Ammonia mg/m$^3$ | 1700 | 460 |
| Hydrogen sulphide mg/m$^3$ | 1119 | 1 |
| Odorant concentration o.u./m$^3$ | 96 090 | 4350 |

planning phase when no landfill is yet present. When in operation, different measurements may be available or possible, ranging from methane measurements to olfactometric emission measurements and sensoric immission measurement by means of a test panel. As it is difficult to estimate or analyse the connected values 'emission concentration' and 'emission volume flow' — where the latter may be most difficult to measure — it is always recommended to do an emission concentration measurement using olfactometry and simultaneously get an idea of the actual immission situation using test persons located off site in a downdraught direction (see Thiele *et al.*, 1986) whilst recording associated meteorological data such as wind velocity, wind direction, date and time, temperature, cloudiness, etc. From the data obtained a dispersion calculation can be made 'backwards' to calculate the emission flow rate and subsequently do an immission prognosis calculation. Actually, some refinements are necessary in order to consider the short-time-effects of odour impressions, for example by using the Högström-model (Högström, 1972) or its derivations or by means of lowering the maximum allowable immission concentration as done by Thiele *et al.*, (1986) using 1/10 of the odour threshold of 1 o.u./m$^3$.

Based on current experiments there are some recommendations to be used at the planning phase, when no real measurements are available yet. If in subsequent stages data are available, then the recommendations should subsequently be replaced by the real data to improve the accuracy.

From the viewpoint of odour emissions, landfill sites may be divided into characteristic parts:

1. The face area, where the waste is tipped
2. The site surface, more or less covered
3. Slopes
4. Leachate stores, passive degassing installations

As a first approach during the planning phase, the following remarks may be relevant:

*Face area.* Odorant flow rate (OFR in o.u./h) is dependent upon hourly waste amount tipped (W in m$^3$/h), extent of the area, characteristics and history of the waste, etc. Assuming a daily coverage of this area, which also leads to a small area and a point source characteristic, as a first step the OFR may be calculated as

$$OFR = k \times OC \times W$$

Several *k*-factors can be found considering diurnal, weekly or annual amount of hours of operation, but it is not yet obvious which one leads to 'true' results, especially when it is considered that the immission concentration also depends upon the dispersion model used. Having no better data, we would recommend a $k$-factor of 1–2 and an odorant concentration (OC) of about 1000 o.u./m$^3$ unless measurement indicates a value differing significantly.

*Covered site surface.* Usually the biggest portion of the landfill, a calculation may be done by using the specific gas loss ($q$ in m$^3$/(m$^2$/h)) and area ($a$ in m$^2$) forming

$$OFR = OC \times q \times a$$

The value of $q$ can be gained from calculations considering the age and thickness of waste layers, biogas formation depending upon waste characteristic, degassing efficiency and others. In an example based on a case study values were in the range of 0.005 m$^3$/(m$^2$/h) and an odorant concentration value of 200 000 o.u./m$^3$ for genuine biogas formed was obtained. If a different gas quality is detected due to different wastes or a different test panel, then this value should be replaced in order to get corresponding immission estimates.

*Slopes.* In principle, the same values and remarks as mentioned before are relevant. But in fact the question has to be answered how to minimize fracture formation with subsequent direct gas losses. According to the reported measurements, the value of $q$ can be estimated perhaps 2–10 times higher than the one relevant for covered surface when having steep slopes and no special covering is used.

*Leachate stores, passive degassing installations.* It is impossible to give overall values for these installations. It is recommended to connect the leachate facilities to the active degassing system and to replace passive degassing by active degassing. If this is not the case, these facilities represent point sources with high OFR values. At the hazardous waste landfill site, a very rough estimation was made leading to an OFR value of 3000–5000 o.u./m$^3$ per hour served by the respective leachate store.

## FINAL REMARKS

The above mentioned remarks show possible points of measurement to be made against odour emissions, but in general, it must be stated that 'odour' is a very complex topic covering different disciplines. The statements given here may be evaluated as an approach from a more technical point of view, and other related aspects such as assessment of immissions with regard to sociology or jurisprudence must always be taken into consideration to get a complete view of the situation.

## REFERENCES

AA (1986a). 'Durchführung der Technischen Anleitung zur Reinhaltung der Luft', 14.10.1986, MBl. NW, No. 88, 17 November 1986, pp. 1658.

AA (1986b). 'Technische Anleitung zur Reinhaltung der Luft', 1.3.1986, GMBl, 27 February 1986, pp. 95.

Frechen, F.–B. and Kettern, J.T. (1986). 'Reduction of odorous emissions

# 6.5 Assessment Techniques for Gas Emission and Dispersion from Waste Landfills

S. CERNUSCHI and M. GIUGLIANO

*Institute of Sanitary Engineering, Polytechnic of Milan,*
*Via Fratelli Gorlini 1, I-20151 Milano, Italy*

## INTRODUCTION

Gas generation in waste landfills emerges from biological processes, waste volatilization and chemical reactions. Control of these emissions has focused principally on explosive hazards due to methane migration and accumulation in man-holes and house basements around the landfill, and on nuisance due to the odorous trace compounds that can lead to justifiable complaints from local residents.

Recently some concern has been expressed regarding the health risks associated with trace constituents of the gas landfill, even if adverse effects on human health have not been documented (Walsh *et al.*, 1988). Trace organic compounds, many of which are considered toxic, carcinogenic, mutagenic and teratogenic, have been identified in landfill gas (Shen, 1981; US-EPA, 1981; Young *et al.*, 1983; Vogt *et al.*, 1984; Rettenberger, 1987). These studies confirm that the air pollution aspects, which until now have received very little attention, can have an important role in the environmental impact assessment of the waste landfill.

Waste disposal in landfills, particularly industrial wastes volatilizing over time, can be considered actual air pollution sources and treated with suitable dispersion models describing the gas distribution around the landfill site.

This chapter considers the problems of this particular source and evaluates the performance of techniques available for the estimation of the emission and the dispersion of landfill gas in the environmental impact assessment.

## CHARACTERIZATION OF LANDFILL GAS

The composition of landfill gas is highly dependent upon the type of waste disposed. The gas produced by the decomposition of municipal waste consists principally of methane (approximately 60–65%) and carbon dioxide (approximately 30–35%) with over 100 trace components, the concentration

_...wastes landfills, reflect obviously the presence of volatile components of the waste disposed.

In Table 1 the characteristics of gas from nine landfills are reported with the corresponding literature references. Table 2 shows the principal gas trace

**Table 1.** Characteristics of the landfill sites whose trace gas components are reported in Table 2.

| Site | Type of waste | Average age of waste | No. of locations on site sampled | References |
|------|---------------|----------------------|----------------------------------|------------|
| A | Domestic (crude) | 5–6 years | 2 | Young et al., 1983 |
| B | Domestic (baled) | 7 months | 1 | Young et al., 1983 |
| C | Domestic (pulverized) | 3 weeks | 1 | Young et al., 1983 |
| D | Domestic (commercial demolition) | 5 years | 1 (3 readings) | Walsh et al., 1988 |
| E | Industrial | Not specified | 4 | Rettenberger, 1988 |
| F | Domestic | 12 years | 2 | Höjlund et al., 1988 |
| G | Industrial | 6 months | 3 | Höjlund et al., 1988 |
| H | Industrial | 15 years | 1 | Höjlund et al., 1988 |
| I | Industrial | 7 years | 2 | Höjlund et al., 1988 |

**Table 2.** Principal trace gas components of domestic and industrial landfills.

| Trace components | Site A | Site B | Site C | Site D | Site E | Site F | Site G | Site H | Site I |
|---|---|---|---|---|---|---|---|---|---|
| Benzene ($mg\ m^{-3}$) | 8 | 42 | 95 | 35.5 | 12–14 | | | | |
| Toluene ($mg\ m^{-3}$) | 36 | 72 | 54 | 13 | 43–95 | | | | |
| Xylenes ($mg\ m^{-3}$) | 69 | 77 | 120 | 27.4 | 15–57 | | | | |
| Propylbenzenes ($mg\ m^{-3}$) | 138 | 94 | 32 | | | | | | |
| Butylbenzenes ($mg\ m^{-3}$) | 21 | 63 | 105 | | | | | | |
| Limonene ($mg\ m^{-3}$) | | 230 | | | | | | | |
| Trichloroethylene ($mg\ m^{-3}$) | 3 | | 30 | 8.6 | 6–13 | 0.5 | | 0.6 | |
| Tetrachloroethylene ($mg\ m^{-3}$) | | | | 6.2 | 3–8 | | | | |
| Vinylchloride ($mg\ m^{-3}$) | | | | 8.4 | 14–50 | | | | |
| Hydrogen sulphide ($mg\ m^{-3}$) | | | | | | 48–52 | | 340 | |
| Methanethiol ($mg\ m^{-3}$) | | | 87 | | 12 | | | | |
| PAH ($\mu g\ m^{-3}$) | | | | | | | 2–136 | | 0.6–1.0 |
| Chlorobenzenes ($\mu g\ m^{-3}$) | | | | | | | 2–63 | | 0.1–0.3 |
| Mercury ($\mu g\ m^{-3}$) | | | | | | 183–280 | 0.9–3.5 | 0.6 | 0.6–1.2 |
| Cadmium ($\mu g\ m^{-3}$) | | | | | | 1.3 | | 0.2 | |
| Lead ($\mu g\ m^{-3}$) | | | | | | 18–68 | | 7 | |
| Zinc ($\mu g\ m^{-3}$) | | | | | | 20–120 | | 8 | |
| $TCDD_e$[1] ($ng\ m^{-3}$) | | | | | | | <0.1 | | <0.2 |
| 7CDD/F–8CDD/F[2] ($ng\ m^{-3}$) | | | | | | | 0.1–0.5 | | <1.0 |

[1] Tetrachlorodibenzodioxins equivalent (Eadon method).
[2] Heptachloro- and octachloro-dioxins and furans.

components which could have environmental significance as toxic and odorous compounds.

The major contribution to the toxicity, confirmed also by other concentration data not reported in Table 2 (Young *et al.*, 1983), derives from the presence of benzene, vinylchloride and mercury vapours in the gas from industrial landfills. Mercury vapours are present in appreciable quantities at landfill sites where measurements were performed, with ambient air concentrations increasing with temperature (Bergvall *et al.*, 1988).

The toxic concerns for domestic landfills are certainly less, but are an important contribution to the nuisance of the odorous compounds for this type of landfill, due to the presence in the gas of thiols, alkylbenzenes and limonene. These substances require individually a dilution of more than 1000-fold to reach the odour threshold.

## EMISSION RATES ESTIMATION

The evaluation of the emission rate of gaseous pollutants from waste sites is a very difficult problem, due to the high number of factors affecting the emission process. Emission rate values can be obtained, in principle, with two different approaches:

1. Utilization of theoretical and/or empirical models of gas generation and migration processes through the landfill
2. Calculation of the emission from the measurement of the pollutant concentration immediately above and/or in the surroundings of the waste site.

For municipal solid waste landfills, the sizeable body of scientific data that has become available over the last few years on the production and migration characteristics of gas has led to the proposal of some empirical interpretation schemes of the processes involved. However, a comprehensive model relating quantitatively the complex interactions between the decomposition processes and the physical environment, and also applicable for the evaluation of the emission of pollutants, has not yet been proposed.

The quantity of gas dispersed to the atmosphere through the top cover of the landfill depends, essentially, on the gas production rate, on its migration properties through the waste deposited and through the top layer of the site, on the collection efficiency of the gas extraction system and on the factors affecting the transfer of the gas from the exposed area to the atmosphere. Gas production rates are strongly dependent on the organic matter content of the refuse and its biodegradability and on the moisture and the temperature inside the landfill, with a significant decrease over time. Theoretical predictions and laboratory scale studies can give only approximate values, whose comparison with data obtained from field studies is made difficult by the problems related to the physico-chemical characterization of the landfill environment and to the age of the refuse disposed in different landfill areas. Total specific gas productions reported in the literature for MSW (Stegmann, 1988; Hoeks, 1983; Shen, 1981) cover a relatively broad range, between $120 \, m^3 \, t^{-1}$ and $300 \, m^3 \, t^{-1}$; specific production rates can be as high as $15$–$20 \, m^3 \, t^{-1} \, year^{-1}$ in the early stages of the decomposition process (up to the first 5 years), with a decrease to values in the range $4$–$8 \, m^3 \, t^{-1} \, year^{-1}$ in the successive $5$–$30$ years. Mean rate values derived from full-scale landfills in U.S.A. (Stegmann, 1988), for a period of $10$–$20$ years, are between 3 and $7 \, m^3 \, t^{-1} \, year^{-1}$.

The migration through the landfill of the gas produced, and its ultimate release to the atmosphere, are direct consequences of the pressure and concentration gradients of the gas inside the landfill. Movement of the gas is dependent on many factors, difficult to include in a simulation model, related to transport properties of the gas itself (diffusivity, viscosity), to physical and chemical characteristics of the waste (permeability, moisture content, temperature) and to the layout and efficiency of the gas collection system. Pressure is exerted by the gas in all directions inside the landfill, so its movement is not restricted to the vertically upward direction. However, the gas generally tends to migrate preferentially upward to the cover layer (Shen, 1981) since the lateral movement will often be restricted by the static pressure within the landfill and the higher flow resistance of the surrounding soils and, eventually, of the bottom and lateral liners.

Release of gas to the atmosphere through the top layer of the landfill is also dependent on some of the factors that influence the migration inside the landfill, like the transport properties of the gas and the permeability, moisture content and thickness of the cover. The release is also affected by some factors related to atmospheric meteorological conditions, mainly the wind speed (increased wind at the surface speeds up diffusion), barometric pressure fluctuations (pumping action from pressure fluctuations enhances diffusion) and air temperature (larger temperature differences between the interior of the landfill and the outside top surface favour thermal diffusion).

As stated previously, a model including all the aspects related to the gas production, migration and emission in MSW landfills is still not available. Emission rates can thus be evaluated, as a first approximation, utilizing mean values for the gas production rate ($m^3$ of gas per t of refuse per year), and assuming a release to the atmosphere, through the top cover of the landfill, of all the gas that is not collected by the extraction system. This requires, in turn, knowledge of the gas collection efficiency of the extraction system, dependent on the characteristics of the system itself and on the particular landfill. Limited data reported in the literature (Esposito, 1984) indicate a range of efficiency between 23% and 56%, with mean values around 40%.

Emission rates $E$ ($m^3\ m^{-2}\ year^{-1}$) can thus be evaluated with the following equation:

$$E = (G)(\gamma_R)(1 - \eta)(L) \tag{1}$$

where $G$ is the mean specific gas production rate ($m^3\ t^{-1}\ year^{-1}$) $\gamma_R$ is the specific weight of the waste ($t\ m^{-3}$), $\eta$ is the gas collection efficiency of the extraction system and $L$ (m) is the depth of the landfill. Utilizing the mean values for the production rate ($4$–$8\ m^3\ t^{-1}\ year^{-1}$) and the extraction efficiency (40%) previously reported, Equation (1) results in emissions between 38 and $76\ m^3\ m^{-2}\ year^{-1}$ ($0.004$–$0.008\ m^3\ m^{-2}\ h^{-1}$) for a refuse density of $0.8\ t\ m^{-3}$ (Findikakis *et al.*, 1979), and a mean landfill depth of 20 m. The values are comparable with those reported in the literature. Emission rates $E_U$ ($g\ m^{-2}\ s^{-1}$) for the compounds of interest can then be evaluated from the concentration $C$ ($g\ m^{-3}$) of the compound in the gas (see first chapter) and the emission rate $E$.

Although the approach outlined gives only approximate values, it should be considered conservative for atmospheric dispersion calculations: neither the gas remaining inside the landfill nor the gas that escapes from its lateral boundaries are substracted from the values obtained.

For industrial waste landfills, the emission of gaseous compounds is primarily determined by the volatilization rate of the compound and its migration properties through the waste and the landfill top cover to the atmosphere.

The rate of the chemical waste volatilization at landfill sites is dependent upon the physical and chemical properties of the waste and the surrounding environment. The processes involved in the volatilization are essentially three (Shen *et al.*, 1980): volatilization of an organic liquid from a pure solution or a mixture of chemicals, volatilization from a water solution and volatilization from soils or solids on which the compound is adsorbed. The first process is strongly influenced by the vapour pressure of the compound, and hence by the temperature, and by its molecular diffusion through the vapour phase close to the chemical or mixture of compounds. For the second process, water solubility also plays an important role: compounds with low vapour pressure but also with limited solubility in water can be vaporized in rates comparable with those of chemicals with vapour pressures several orders of magnitude higher but readily soluble in water. In the third process, the adsorption capacity of the solid, the strength of the adsorption between the compound and the solid and the effective surface area for the desorption process are all important in determing the volatilization rate.

In a chemical waste landfill, a complex combination of the three processes mentioned is involved. The rate of diffusion of the vaporized compound

where $C_m(z)$ is the average concentration obtained during sampling at the different heights $z$ and $C_0$ is the concentration at surface. Equation (6) contains two unknowns $C_0$ and $b$, whose values can be determined by linear regression, provided $C_m(z)$ has been measured at three or more heights. The height of the plume $Z_b$ can be estimated from the vertical dispersion parameter $\sigma_z$:

$$Z_b = 2.15\sigma_z \tag{7}$$

where $\sigma_z$ is a function of the atmospheric stability, determined from surface observations of wind speed and insolation, and of the maximum downwind distance between the leading edge of the source and the sampling path.

Substitution of Equations (5) and (6) in Equation (4) leads to the following expression:

$$Q = C_o W U_o \int_0^{Z_b} (z/10)^p (1 - z/Z_b)^b \, dz \tag{8}$$

which is readily solvable for $Q$.

The technique described has been applied successfully to the evaluation of total reduced sulphur emission rate from an aerated effluent lagoon, yielding values with a standard deviation of 15% over 4 different tests, conducted with significant variations in the wind speed ($1.0$–$2.7 \, \mathrm{m \, s^{-1}}$) and direction and in the estimated plume boundary height ($60$–$400 \, \mathrm{m}$). In comparison with other methods that rely on field measurements, the technique appears to be relatively simple and reasonably accurate, although it requires a careful measurement of the wind speed and direction during sampling and the availability of the atmospheric stability that cannot be readily obtainable.

## ATMOSPHERIC DISPERSION MODELLING

The easiest way to estimate pollutant ground level concentrations downwind of area sources is by the use of the Gaussian model for a point source, considering all the emission concentrated in the centre of gravity of the area:

$$C(x,y) = Q(\pi \sigma_y \sigma_z u)^{-1} \exp\left[-\tfrac{1}{2}(y/\sigma_y)^2\right] \tag{9}$$

where $C(x,y) \, (\mathrm{g \, m^{-3}})$ is the ground level concentration at downwind distance $x$ and crosswind distance $y$ from the area centre of gravity, $Q$ is the mass emission rate ($\mathrm{g \, s^{-1}}$), $\sigma_y$ (m) and $\sigma_z$ (m) are the horizontal and vertical standard deviation of the plume concentration and $u$ ($\mathrm{m \, s^{-1}}$) is the wind speed. Equation (9) further considers the pollutant emitted at ground level: the assumption is quite reasonable for landfill emission, characterized by gas velocities and temperatures which result in plume rise values that can be neglected in dispersion calculations.

The approximation of an area source with a ground level point source, with the total emission located in the centre of gravity, results in significant overpredictions of the concentration, especially for receptors placed near the area boundaries. Area sources have lateral dimensions and, consequently, give rise to a horizontal spreading in the emission plume already at the source that is not taken into account by simply considering all the source concentrated in the area centre of gravity.

The approach commonly utilized for the inclusion of the initial concentration spreading still considers the area source as a point source, but locates it at a distance upwind of the area boundary. This virtual distance is then added to the real source–receptor distance for the evaluation of the horizontal and vertical dispersion coefficients $\sigma_y$ and $\sigma_z$ utilized in Equation (9). The area source is, in practice, approximated with a virtual upwind point source, whose distance from the real source is chosen to give to the plume of the virtual point source the lateral and vertical dispersion that it already has at the effective source location. The distance is dependent on atmospheric stability, and is evaluated from the relationships that apprxomately fit the Pasquill–Gifford curves for $\sigma_y$ (m) and $\sigma_z$ (m) (Turner, 1969):

$$\sigma_y = px^q \tag{10}$$

$$\sigma_z = ax^b \tag{11}$$

where $p$, $q$, $a$ and $b$ are stability dependent coefficients and $x$ is the distance (km). The lateral and vertical distances, $x_y$ and $x_z$ respectively, are then given by:

$$x_y = (\sigma_{y_0} \rightleftarrows p)^{1/9} \tag{12}$$

$$x_z = (\sigma_{z_0} \rightleftarrows a)^{1/6} \tag{13}$$

where $\sigma_{y_0}$ (m) and $\sigma_{z_0}$ (m) are the standard deviations of the initial lateral and vertical plume concentration distributions at the real source. For a surface based source the initial lateral dispersion $\sigma_{y_0}$ is set equal to the length $S$ (m) of the side of the source divided by 4.3 (U.S. EPA, 1987):

$$\sigma_{y_0} = S/4.3 \tag{14}$$

and the initial vertical dispersion $\sigma_{z_0}$ is assumed to be equal to the vertical dimension $z$ (m) of the source divided by 2.15 (U.S. EPA, 1987):

$$\sigma_{z_0} = z/2.15 \tag{15}$$

The area source is required to be square. Sources with irregular shape, that constitute the most frequent situations, can be simulated by subdividing them into multiple squares that approximate the geometry of the total area. However, the evaluation of the ground concentration must then be conducted for every single source. A more simple, if less exact, approximation utilizes the square root of the area as the length $S$ of the side, thus requiring the evaluation of the dispersion from just a single source.

The concept of a virtual upwind point source is included also in a dispersion equation for a surface-based area source recommended by the U.S. EPA (Baker *et al.*, 1985) for assessments of air pollution from landfills. The equation is derived on the basis of a uniform horizontal concentration distribution in every wind-rose sector $\theta$, commonly utilized in long term climatological models, with a virtual distance $L$, equal to 2.15 times the crosswind width of the landfill, added to the real downwind distance $x$ from the landfill centre to the receptor:

$$C(x, \theta) = 16Q[2^{1/2} \pi^{3/2}(x + L) \sigma_z u]^{-1} \qquad (16)$$

where 16 is the number of sectors in which the wind-rose is usually divided.

The initial area source plume dispersion can also be taken into account by considering the source as a finite crosswind line source (U.S. EPA, 1987). Initial lateral plume spreading is included by considering the source as a line source, where the initial vertical spreading is simulated with the addition of a vertical virtual distance $x_z$, equal to the length $S$ of the side of the area source, in the $\sigma_z$ evaluation with Equation (11). Ground level concentration $C(x, y)$ for a surface-based source can then be calculated with the following equation:

$$C(x, y) = QS[(2\pi)^{1/2} u\sigma_z] \\ \times \{\mathrm{erf}[(0.5S' + y)(2^{1/2} \sigma_y)^{-1}] + \mathrm{erf}[(0.5S' - y)(2^{1/2} \sigma_y)^{-1}]\} \qquad (17)$$

where erf is the error function (Burington *et al.*, 1970) and $S'$ is the effective crosswind width of the source, defined as the diameter of a circle whose area is the same as the area of the source:

$$S' = 2S(\pi)^{-1/2} \qquad (18)$$

The approach also requires the area source to be square: for irregular sources, the same considerations previously outlined can be applied.

Virtual point source models appear to be primarily indicative for the evaluation of ground-level air pollution from landfills. A comparative study conducted on a hazardous waste landfill (Baker *et al.*, 1985) showed a more accurate estimation of the ambient concentrations monitored in the surrounding of the site when a virtual source model was applied. Average concentrations, obtained over a 10 day period, resulted in over-estimation in the range 8%–97%, compared with an overprediction of 374% obtained with the simple point source model. Virtual source models also appear to be more precise, with the smallest values of scatter around the mean values.

As stated before, emissions from landfills typically contain compounds characterized by short-term, acute effects (i.e. odorous substances) and by long-term cumulative effects (i.e. toxic trace substances). The assessment of the air quality impacts of the two categories of substances must then be conducted utilizing different approaches in the application of the area source dispersion models.

For compounds with acute effects, short-term versions of the models should be applied. This models calculate hourly concentration values and, consequently, require meteorological data available on the same time basis. For toxic compounds with cumulative effects, long term models, which calculate mean annual concentrations, must be applied. The models utilize statistical summaries of meteorological data in the form of joint frequency tables of wind speed, wind direction and atmospheric stability: each entry in the table represents the time fraction with which the considered combination of wind speed, wind direction and stability occurs in a year. Mean annual concentrations at the receptor are then simply evaluated by summing up the concentrations corresponding to the particular wind speed and direction and stability combination multiplied by the corresponding frequency of occurrence of the combination.

## EXAMPLE OF IMPACT ASSESSMENT OF GAS LANDFILL EMISSIONS ON THE ATMOSPHERIC ENVIRONMENT

The evaluation of the impact on air quality due to emissions arising from landfills makes use, as previously outlined, of a multidisciplinary approach.

The identification and quantification of the emission of compounds developing short term effects (typically odorous substances like thiols, alkylbenzenes, limonene, etc.) is carried out from the type of the wastes disposed of. In the same way, the emission of trace substances with long term effects (typically toxics like benzene, organohalogenated compounds, mercury vapours, etc.) should also be evaluated, particularly for industrial wastes disposal sites.

The atmospheric transport and diffusion of pollutants with long term effects is described with mathematical dispersion models in climatological versions, suitable for area sources. The concentrations are calculated on a long term basis (for example, annual mean values), with the meteorological data usually required in input as the joint frequency of occurrence of the several combinations of wind velocity, wind direction and atmospheric stability. The influence of the mixing layer on the vertical diffusion can normally be neglected since the source is located at ground level (U.S. EPA, 1987). In Fig. 1 the mean annual concentration contours of benzene in the surroundings of a MSW landfill are reported. The estimation of atmospheric

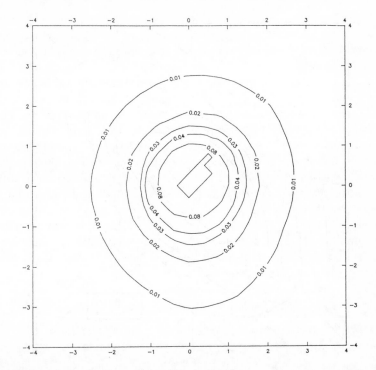

**Figure 1.** Mean annual concentration contours of benzene ($\mu$g m$^{-3}$) in the surroundings of a MSW landfill. Map coordinates are in km; the landfill boundaries are reported in the centre of the graph.

diffusion was conducted utilizing the ISCLT dispersion model of U.S. EPA (U.S. EPA, 1987); the emission of benzene was evaluated by Equation (1), on the basis of a 60% efficiency of the gas collection system and a gas concentration of benzene of 35.5 mg m$^{-3}$ (Walsh *et al.*, 1988).

For pollutants with short term effects, the atmospheric dispersion should be evaluated with models resulting in short term concentration estimates (for example, mean hourly values). A screening model can be utilized in advance for the recognition of the critical combinations of wind velocity and atmospheric stability for the area of concern. The short term dispersion model, applied with such critical meteorological data as input, gives the concentration contours determined by every critical combination recognized. The number of hours per year in which the estimated concentration is expected to occur can then be obtained from the joint frequency table of meteorological data, already utilized for long term evaluations, determining the occurrence of the critical

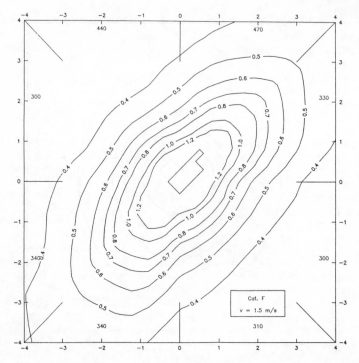

**Figure 2.** Mean hourly concentration contours ($\mu$g m$^{-3}$) of propylbenzene (tracer of odorous compounds) in the surroundings of a MSW landfill, for a wind velocity of 1.5 ms$^{-1}$ and an F stability category. Map coordinates are in km; the landfill boundaries are reported in the centre of the graph.

wind velocity and stability combinations for every wind direction considered. In Fig. 2 the concentration of propylbenzene (considered as the tracer of odorous compounds) around a MSW landfill for a critical combination of wind velocity and atmospheric stability, is reported in terms of mean hourly concentration contours. The number appearing in each one of the eight sectors in which the wind direction has been divided gives the hours per year of occurrence of the concentration reported in every sector. The atmospheric dispersion calculations were conducted with the ISCST model of U.S. EPA (U.S. EPA, 1987). The emission of propylbenzene from the landfill was evaluated with the same approach utilized for the long term pollutant case previously illustrated, utilizing a concentration value in the gas of 69 mg m$^{-3}$ (Young *et al.*, 1983).

## REFERENCES

Baker, L.W. *et al.* (1985). 'Screening models for estimating toxic air pollution near a hazardous waste site'. *JAPCA*, **35**, 1190–1195.
Bergvall, G. *et al.* (1988). 'Measurement of mercury vapour emissions from Swedish waste landfills'. In: Proc. 'ISWA '88 – 5th. International Solid Waste Conference', Copenhagen (Denmark), 11–15 September, 56–60. Academic Press.
Burington, R.S. *et al.* (1970). 'Handbook of probability and statistics with tables'. 2nd ed., McGraw-Hill Book Co., New York (USA).
Esplin, G.J. (1988). 'Boundary layer emission monitoring'. *JACPA*, **38**, 1158–1161.
Esposito, A. (1984). 'Recupero energetico del biogas prodotto da una discarica controllata di rifiuti solidi urbani'. Proc. of CISPEL Course 'Il recupero energetico dai rifiuti', Gardone Riviera (Bs), 8–12 October.
Findikakis, A. *et al.* (1979). 'Numerical simulation of gas flow in sanitary landfills'. *Journal ASCE*, **105**, 927–945.
Hoeks, J. (1983). 'Significance of biogas production from waste tips'. *Waste Management and Research*, **1**, 323–335.
Höjlund, A. *et al.* (1988). 'Environmental impact of landfill gas in Sweden'. Proc. ISWA Seminar 'Can landfill technology be improved? — Current technology and environmental aspects', ENVIRO 88, Amsterdam (Netherlands), 19–23 September.
Rettenberger, G. (1987). 'Composti in tracce nel gas da scarichi controllati'. In: Proc. of 'Simposio internazionale sullo scarico controllato', Cagliari (Italy), 19–23 October, IX-1–IX-14.
Shen, T. *et al.* (1980). 'Air pollution aspects of land disposal of toxic wastes'. *Journal ASCE*, **106**, 211–226.
Shen, T. (1981). 'Control techniques for gas emissions from hazardous waste landfills'. *JAPCA*, **31**, 132–135.
Stegmann, R. (1988). 'Landfill gas as an energy source'. Proc. Int. ISWA Conf. 'Valorization des dechets – Aspects economiques', Paris (France), 22–24 April, 311–322.
Turner, B. (1969). 'Workbook of atmospheric dispersion estimates'. U.S. Public Health Service, Cincinnati, USA.
U.S. EPA (1981). 'Evaluation guidelines for toxic air emissions from land disposal facilities'. Office of Solid Waste, Washington, D.C. (U.S.A.).
U.S. EPA (1987). 'Industrial Source Complex (ISC) dispersion model user's guide'. 2nd ed. (revised), U.S. EPA 450/ 4-88-002a.
Vogt, W.G. *et al.* (1984). 'Volatile organic compounds in gases from landfill simulators'. GRI-84/0144, Gas Research Institute, Chicago, USA.
Walsh, J.J. *et al.* (1988). 'Control of volatile organic compound emissions at a landfill site in New York: a community perspective'. *Waste Management and Research*, **6**, 23–34.
Young, P.J. *et al.* (1983). 'The identification and possible environmental impact of trace gases and vapours in landfill gas'. *Waste Management and Research*, **1**, 213–226.

## 6.6   Unsaturated Zone Attenuation of Leachate

HOWARD ROBINSON

*Aspinwall and Company Ltd, 5 Swan Hill Court,*
*Shrewsbury, SY1 1NP, UK*

### INTRODUCTION

The United Kingdom is at odds with many of its European and international neighbours in continuing to place great reliance on landfills where waste degradation products, including leachates and gases, are allowed to percolate downwards, and out of landfill sites. Success of this strategy requires that in passing through the unsaturated zone beneath the site, and on mixing with ground-water, natural processes act to improve the quality of leachate to an extent where any contamination of the ground-water is minimal and acceptable.

In the past, these processes were known collectively as 'dilute and disperse', but this term has been replaced in recent years by the phrase 'dilute and attenuate', as the importance of processes other than dilution has been recognized. Such processes include both biological and physicochemical reactions, and can occur most effectively where leachates seep slowly through intergranular unsaturated zones beneath sites (Anon., 1978).

### TECHNICAL BASIS FOR THE 'DILUTE AND ATTENUATE' CONCEPT

Although historically these processes of attenuation may well have been fortuitous in many instances, since the mid-1970s the philosophy has been based on an increasing body of technical research and experience. The period from 1973 to 1977 witnessed the most intensive programme of landfill research ever conducted in Britain. The Department of the Environment's study 'Co-operative Programme of Research on the Behaviour of Hazardous Wastes in

Landfill Sites', published in 1978 as the report now termed 'The Brown Book' (Anon., 1978), proposed the benefits of intergranular unsaturated zones most clearly. Its main conclusion was that pollution plumes beneath and around landfill sites were in general extremely limited in extent by attenuating processes, even beneath sites which by today's standards were poorly managed, or in many instances overloaded with hazardous wastes. Nineteen landfill sites were investigated, on various strata and containing a variety of wastes. Innovative site investigation techniques were developed which are widely used today. Such detailed site investigations, monitoring and assessment have become standard practice, both at proposed new sites and at existing landfills. It is true to say that, in spite of this much higher level of site monitoring, which waste disposal and water authorities have required during the last ten years, and increasing analytical expertise, serious ground-water pollution by leachate has no more been detected now than it was then.

However, it is a fact that in Britain, as sites contain much larger, more compacted accumulations of waste, with higher potential for pollution, and as demands on ground-water increase, more new sites are being designed on a containment basis. This is not to say that all such sites will be necessarily operated satisfactorily. Control of leachate, gas, restoration, and settlement are all still required. In Britain, there is a wide range of geological formations which have been quarried and may be suitable to receive wastes. These include sands, gravel, silts, chalk, clay, limestones and granites. The wide range of different problems which these pose means that there is no single landfill design which can be universally applied. What is essential in each instance is detailed and professional site assessment, followed by a high quality, site-specific engineered landfill design. This must be good enough to demonstrate with a high degree of confidence that the site will not have an adverse impact on the environment. It is not an acceptable strategy to dismiss the philosophy of 'dilute and attenuate' for political rather than for sound technical reasons at any specific site. However, what such assessment requires are scientific data which can be applied and used for decision-making.

## ATTENUATION PROCESSES

Results from site investigations and assessments continue to demonstrate the remarkable attenuation capacity of many natural materials (e.g. Tester and Harker, 1982; DoE, 1978), and the ability of physical, chemical and biological processes (including dilution) both within and below landfills, to reduce levels of contaminants. It is, however, more of a problem to quantify and assess the efficiency and reliability of these processes, and allow decisions to be made regarding assessment of potential landfills in varying geological conditions.

Some workers have identified that particular strata are less well suited to provide attenuation of contaminants. For example, cation exchange and buffering capacity are limited in the triassic sandstone strata of the UK Midlands (Harris and Lowe, 1984). This can mean that dissolution of carbon dioxide from landfill gas, and generation of volatile fatty acids in leachate, can reduce pH-values such that microorganisms which might effect biodegradation are inhibited. Despite this, significant degradation of volatile fatty acids has been observed in recent work (Blakey and Towler, 1987) beneath a major landfill on this strata, although it seems certain that a much greater depth of unsaturated zone would be needed in this situation to attenuate a given flux of contaminants, than in other more suitable environments.

A significant futher benefit of an unsaturated zone is in controlling the seepage of leachate into the saturated zone, such that the residual contaminant load is balanced by a continuous supply of oxygen and dilution which may well be available in the groundwater underflow. Unsaturated zone calculations, together with estimates of flow rates, must therefore be coupled with determinations of the supply of oxygen, dilution and addition attenuation available within the saturated zone, to allow a total estimate to be made of the overall degree of protection afforded to adjacent aquifers.

If a landfill is to be allowed, it must be possible to demonstrate with a reasonable degree of confidence that both unsaturated and saturated zones possess properties which can be measured, and which are at high enough values to indicate that the likelihood of a detectable impact on ground-water resources occurring is minimal and acceptable. However, leachate/rock interactions are complex and variable, so that there can be no simple means of assessing the degree of attenuation which will occur at a specific site. In spite of this, the following properties are important, and can be determined by field or laboratory tests.

## Measurable Properties

### Particle Size Distribution

Medium-to-fine-grained strata, where flow is primarily intergranular, have high surface areas for reactions to take place, and allow filtration. Major fissure flow systems are clearly disadvantageous.

### Clay Content

Clays have very large surface areas, reduce permeability (increasing contact time for reactions to occur), and different clay minerals have significantly different exchange capacities and hence potential for attenuation.

## Cation Exchange Capacity

Ion exchange is a major attenuation mechanism for cations in particular ammonium (ammonium acetate is in fact used for the laboratory determination of the cation exchange capacity, CEC).

## Carbonate and Oxide Content

Neutral or slightly alkaline conditions reduce the mobility of many pollutants, for example several metals, and provide an optimum environment for both aerobic and anaerobic microorganisms. It has been demonstrated (Harris and Lowe, 1984; Robinson and Lucas, 1984) that landfill gases diffuse into the unsaturated zone beneath landfills in advance of a leachate front. This has two consequences; firstly that all subsequent processes will occur under essentially anaerobic conditions; secondly, that carbon dioxide in the gas will dissolve in pore waters to form carbonic acid ($H_2CO_3$). This is a weak acid, but in the absence of sufficient carbonate minerals and at typical partial pressures of carbon dioxide beneath landfills, pH values as low as 5 can occur, and can act to inhibit beneficial microorganisms (Tester and Harker, 1982). Carbonates are not the only means by which acids can be neutralized, but they are generally much more important than other (usually surface) processes.

Oxides of iron can also reduce pollutant mobility by sorption.

## Zone Thickness

Unsaturated zone thickness has a direct relationship with the attenuating capacity of a given stratum, and influences the contact time between it and the leachate during which reactions can take place.

## Ground-water Underflow

Ground-water underflow can provide a constant supply of well-oxygenated ground-water, allowing aerobic degradation processes to take place subsequently, and giving potential for dilution of more-resistant contaminants.

## Biological Degradation of Contaminants

The unsaturated zone underlying a landfill generally becomes anaerobic within a matter of months in the regions where attenuation will occur. Production of an anaerobic microbial population will therefore depend on a supply of nutrients (principally organic material), pH values, redox potential, temperature and inoculum, which may well come from the wastes themselves. Unlike the earlier parameters, biodegradation rates cannot be measured in advance and although laboratory studies (Hoeks and Borst, 1982; Campbell *et al.*, 1983) can help in predictions, the most useful data come from detailed monitoring of existing sites, described below.

With a knowledge of these characteristics at a given location, obtained as part of a professional and thorough site investigation, it will be possible to estimate a likely minimum degree of attenuation, which can be used in conjunction with a detailed landfill design to arrive at a sound assessment of pollution risk. A vital factor to remember is that operation of an acceptable attenuation landfill design depends on matching the overall flux of contaminants to the available attenuation capacity. Good landfill practices, such as cellular tipping, progressive restoration, or even removal of leachate as it arises, all assist in reducing the quality of leachate entering the underlying strata and requiring to be attenuated.

## CASE STUDY OF AN ENGINEERED UNSATURATED ZONE

The monitoring of the unsaturated zone beneath the Stangate East Landfill site, has allowed the processes of attenuation to be observed in situ since before wastes began to be emplaced in early 1982. Details of the landfill investigation, design, instrumentation and monitoring have been published elsewhere (Robinson and Lucas, 1984, 1985a, 1985b), but the work is summarized below, and examples of the most interesting results are presented.

### Stangate East Landfill

The investigations began in 1979 at the site which was originally excavated into an alternating sequence of limestones for which the quarry was worked, and a soft calcareous sand and silts (known as 'hassock') which were a waste product.

The strata are permeable, and contain a minor aquifer. A high quality river flows within 100 m of the site, suitable for potable supply. A full hydrogeological investigation confirmed that this river receives spring flows from beneath the quarry area, and calculations indicated that ground-water underflow would provide sufficient dilution although additional attenuation would be needed to reduce ammonia and COD to acceptable levels. The characteristics of the underlying strata, and the waste hassock and silt available for site preparation, were therefore investigated. Laboratory determinations were made of clay content and mineralogy, cation exchange capacity and carbonate content, and of the permeability of various hassock/silt mixtures.

The landfill design included an engineered, semi-permeable, 6 m thick attenuation layer which has sufficient attenuating characteristics to retard the movement of more readily attenuated pollutants, and which would reduce seepage through the floor such that other pollutants would be adequately diluted by ground water. The design is shown in Fig. 1. Migration of leachate or gas through the quarry side walls is prevented by 1 m of silt. The landfill receives up to 1300 tonnes per day of mainly road-borne domestic wastes from south-east London.

## Instrumentation

A programme was established to monitor in situ the behaviour of the landfill design by means of readings and samples obtained from within and beneath the site, achieved by installing more than 100 instruments and sampling devices in the unsaturated and saturated zones before waste disposal began, and by further instrumentation of the landfill when 30 m of wastes had been placed, within 18 months (see Fig. 2).

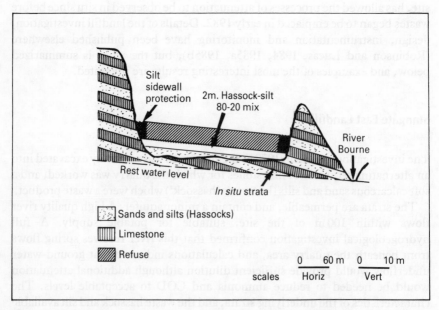

**Figure 1.** Schematic cross-section through Stangate East Landfill.

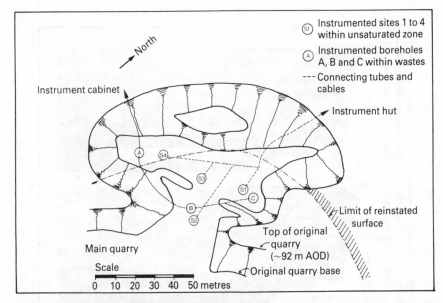

**Figure 2.** Plan of experimental area of the site at Stangate, showing instruments within wastes and the attenuation blanket.

The original instruments were installed in four areas of the site, and had to be of a specification such that they could continue to function within the aggressive environment of the landfill for many years, without any possibility of being repaired, and provide samples or readouts at a location up to 200 m (and up to 30 m higher) from where they were installed. These provide water samples from unsaturated or saturated strata and wastes, gas samples, water table levels, in situ electrical conductivity, moisture content data and temperature results. Monitoring initially took place every three months and, more recently, every four months. A very high proportion of the emplaced samplers continue to function well after more than four and a half years, and have provided a unique record of changes occuring in both the unsaturated and saturated zones beneath the 30 m depth of waste now placed in the landfill.

**Results**

Movement of landfill gases through the unsaturated zone at one of four sites is shown in Fig. 3. Initial depletion of oxygen is followed by anaerobic

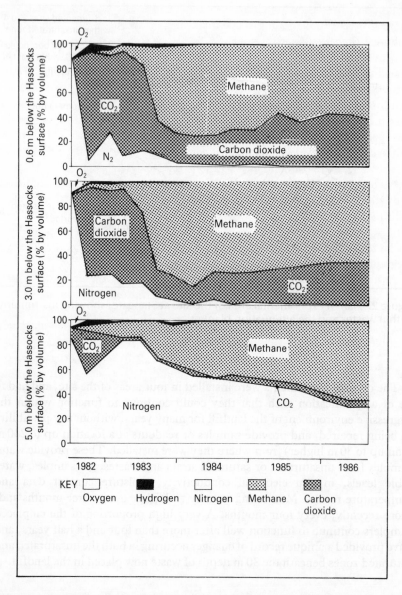

**Figure 3.** Movement of landfill gases through the unsaturated zone at Stangate East; results from site 4.

production of carbon dioxide, with methanogenesis becoming established more slowly. Production of minor quantities of hydrogen (maximum 6% by volume) occured at an early stage.

Movement of landfill gases through the unsaturated zone is characterized by displacement of nitrogen and dissolution of carbon dioxide, which is much more soluble than methane. The most important finding is that anaerobic conditions rapidly became established within the full 6 m depth of the unsaturated zone beneath the landfill, by diffusion of landfill gases well in advance of any contaminated liquors percolating from the wastes. This has an important bearing on the nature of processes which might subsequently act to attenuate leachate contaminants within this zone.

At Stangate, dissolution of carbon dioxide did not result in significant reductions in pH value at any site, due to the high buffering capacity of the calcareous hassock. This contrasts with results from sites on poorly buffered strata (e.g. some triassic sandstones) where production of carbonic acid may make the strata less hospitable to anaerobic bacteria, which could otherwise degrade organic materials in leachate.

As a result of high waste inputs, relatively low effective rainfall at Stangate, and good landfill practices, only small quantities of leachate have been generated to date. However, shallow accumulations of leachate (from 0.1 to 0.4 m) have been shown to exist at the base of wastes in the experimental area. Samples have been recovered, and shown to vary from high strength liquors (COD values > 50 000 mg/l, ammonia from 920 to 1900 mg/l) in two locations, to a more methanogenic leachate (COD 4000 mg/l; ammonia 450 mg/l) in a third. High concentrations of chloride (1400 to 3000 mg/l) were present in all leachate samples.

Migration of these leachates into the unsaturated zone has been noted at each of the four instrumental sites. The in situ water samplers have performed well (only two failures in four and half years from a total of 33 installed), and allowed detailed monitoring of attenuation processes. Results from site two (see Fig. 4) are typical, with initial movement of high concentrations of volatile fatty acids, and COD, into the Hassock. Movement of ammonia has been minimal, restricted by cation exchange processes, and Fig. 4 demonstrates that as anaerobic microorganisms have been established, fatty acids have been degraded rapidly, despite continuing inputs of leachate from above (shown by the continuing increase in concentrations of chloride within the Hassock; Fig. 5). At present no contamination of the water table has been observed. The saturated zone provides a further safeguard to the local water resources and biological degradation of many organic components is likely to be more rapid in this zone, the residual contaminant load being balanced by the continuous supply of oxygen and dilution available in the ground-water underflow.

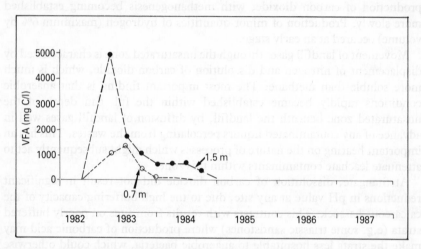

**Figure 4.** Migration and degradation of volatile fatty acids (VFA) in the unsaturated zone at site 2, at depths 0.7 and 0.5 m below Hassock surface.

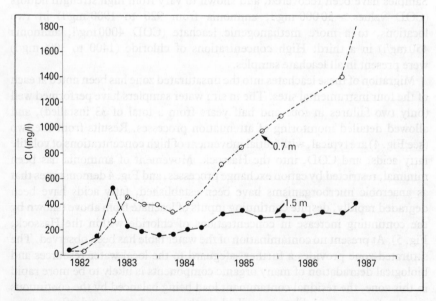

**Figure 5.** Migration of chloride through the unsaturated zone at site 2 at depths 0.7 and 1.5 m below Hassock surface.

## Future Monitoring

Regular monitoring of both the unsaturated and saturated zones, including local springs, and also of the wastes themselves, will continue for many years, even though the site is now completed and being restored to agricultural use. A high standard of restoration is important, since there are two aspects to the site design; that is, the minimization of leachate quantity and the safe attenuation of that which is inevitably generated.

This continued monitoring will provide a unique assessment of the performance of the engineered design. It is only be means of investigations such as this that the behaviour of future such 'dilute and attenuate' landfills can be quantitatively predicted with confidence, and that sites can be designed to meet the high standards which environmental pressures demand.

## CONCLUSIONS

Based on the presented investigations several conclusions can be reached.

Firstly, the dilute and attenuate philosophy of landfill design depends for its effectiveness on chemical, physical and biological processes which operate within the wastes themselves, and in the underlying rock strata. In the UK many landfills continue to operate safely as a result of these principles, and significant case histories of major ground or surface water pollution are rare, despite great increases in monitoring activities in the last ten years.

Secondly, if a dilute and attenuate landfill is to be constructed, it must be possible to demonstrate with a reasonable degree of confidence that the likelihood of a detectable adverse impact on local ground-water resources is minimal and acceptable. To allow this, relevant properties of the strata must be measured and assessed. These include particle size distribution, clay content, cation exchange capacity, unsaturated zone thickness, and carbonate content. The latter is very important, in providing capacity for buffering of pH values, as carbon dioxide and fatty acids migrating from the wastes tend to lower these to values which may be hostile to beneficial microorganisms.

Thirdly, a significant further benefit of an unsaturated zone is in controlling the seepage of leachate into the unsaturated zone, such that the residual contaminant load is balanced by the continuing supply of oxygen and dilution which is commonly available from ground-water underflow.

Finally, it is important to remember that there are two aspects to consider in site design. One is the minimization of leachate quantity, by good landfill practice, and the other the safe attenuation of that leachate which is generated and released.

## ACKNOWLEDGEMENTS

This chapter is presented by permission of ARC South Eastern who operate Stangate East Landfill, and the Department of the Environment who are funding the research programme; the help and co-operation of each are gratefully acknowledged. The views expressed are those of the author, and do not necessarily represent those of either of the above parties.

## REFERENCES

Anon. (1978). 'Co-operative Programme of Research on the Behaviour of Hazardous Wastes in Landfill Sites'. Final Report of the Policy Review Committee, UK Department of the Environment, HMSO, London.

Blakey, N.C. and Towler, P.A. (1987). 'The effect of unsaturated/saturated zone property upon the hydrogeochemical and microbiological processes involved in the migration and attenuation of landfill leachate components'. Proceedings of an IAWPRC International Symposium, Groundwater Microbiology; Problems and Biological Treatment, Kupio, Finland, 4–6 August 1987.

Campbell, D.J.V., Parker, A., Rees, J.F. and Ross, C.A.M. (1983). 'Attenuation of potential pollutants in landfill leachate by lower greensand', *Waste Management and Research* 1, 31–52.

Department of the Environment (1978). 'Investigation of a landfill at Greenoakhill', *WLR Technical Note* 30, by I.B. Harrison, HMSO London.

Harris, R.C. and Lowe, D.R. (1984). 'Change in the organic fraction of leachate from two domestic refuse sites on the Sherwood Sandstone, Nottinghamshire'. *Quarterly Journal of Engineering Geology* 17, 57–69.

Hoeks, J. and Borst, R.J. (1982). 'Anaerobic digestion of free volatile fatty acids in soils below waste tips'. *Water, Air, and Soil Pollution* 17, 165–173.

Naylor, J.A., Rowland, C.D., Young, C.P. and Barber, C. (1978). 'The Investigation of Landfill Sites'. UK Water Research Centre Technical Report TR91.

Robinson, H.D. and Lucas, J.L. (1984). 'Leachate attenuation in the unsaturated zone beneath the landfill: instrumentation and monitoring of a site in Southern England, *Water Science and Technology* 17, 477–492.

Robinson, H.D. and Lucas, J.L. (1985a). 'The behaviour and attenuation of leachate from domestic wastes in a landfill with a designed and engineered unsaturated zone. Instrumentation and research at Stangate East Landfill, in Kent, England'. Proceedings of the International Conference, New Directions and Research in Waste Treatment and Residuals Management, vol. 1, 31–49, the University of British Columbia, Vancouver, BC, Canada, 23–28 June 1985.

Robinson, H.D. and Lucas, J.L. (1985b). 'Attenuation of leachate in a designed, engineered and instrumented unsaturated zone beneath a domestic waste landfill', *Water Pollution Research Journal of Canada* 20(3), 76–91.

Tester, D.J. and Harker, R.J. (1982). 'Groundwater pollution investigations in the Great Ouse Basin', *Solid Waste Disposal Water Pollution Control* 81, 308–328.

# 6.7 Behaviour of Leachate Pollutants in Groundwater

THOMAS H. CHRISTENSEN, PETER KJELDSEN, JOHN LYNGKILDE,
JENS CHRISTIAN TJELL

Department of Environmental Engineering, Technical University,
Bldg. 115, DK-2800 Lyngby, Denmark

## INTRODUCTION

Pollution of groundwater causing risk to the quality of surface waters and water wells is the major environmental concern related to sanitary landfills. Although measures are taken to collect leachates at most new landfills, knowledge and documentation of the behaviour of landfill leachate pollutants in groundwater are needed for evaluation of environmental risks at existing and planned landfills, for establishing cost-effective groundwater quality monitoring programmes, and for evaluating the consequences of unexpected leakage of leachate and defining appropriate remedial action. The latter aspect is currently a major task with regard to the old landfill sites with no or inappropriate leachate collection systems.

The current knowledge about leachate behaviour in groundwater is relatively scarce (Farquar and Constable, 1973; Griffin et al., 1976; Hoeks et al., 1979; Hoeks et al., 1984; Kjeldsen and Christensen, 1984; Kjeldsen, 1986; Loch et al., 1981 and Nicholson et al., 1983) and covers only a few of the many important aspects involved in the fate of leachate leaking into an aquifer.

Focusing on the common type of sanitary landfill receiving a mixture of municipal and commercial wastes but excluding significant amounts of concentrated specific industrial wastes, the landfill leachate pollutants may be categorized as:

1. Common inorganic cations: calcium ($Ca^{++}$), magnesium ($Mg^{++}$), sodium ($Na^+$), potassium ($K^+$), ammonium ($NH_4^+$), iron ($Fe^{++}$), manganese ($Mn^{++}$) and anions: chloride ($Cl^-$), sulphate ($SO_4^{--}$) and hydrogen-carbonate ($HCO_3^-$).

2. Heavy metals: cadmium (Cd), zinc (Zn), lead (Pb), copper (Cu), nickel (Ni) and cobalt (Co).
3. Organic matter, expressed as Chemical Oxygen Demand (COD), including volatile fatty acids (in particular in the acid phase of the waste degradation) and more refractory compounds (fulvic like).
4. Specific organic compounds originating from household or industrial chemicals and present in relative low concentrations in the leachate (usually less than 0.1 mg/l). These compounds include a variety of aromatic hydrocarbons, phenols and chlorinated aliphatics.

Other compounds may be found in leachate from landfills: e.g. borates, sulphides, arsenates, selenate, mercury, barium and oil compounds. But in general these compounds are only of secondary importance.

Discussing the fate of leachate pollutants in groundwater, it is important to realize the dynamic development of various redox zones in the groundwater. This is illustrated in Fig. 1 showing the theoretical zones in a horizontal and vertical plane downstream of a leachate leakage. Close to the leakage point a strongly anaerobic zone will develop with a redoxpotential suitable for methane generation and redoxprocesses feasible at higher redoxpotentials. With increasing redoxpotentials further apart from the landfill (the leachate is titrating the aquifer redox–buffer system) inorganic redox–pairs (although microbially mitigated) will be dominating. This zone is here called the anoxic zone and is characterized by the absence of methane formation and the absence of free, dissolved oxygen. In this zone nitrate, manganate, ferric compounds and sulphate are being reduced in the stipulated order. In many cases, e.g. in areas of extensive agricultural activities nitrate reduction is probably the dominating redoxprocess, since nitrate is continuously diffusing into the anoxic zone. Finally the diluted and probably attenuated leachate may enter into an aerobic zone, where free, dissolved oxygen is present although in modest concentrations. Since landfills often are polluting surfaces near secondary aquifers, the presence of an aerobic zone is not unlikely.

For a continuous leakage these zones may develop over many years (up to a couple of decades) and may never stabilize. The change from acid phase to methane phase degradation in the landfilled waste highly affects the quality of the leachate (Ehrig, 1983) and hence the input to the aquifer. As the content of degradable organic matter in the leachate decreases, the reductive capacity of the leaking leachate is diminished and the reduced zones will slowly be eliminated. Not much is known about the time frames involved, but the dynamics are slow and for an ordinary landfill with a continued leakage this impact on the groundwater may last for more than 100 years. Hopefully the effect of the plume is modest considering the long time span.

The actual shape of the three dimensional leachate plume strongly depends

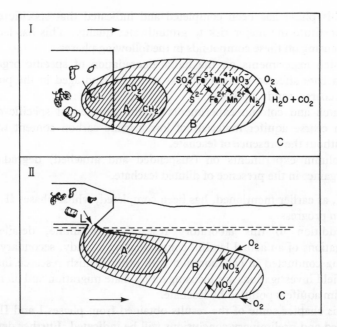

**Figure 1.** Illustration of theoretical redoxzones down stream from a leachate leakage in a landfill. I: Horizontal plane. II: Vertical plane. A: Anaerobic zone (methane formation). B: Anoxic zone, C: Aerobic zone. L indicates leakage point.

the local hydrogeological conditions. The movement of the leachate as a conservative tracer into an aquifer is in itself a very complicated issue, which is beyond the scope of this chapter. Here, the movement of the reactive pollutants being subject to sorption, ion exchange, precipitation/dissolution redox processes or microbial degradation is discussed relative to the movement of water.

## THE TECHNICAL UNIVERSITY OF DENMARK RESEARCH PROGRAM ON LEACHATE BEHAVIOUR IN GROUNDWATER

A research programme has been established at the Technical University of Denmark focusing on the behaviour of landfill leachate pollutants in groundwater. The laboratory part basically consists of four phases:

(I)   Column studies of the strongly anaerobic zone beneath the landfill.

This phase has been completed and indicated that specific organics constitute the major risk to groundwater quality. This has led to the focusing on these compounds in the following phases.

(II)   Batch experiments on the biotic degradation of specific organics in leachate affected groundwater as could be envisaged in the porewater of coarse, sandy aquifers.

(III)  Batch and column experiments on the sorption of specific organics on coarse aquifer materials of low organic carbon content with and without the presence of leachate.

(IV)   Column experiments on (suspended and attached) degradation of organics in the presence of diluted leachate.

Phase I, as earlier mentioned, has been completed, whilst phases II, III and IV are in progress.

In addition to this laboratory research programme, detailed field investigations of an actual landfill contaminating a sandy, secondary aquifer are being conducted in collaboration with other Danish research institutes. These field investigations are focusing on leachate migration and on methods to reclaim aquifers polluted by leachate.

In this chapter some of the results obtained from phases I and II will be presented and preliminary conclusions will be indicated. Further details can be found in Kjeldsen and Christensen (1984), Lyngkilde *et al.* (1988) and Larsen *et al.* (1989).

## EXPERIMENTS

The purpose of the column experiments of phase I was to simulate in the laboratory what would happen in the groundwater zone near a landfill with a sudden release of concentrated leachate. A broad spectrum of pollutants representing all four previously presented groups of leachate pollutants were studied. Based on the results of the first phase the second phase focused on the degradability of 22 specific organics which as trace compounds may migrate with the leachate into coarse sandy aquifers. The experimental set-up and procedures are presented in the following paragraphs.

### Column Experiments of Phase I

The basic principles of the column experiments are shown in Fig. 2. Actual landfill leachate, stored at 1°C under a $CO_2$ atmosphere, is by a carbondioxide

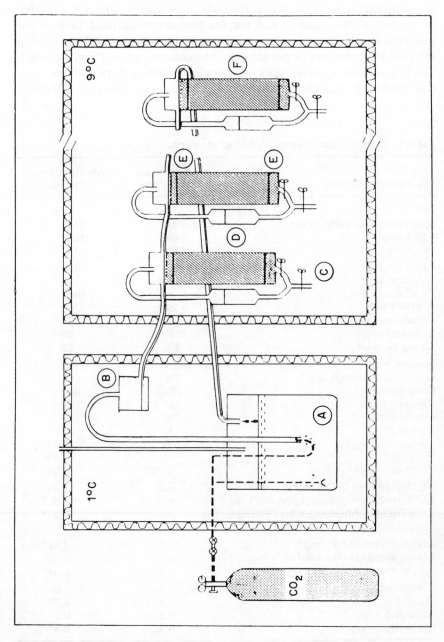

**Figure 2.** Experimental set-up of column experiments of phase I: A: Leachate reservoir, B: Splash reservoir, C: Sampling gate, D: Sample reservoir, E: Gravel layers, F: Soil/aquifer material.

lift pump led to a splash reservoir, from which the leachate continuously is fed to the top of the columns and surplus leachate recirculated back to the reservoir. Leachate migrates through the packed columns and effluent is being sampled at the bottom of the column. The columns were of varying length (30–90 cm) and various types of sampling gates and procedures were employed allowing for hydraulic retention time ranging from 10 to 150 days.

**Table 1.** Composition of employed leachate in phase I.

| Leachate component | Acid phase leachate | Methane phase leachate |
|---|---|---|
| pH | 6.5 | 6.5 |
| Specific conductivity, mS/m | 1000 | 1140 |
| COD, mg/l | 7400 | 1460 |
| BOD, mg/l | 5700 | 300 |
| Chloride, mg/l | 1160 | 2120 |
| Sulphate, mg/l | n.a. | 116 |
| Bicarbonate, mg/l | n.a. | 6890 |
| Acetate, mg/l | 1580 | 167 |
| Propionate, mg/l | 800 | 87 |
| Butyrates, mg/l | 760 | 39 |
| Valerates, mg/l | 290 | 12 |
| Sodium, mg/l | 850 | 2320 |
| Ammonia–Nitrogen, mg/l | 730 | 685 |
| Potassium, mg/l | 750 | 640 |
| Calcium, mg/l | 725 | 103 |
| Magnesium, mg/l | 280 | 144 |
| Iron, mg/l | 130 | 20.6 |
| Manganese, mg/l | 6 | 0.33 |
| Zinc, mg/l | 2 | 1.15* |
| Cadmium, mg/l | 0.007 | 0.29* |
| 2-Chlorophenol (2-CP), mg/l | n.a. | 1.16* |
| 4-Chloro-2-methylphenol (2M-4CP), mg/l | n.a. | 2.86* |
| 2,4,5-Trichlorophenol (2,4,5-TCP), mg/l | n.a. | 1.04* |
| Pentachlorophenol (PCP), mg/l | n.a. | ~1* |
| 2-Nitrophenol (2-NP), mg/l | n.a. | 0.51* |
| 4-Nitrophenol (4-NP), mg/l | n.a. | 0.22* |
| 2,4-Dinitrophenol (2,4-DNP), mg/l | n.a. | 0.34* |
| Dimethoate, mg/l | n.a. | 1.0* |
| Malathion, mg/l | n.a. | 0.7* |
| Sulfotepp, mg/l | n.a. | 0.2* |
| Fenitrothion, mg/l | n.a. | 0.3* |

* The concentration is elevated due to spiking of the leachate.
n.a. = not analysed.

**Table 2.** Physical and chemical composition of the experimental soils.

| | Experimental soils | | | | |
|---|---|---|---|---|---|
| | A | B | C | D | E |
| Clay, <0.002 mm, % | 2.4 | 8.1 | 13.4 | 1.0 | 1.0 |
| Silt, 0.002–0.02 mm, % | 0.8 | 8.0 | 10.8 | 0.5 | 0.5 |
| Fine sand, 0.02–0.2 mm, % | 16.3 | 49.3 | 49.3 | 8.7 | 2.7 |
| Coarse sand, 0.2–2 mm, % | 80.1 | 34.5 | 25.7 | 89.2 | 84.3 |
| Gravel, >2 mm, % | 0.0 | 0.0 | 0.0 | 0.6 | 11.5 |
| Organic matter, mg C/g soil | 2.1 | 1.6 | 1.5 | <1 | n.a. |
| Cation exchange capacity, meq/100 g | 2.2 | 4.0 | 7.3 | 0.7 | 0.3 |
| Calcium, exchangeable, meq/100 g | 0.87 | 1.48 | 5.75 | n.a. | n.a. |
| Magnesium, exchangeable, meq/100 g | <0.01 | 0.37 | 0.29 | n.a. | n.a. |
| Sodium, exchangeable, meq/100 g | 0.06 | 0.07 | 0.15 | n.a. | n.a. |
| Potassium, exchangeable, meq/100 g | 0.07 | 0.27 | 0.12 | n.a. | n.a. |
| Manganese, exchangeable, meq/100 g | 0.004 | 0.017 | 0.004 | n.a. | n.a. |
| pH–CaCl$_2$ | 5.3 | 4.4 | 6.4 | n.a. | n.a. |
| Cadmium, $\mu$g/g | 0.01 | 0.04 | 0.04 | 0.06 | n.a. |
| Zinc, $\mu$g/g | 8.34 | 20.16 | 29.31 | 9.85 | n.a. |
| Iron, $\mu$g/g | 3810 | 9440 | 16 500 | n.a. | n.a. |
| Manganese, $\mu$g/g | 170 | 201 | 212 | 116 | n.a. |
| Calcium, $\mu$g/g | 262 | 454 | 897 | 35 700 | n.a. |

n.a. = not analysed.

Two types of leachate were employed: An acid phase leachate with a high concentration of organic matter (7400 mg COD/1) and a methane phase leachate with a low content of organic matter (1460 mg COD/1) of low degradability (BOD$_5$/COD = 0.21). The composition of the leachates is given in Table 1.

Five soils and aquifer materials where included in the experiments ranging from a very coarse sand to a sandy loam. Physical and chemical characteristics are shown in Table 2.

The soil columns were monitored in terms of effluent breakthrough curves (effluent concentration as function of number of pore volumes passing through the column) and final pollutant distribution in soil columns. For further details on experimental procedures and the principles of interpreting column experiments see Kjeldsen and Christensen (1984) and Kjeldsen (1986).

## Batch Experiments of Phase II

Large batches, as shown in Fig. 3, were obtained by dilution of actual methane phase leachate established with levels of organic matter corresponding to 10, 50 and 200 mg COD/1. The batches were at temperature 25°C and 10°C adjusted to appropriate redoxpotentials representing anoxic and aerobic conditions by addition of nitrate (to ensure anoxic conditions) and oxygen (to ensure aerobic conditions). The batches were spiked with 22 specific organics

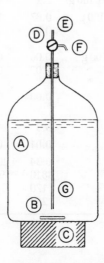

**Figure 3.** Experimental setup of batch experiments of phase II: A: Glass-container for diluted leachate, B: Rod for stirring, C: Magnetic stirrer, D: 3-way valve, E: Input of oxygen (air) or nitrogen for sampling, F: Sampling gate.

in concentrations of 100 μg/1. The specific organics studied represent various groups of common, rather mobile organic trace compounds in landfill leachate. The organics are listed in Table 3. At various time intervals, 50 ml of the batches were sampled and extracted with pentane and ether for gas-chromatographic analysis (FID/ECD). ATP-activity measurements, as an expression of microbial activity in the batches, were also made. Decreasing concentrations of a specific organic compound as compared to inhibited controls were taken as microbial degradation.

## RESULTS AND DISCUSSION.

### Column Experiments of Phase I

The column experiments involved a total of 13 experiments monitored in terms of effluent breakthrough curves and 4 experiments monitored in terms of final soil pollutant profiles. Figures 4, 5 and 6 show typical breakthrough curves for a variety of pollutants for a column with soil type A (coarse sand) exposed to methane phase leachate. From the shape and location of the breakthrough curves, as compared to the breakthrough curves of chloride

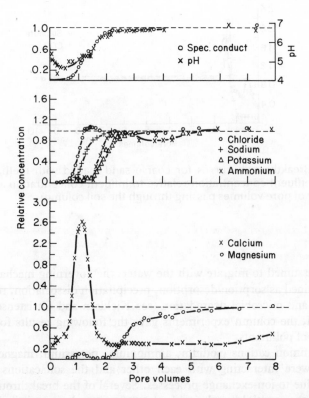

**Figure 4.** Breakthrough curves for coarse sand loaded with methane phase leachate (effluent concentration relative to influent concentration as function of number of pore volumes passing through the soil column).

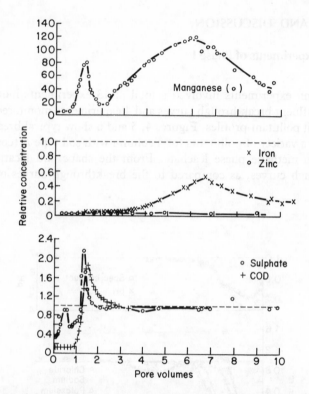

**Figure 5.** Breakthrough curves for coarse sand loaded with methane phase leachate (effluent concentration relative to influent concentration as function of number of pore volumes passing through the soil column).

which is assumed to migrate with the water, the governing mechanisms can be determined as sorption/desorption, precipitation/dissociation, reductions/oxidation and microbial degradation (see Kjeldsen and Christensen, 1984). In general, the column experiments gave the following results for the four categories of pollutants.

The common cations (sodium, ammonium, potassium, magnesium and calcium) were interacting with each other and the soil cations exchange complex due to ion-exchange processes. Several of the breakthrough curves exhibited the multiple peaks and plateau zones characteristic for multi-component ion-exchange. Generally sodium, ammonium and potassium were retarded in increasing order mainly by exchange of calcium and protons. The release of calcium and protons were observed in nearly all columns, as

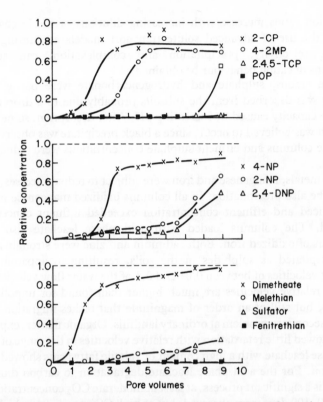

**Figure 6.** Breakthrough curves for coarse sand loaded with methane phase leachate (effluent concentration relative to influent concentration as function of number of pore volumes passing through the soil column). 2-CP: 2-chlorophenol; 4-C-2MP; 4-chloro-2-methylphenol; 2,4,5-TCP: 2,4,5-trichlorophenol; PCP: pentachlorophenol; 4-NP: 4-nitrophenol; 2-NP: 2-nirophenol; 2,4-DNP: 2,4-dinitrophenol.

calcium peaks above inlet concentrations and pH drops at the moments of leachate breakthrough. The retardation of sodium, ammonium and potassium was increasing with increasing clay content. Relative solute transport velocities were in the range of 0.2–0.97 (relative to the water flow velocity). Soil specific distribution ratios (calculated via the linear adsorptions model) for the various cations cannot be determined because they varied with leachate composition (the distribution ratios were twice as high for methane phase leachate as compared to acid phase leachate), calcium and maybe magnesium is also subject to precipitation (calcite, dolomite) and soluble complex formation, the soil/groundwater system is not constant in terms of pH and the observed distribution ratios increase with increasing dilutions of the leachate. Only through the use of advanced solute transport models accounting for ion-exchange selectivity, precipitation and complexation, can acceptable predictions of cation behaviour be obtained.

Of the anions, sulphate and hydrogencarbonate were partly affected. Sulphate was desorbed from the subsoils probably due to a drop in anion-exchange capacity caused by increasing pH. After desorption, some sulphate reduction was believed to occur, since a black precipitate was observed in the top of the columns and effluent sulphate concentrations were reduced by up to 20%.

Of the metals, manganese and iron were subject to redoxprocesses, sorption and maybe also precipitation. In all columns oxidized manganese in the soil was reduced and effluent concentration exceeded influent concentrations multifold. The columns loaded with acid phase leachate also revealed reductions of oxidized iron. Both cadmium and zinc were strongly adsorbed and precipitated as sulphides in the soils, resulting in apparent relative transport velocities of between 0.01 and 0.1 of the water flow velocity.

These relative velocities are much higher than found for unpolluted soil solutions, but still of an order of magnitude that makes migration of heavy metals a secondary problem at ordinary landfills. Organic matter, expressed as COD, showed little retardation with relative velocities in the range of 0.7–1.0. Acid phase leachate with a high content of volatile fatty acids showed the least retardation. For the acid phase leachate degradation to carbon dioxide and methane is a significant process, at least at moderate $CO_2$ concentrations. Half lives of 30–100 days were observed, but at high $CO_2$ pressures, which may be found in landfills, inhibition may take place. For the methane phase leachate, only degradation of the remaining fatty acids content was observed, indicating that the leachate leaving the anaerobic zone will contain a substantial content of refractory (for anaerobic processes) organic matter of the order of 1000 mg COD/1 (excluding effects of dilution).

The specific organics studied in the first phase columns showed varying degrees of retardation. Of concern is the relative high velocities of e.g. 2-chlorophenol, 2-methyl-4-chlorophenol, 4-nitrophenol and the pesticide dimethoate. The latter moving with a velocity of 70% of the water flow. The retardation of the specific organics were found to depend on, as expected, the organic content of the soil and the water solubility of the compound or its octanol–water partition coefficient. For the phenols also their dissociation constants (pKa) and the actual soil-water pH was of importance. For example, the velocity of pentachlorophenol was relatively high due to dissociation. Most of the specific organics were not significantly degraded in the anaerobic environments with the employed retention times. The compounds 2,4,5-trichlorophenol, 4-chlorophenol and the pesticide Malathion showed degradation corresponding to half-life values of 1–20 weeks.

## Aerobic and Anoxic Batch Experiments of Phase II

The batch experiments involve 9 batches representing 3 levels of dilute leachate in aerobic environments and 2 levels of dilute leachate in anoxic (with nitrate) environments plus inhibited references. Figure 7 presents as an example the fate of 1,2-dichlorobenzene (1,2-DCB) in the aerobic zone in the presence of 10, 50 and 200 mg COD/1 and in the anoxic zone (nitrate) in the presence of 50 and 200 mg COD/1. In an aerobic environment 1,2-DCB is degraded easily when a primary substrate is present in sufficient amounts. At the first spiking, the compound is degraded within few days except where only

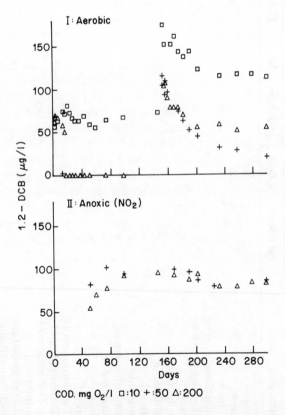

**Figure 7.** Fate of 1,2-dichlorobenzene in completely mixed reactor of varying dilutions of leachate (corresponding to 10, 50 and 200 mg COD/l) at aerobic conditions (I) and anoxic (denitrifing conditions) (II). The aerobic reactors have ben spiked with 1,2-DCB after 155 days.

**Table 3.** Preliminary evaluation of compound degradation in batches simulating aerobic and anoxic environments with a primary substrate present at 50 mg COD/l originating from methane phase leachate.

| Compound | Aerobic/oxygen | | | | Anoxic/nitrate | | | |
|---|---|---|---|---|---|---|---|---|
| | Excellent | Good | Medium | No | Excellent | Good | Medium | No |
| Benzene | | × | | | | | | × |
| Toluene | | × | | | | | | × |
| o-Xylene | | | × | | | | | × |
| 1,4-Dichlorobenzene | | × | | | | | | × |
| 1,2-Dichlorobenzene | | | × | | | | | × |
| Indene | × | | | | | | | × |
| Nitrobenzene | | | × | | | | × | |
| Naphthalene | × | | | | | | | × |
| Biphenyl | × | | | | | | | × |
| Flourenone | × | | | | | | | × |
| Dibenzothiophene | × | | | | | | | × |
| Phenanthrene | × | | | | | | | × |
| Phenol | × | | | | | × | | |
| o-Cresol | × | | | | | | | × |
| 2,4-Dichlorophenol | | × | | | | | | × |
| 2,6-Dichlorophenol | | | × | | | | × | |
| 4,6-o-Dichlorocresol | | | (×) | | | | | × |
| o-Nitrophenol | | | × | | | | | × |
| 1,1,1-Trichloroethane | | | | × | | | | |
| Tetrachloromethane | | | | × | | | | |
| Trichloroethylene | | | | × | | | | × |
| Tetrachloroethylene | | | | × | | | | × |

( ): Abiotic degradation. Comparable degradation in sterile batch.

10 mg COD/l initially was present. Apparently the bacterial growth here is limited by a low primary substrate concentration. After 150 days the batch is fed a new load of specific organics and also here degradation is observed, although the degradation pattern is not identical with that first observed. This is due to the fact that the microbial activity now is reduced, expressed as ATP-activity, since most of the primary substrate has been utilized during the first period (primary substrate was not added at the second loading). The aerobic degradation patterns of 1,2-DCB resemble very well the general trends for biodegradable compounds investigated. The degradation is generally slower and has longer lag phases in batches with decreasing content of leachate. In contrast to the aerobic degradation, degradation of 1,2-DCB has not been observed in the anoxic environment having nitrate as primary electronacceptor (denitrification). Table 3 presents the preliminary results for the studied 22 compounds in the aerobic and anoxic environment. The degradation potential is large in the aerobic environment for all studied compounds except for the chlorinated aliphatic compounds, where no degradation was observed within 280 days. Some compounds even degrade within hours after spiking, e.g. phenol and *o*-cresol. In contrast to this, most of the compounds have shown no degradability under denitrificating conditions. This is in spite of observed microbial activity in the batches. Exceptions are nitrobenzene and tetra-chloromethane, the latter being of special interest because it belongs to the aerobically persistent aliphatics.

## CONCLUSION

The results of the first phases of the laboratory research programme through column breakthrough curves and final soil pollutant profiles and through degradation experiments, has revealed that several mechanisms are involved in the retardation and degradation of leachate pollutants in groundwater.

Only very few of the inorganic leachate pollutants were governed solely by sorption processes (sodium, ammonium, and for some soils potassium and iron), while most pollutants were governed by sorption together with precipitation/dissolution (calcium, magnesium, cadmium and zinc) and redox processes (manganese and in some cases iron). Ranking the pollutants with respect to their velocity showed that chloride, specific conductivity, total solids were very mobile (0.8–1.0), sodium, ammonium, potassium, magnesium and iron were moderately mobile (0.3–0.8) and zinc, cadmium, iron and in most cases also manganese were only partly mobile (< 0.3). In particular the heavy metals zinc and cadmium showed very restricted mobility in the anaerobic zone of even very coarse aquifer materials. Differences in the retardation were found with respect to both leachate composition, degree of dilution and soil types, and exceptions in the above mentioned mobility order were observed. However, in general the mobility decreased with increasing clay content or cation exchange capacity. The extensive interaction between various processes, e.g. formation of dissolved complexes and precipitation/dissociation makes traditional one component modelling a dubious method in predicting leachate migration in aquifers. Only multicomponent models including several simultaneous processes will be able to yield acceptable predictions of leachate fate in acquifers.

The organic matter in the leachate, expressed as COD, show little retardation through sorption. In particular the acid phase leachate with a high content of volatile fatty acids is very slightly retarded. However, the fatty acid content of the leachate is readily degraded anaerobically at 10°C, at least when the $CO_2$ content is not to high, exhibiting half-lives of the order of 25–100 days. In contrast, the non-fatty-acid fraction of the organic matter which typically amounts to 1000 mg COD/l showed no degradation in the anaerobic environment at 10°C. This refractory organic matter may, when leaving the anaerobic zone, eventually be subject to dilution or degradation in denitrifying aerobic zones of the groundwater.

The fate of specific trace organic compounds leaching into an aquifer is still only rudimentarily known. The specific organics may, migrating with the leachate plume, be subject to varying microbial degradation processes. In a leachate contaminated groundwater, trace organics may be subject to co-metabolism, having the unspecified organic matter to act as primary substrate supporting the microbial activity. In the aerobic zone, if present, many of the studied compounds are subject to degradation (of the 22 specific compounds studied, only the chlorinated aliphatics showed no sign of degradation within 280 days). The importance of the concentration of primary substrate was clearly demonstrated, since degradation was slow at low levels (10 mg COD/l). The degradability of the specific compounds apparently is very low in anoxic zones with nitrate as a potential oxidizing agent. The strongly anaerobic zone may be a suitable environment for degradation of some of the organics, and the migration through various redox zones may be beneficial for degradation of various groups of critical organic compounds, but this is still too early to be supported by experimental data.

## REFERENCES

Ehrig, H.-J. (1983). 'Quality and quantity of sanitary landfill leachate'. *Waste Management and Research*, 1, 53–68.

Farquhar, G.J. and Constable, T.W. (1973). 'Leachate contaminant attenuation in soil'. University of Waterloo, Ontario, Canada.

Griffin, R.A., Cartwright, K., Shimp, N.F., Steele, J.D., Ruch, R.R., White, W.A., Hughes, G.M. and Gilkeson, R.H. (1976). 'Attenuation of pollutants in municipal landfill leachate by clay minerals. Part 1 – Column leaching and field verification'. Illinois State Geological Survey, Urbana, Ill., USA (Environmental Geology Notes No. 78).

Hoeks, J., Beker, D. and Borst, R.J. (1979). 'Soil column experiments with leachate from a waste tip. II. Behaviour of leachate components in soil and groundwater'. Institut voor Cultuurtechniek en Waterhuishouding, Wageningen, The Netherlands (NOTA 1131).

Hoeks, J., Hoekstra, H. and Ryhiner, A.H. (1984). 'Kolomproeven met percolatiewater een afvalstort III. Gedrag van niet-vezuurd, gestabiliseerd, percolatiewater in de bodem'. (Soil column experiments with leachate from a waste tip. III. Behaviour of non-acidic, stabilised leachate in soil, in Dutch). Institut voor Cultuurtechniek en Waterhuishouding, Wageningen, The Netherlands (NOTA 1530).

Lyngleilde, J., Tjell, J.C. and Foverskov, A. (1988). Degradation of specific organic compounds with landfill leachate as a primary substrate. K. Wolf, W.J. van der Brink, F.J. Colon (eds), Contaminated Soil '88, pp. 91–100, Kluwer, Academic Press.

Kjeldsen, P. and Christensen, T.H. (1984). 'Soil attenuation of acid phase landfill leachate'. *Waste Management and Research*, 2, 247–263.

Kjeldsen, P. (1986). 'Attenuation of landfill leachate in soil and aquifer material'. Ph.D. Thesis, Department of environmental engineering, Technical University of Denmark, Lyngby, Denmark.

Larsen, T., Kjeldson, P., Christensen, T.H., Skov, B. and Refstrup, M. (1989). Sorption of specific organics in low concentrations on aquifer materials of low organic carbon content: Laboratory experiments. H.E. Kobus, W. Kinzelbach (eds), Contaminant transport in groundwater, pp. 133–140, A.A. Balkema, Rotterdam.

Loch, J.P.G., Lagas, P. and Haring, B.J.A.M. (1981). 'Behaviour of heavy metals in soil beneath a landfill; Results of model experiments'. *The Science of the Total environment*, 21, 203–213.

Nicholson, R.V., Cherry, J.A. and Reardon, E.J. (1983). 'Migration of contaminants in groundwater at a landfill: a case story. 6. Hydrogeochemistry'. *Journal of Hydrology*, 63, Special issue, 131–167.

# 6.8 Soil and Refuse Stability in Sanitary Landfills

ANDREA CANCELLI

*Department of Earth Sciences, University of Milan, Via Mangiagalli, 34 - I 20133 Milan, Italy*

## INTRODUCTION

Sanitary landfills present mulifarious implications: environmental, technical, economical and legal. Among the environmental and technical aspects, surface and ground-water pollution, and the technical solutions to reduce such undesired impacts, are generally thoroughly investigated; in contrast, soil and refuse stability receives comparatively less attention than it deserves.

Reasons for an increase of stability problems in sanitary landfills are many. First of all, present technological trends lead to placing landfills in areas that present stability problems themselves (open pit mines and quarries, sloping ground, recently reclaimed areas, etc.). In addition, anaerobic degradation, which is the most commonly adopted process in modern sanitary landfills, causes generation of combustible biogas, therefore inducing explosion hazard and introducing a further instability factor (Cancelli and Cossu, 1984). Last, but not least, sanitary landfills are frequently located in areas of future urbanization; the possibilities of creating spaces and support for future sporting areas and urban facilities, after the final landfill reclamation, should be carefully evaluated (Sembenelli and Ueshita, 1981; Cartier and Baldit, 1983).

## SLOPE CLASSIFICATION

The source of whatever stability problem is sloping ground surface. However, different slopes can present different problems and can require different analysis, investigation and design philosophies. Therefore, for a better

483

understanding and arrangement of the subjects to be dealt with, slopes can be classified according to the following two criteria (Cancelli *et al.*, 1987; Fig. 1).

The first criterion is based on slope genesis, being either 'natural' or 'human made'. In natural slopes, the stability is essentially governed by long-term conditions; in cut slopes, as in embankments and in landfills, both short-term and long-term stability should be ensured by proper design and construction procedures.

The second criterion is based on the types of slope-forming materials, which can be 'in situ soils', 'compacted soils' and 'wastes'. In natural soils, the geotechnical behaviour essentially depends on geological factors, including depositional conditions, stress history and weathering processes; investigation results are always affected by a certain degree of uncertainty, and unpredictable events can always occur. In compacted soils, grain size composition and method of emplacement are human-controlled; failures are generally related to improper decision or careless construction procedures. In wastes, origin, chemical composition, method of emplacement and time elapsed from emplacement are the dominant factors influencing their geotechnical behaviour.

A further subdivision should be introduced between 'tailings dams' and 'landfills', the former being obtained by solid wastes disposal in a wet stream, and the latter in a dry stream (Morgenstern, 1985). In turn, waste landfills include various kinds of mine and industrial dumps, urban demolition dumps, and sanitary landfills.

Bearing in mind the stability problems of sanitary landfills, it is also convenient to classify them according to the site morphology (Cancelli and Cossu, 1984):

1. 'depression' landfills, into natural gullies, shallow human-made trenches, or abandoned quarries and open pit mines;

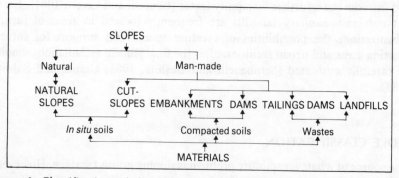

**Figure 1.** Classification of slopes, according to the slope genesis and to the slope-forming materials (from Cancelli *et al.*, 1987).

2. 'area' landfills, as embankments on relatively flat natural ground;
3. 'slope' landfills in adjacent to natural or human-made slopes.

Abandoned quarries of clay material for bricks and tiles industry are generally preferred for depression landfills, due to the low hydraulic conductivity of natural clay deposits. However, in many countries it may become necessary to fill quarries of coarser aggregates (such as sand and gravel for concrete constructions), or other kinds of open pit mines. In these cases, the environmental protection is ensured by artificial liners on the bottom and the slopes of the depression. Area landfills are often adopted, in preference to depression ones, provided the final landscape and the future land use be acceptable. Moreover, this kind of sanitary landfill is often an obliged solution, for surficial or shallow ground-water table. In any case, bottom watertightness is required. Slope landfills generally concern areas that had been previously degraded (e.g. following erosion processes), or in areas without any landscaping importance. This kind of waste disposal is often associated to serious stability problems.

## STABILITY PROBLEMS

According to many codes (e.g. the Italian Code, CI, 1984), all kinds of landfills should be settled on, or in, foundation soils, the stability of which being sufficient to avoid risks of landslides, failures and settlements, with reference both to bottom and sides of the landfill, and to all water drainage systems. These requirements correspond to the recently issued Eurocode (EC, 1987), according to which every geotechnical structure must fulfil fundamental performance criteria of stability and stiffness during construction and throughout its life, without reaching a 'limit state'. In turn, limit states may include:

1. 'ultimate limit states', at which a mechanism is formed in the ground and/or in the landfill structure;
2. 'serviceability limit states', at which deformations in the ground and/or in the landfill can cause loss of serviceability in the landfill itself.

In other words, bearing in mind the above reported morphological classification of sanitary landfills, the following stability problems have to be considered:

1. stability of natural slopes, or of cut slopes in abandoned quarries.
2. stability and settlement of the waste landfill itself;
3. stability and settlement of the 'landfill–foundation soil' complex.

## Stability of Natural and Cut Slopes

It is obvious that active landslides and landslide prone areas should be avoided as waste disposal sites. A preliminary assessment of the stability of natural slopes is usually carried out on the basis of geological, hydrogeological and geomorphological surveying. Ground-water levels should be measured over long periods in order to check the most unfavourable conditions; surface and subsurface movements should be monitored, by means of topographic surveying and inclinometer measurements. Geotechnical investigations include borings, in situ tests, sampling and laboratory tests. The factor of safety is evaluated according to the well-established principles and methods of soil and rock mechanics.

In Italy, most waste landfills are placed on highly consistent to hard, overconsolidated, marine silty clays (e.g. the Plio-pleistocenic 'Blue clays', which are widespread along the peri-adriatic regions of the Apennines, or on highly consistent, overconsolidated, lacustrine silty clays, extending at the bottom of several inner valleys of the Italian peninsula). Abandoned quarries for clay material, which is used in bricks and tiles industry, are generally preferred, in order to obtain a satisfactory final reclamation of compromised areas.

For both natural and human-made slopes in overconsolidated clay soils, failures can occur according to different mechanisms; among them, the most common are (Fig. 2):

1. Rotational slides, in homogeneous clay deposits, induced by toe excavation or following processes of natural softening and progressive failure (Fig. 2a);
2. combined, mainly translational slides along sand seams, following insurgence of high porewater pressures (Fig. 2b).

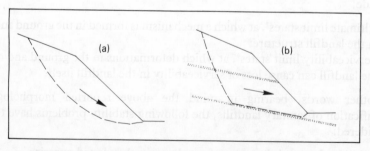

**Figure 2.** Typical examples of failures in overconsolidated clay slopes: (a) rotational; (b) combined, mainly translational (from Cancelli, 1987).

The degree of slope failure hazard is maximum at the beginning of waste landfilling, in connection with the maximum height of slopes. Re-profiling the slope is always necessary; the average slope angles are ranging from 30 to 35°, depending on slope height and shear strength of clay soils (Cancelli *et al.*, 1987). In order to inhibit or reduce the hazard of progressive failures in long-time abandoned cut slopes, the definitive re-profiling should be carried out by steps, each 3–5 m high, and each step should be excavated only after completion of the underlying waste layers.

In the northern part of River Po basin, many quarries for the exploitation of coarse aggregates have been abandoned, and converted to sanitary landfill sites, provided a pollution control system be realized (Cancelli and Francani, 1984). Sandy and gravelly deposits are often slightly cemented, and quarry exploitation slopes may be close to the vertical (Fig. 3); however, water seepage can leak and dissolve cementing substances, so decreasing cohesive bonds between solid particles. In addition, vertical cracks can develop, parallel to the cliff, following lateral decompression and long-time exposures; as a consequence, sudden failures can occur, according to one of the following mechanisms (Fig. 3):

1. falling of blocks from competent conglomerate layers, underlain by more easily eroded sandy layers (Fig. 3a);
2. tilting and toppling of prismatic masses, caused by freezing or by excess cleft water pressures (Fig. 3b);
3. sliding of soil masses along a slightly curved surface out-cropping at the toe of the cliff, sometimes favoured by excess water pressures into vertical cracks (Fig. 3c).

In any case, the slope must be flattened in order to improve the initial stability of the disposal site, and also because of the necessity to put in place liner

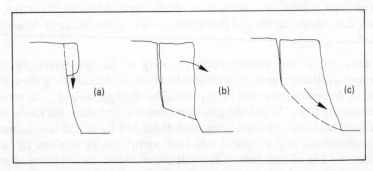

**Figure 3.** Typical examples of failures in subvertical slopes in slightly cemented sandy and gravelly deposits: (a) falls; (b) topples; (c) slides (from Cancelli, 1987).

materials (compacted clays, sand–bentonite mixtures, synthetic geomembranes); in such a way, inert material for daily covering of wastes and a
supplementary volume for waste disposal can be obtained. According to the
height of the slope and to the nature of slope forming soils, the common
practice suggests average slope angles ranging from 35° to 45° (Cancelli *et al.*,
1987).

## Stability and Settlement of Compacted Wastes

The stability of the waste landfill (considered independently on its subgrade) is
influenced by several, interacting factors; the most important are:

1. the geotechnical properties of in situ wastes (in turn, depending on initial
   composition, methods of putting in place and compaction, chemical and
   physical conditions of waste mineralization, settlements and elapsed time
   since waste deposition);
2. the geotechnical properties of intermediate, and final, covering soils;
3. the inclination and height of slopes;
4. the presence of biogas in the waste landfill;
5. the presence of water seeping into the waste landfill, and the pore pressure
   regime.

When theoretical concepts and experimental procedures of classical soil
mechanics are extended to wastes, the following important difference must be
kept in mind:

1. mineral components of natural soils have been subjected to weathering
   processes for periods longer than thousands of years, and, consequently
   they can change only at very low rates, in geological times.
2. in contrast, most waste materials are chemically and biologically active, and
   they can change nature and properties in short times (order of decades, or
   less).

The most significant characteristic property of in situ wastes (in turn
governing all other properties, including hydraulic and mechanical ones) is the
degree of denseness, that can be expressed by the unit weight $\gamma$. In general,
values ranging from 150 to 350 kg/m$^3$ correspond to uncompacted wastes (as in
many slope landfills), values ranging from 350 to 550 kg/m$^3$ can be obtained by
light compaction, and values of 800–1000 kg/m$^3$ can be attained by heavy
compaction. The in situ density depends above all, on the thickness of waste
layers, and also on the depth below the upper landfill surface; the best results
are obtained by the so-called 'onion-skin' compaction technique, with values

of $\gamma$ that can be higher than 1000 kg/m$^3$ (Wiemer, 1982; Cancelli, 1987). Other properties, strictly related to $\gamma$, are:

1. the void ratio $e$, ranging from about 1 (well compacted wastes) to values as high as 15 (in the absence of compaction).
2. the humidity $w$ (expressed as percentage of water by volume of wastes), varying from 10–15% (loose wastes) to about 40% (well-compacted wastes);
3. the (field) retention capacity $w'$, corresponding to the maximum humidity than can be retained by wastes without gravity percolation.

The difference

$$w'' = w' - w \qquad (1)$$

is called adsorption capacity and represents the maximum quantity of water that can be absorbed by wastes without forming leachate. From field measurements, it seems that $w''$ reaches a maximum value for a degree of compaction, corresponding to a unit weight of about 600–800 kg/m$^3$; besides, mechanical shredding of wastes (before putting in place and compacting them) greatly increases $w''$, also reducing the quantity of leachate that can be produced by a sanitary landfill (Franzius, 1977; Cancelli *et al.*, 1987).

The state of compaction directly influences hydraulical and mechanical properties, including shear strength and compressibility parameters.

Heterogeneity, nature and size of different components of urban wastes do not allow easy and reliable determinations of 'drained' shear strength parameters, to be introduced into stability analyses. The few data collected from literature are plotted and compared in Fig. 4; on this basis, the following values of the effective shear strength parameters may be assumed for routine design (Cancelli *et al.*, 1987): friction angle $\phi' = 25\text{–}26°$; intercept cohesion $c'$ limited to a maximum of 30 kPa.

The bearing capacity of wastes, even if thoroughly put in place, is generally low. The surficial crust can play an important role, despite of its reduced thickness; below the crust, values of the bearing capacity ranging from 25 to 100 kPa have been evaluated by means of static penetrometer and pressuremeter tests (Cartier and Baldit, 1983). Nevertheless, it is generally recommended to assume values not greater than 25–40 kPa (Sargunan *et al.*, 1986).

If the most significant refuse parameters ($\gamma$, $\phi'$, $c'$) are known with sufficient accuracy, designing height and slope angle of waste landfills can be accomplished by the classical methods of soil mechanics. The low unit weight of loose wastes gives reason for the high values of the slope angle, as it is sometimes obtained for dry waste landfills (up to about 60°). However, the most unfavourable conditions must be taken into account (pore pressures given by leachate circulation and by biogas developing with the landfill); the

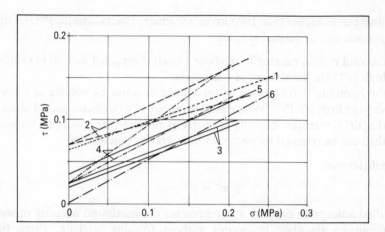

**Figure 4.** Summary of effective stresses envelopes for refuse. Sources of data: (1) from laboratory tests (Fang *et al.*, 1977); (2) extreme values from laboratory tests (*Fang*, 1983); (3) suggested values for sanitary landfill design (Oweis *et al.*, 1985); (4) extreme values suggested by STS (1985); (5) suggested values (Chen, 1986); (6) back-computed values (Tonteri and Lindroos, 1987); (2), (3), (4) after Oweis and Khera (1986).

actual value of cohesion, and the evolution of $c'$ with time, are doubtful; a minimum value of the safety factor ($F \geq 1.3$) should be verified for every earth structure. Therefore, a maximum slope angle of 30° is normally adopted when properly designing waste landfills: in addition, bearing in mind the presence of daily cover material interbedded into refuse, the slope profile is interrupted by horizontal beams, and the average slope inclination will be lower than 30° (Cancelli *et al.*, 1987).

In modern sanitary landfills, the prominent care is devoted to collecting leachate and taking it out of the refuse mass. Therefore, the cover layers (about 0.3 m thick) have also a function of hydraulic separators between two successive waste layers (each 2–3 m thick); they are slightly inclined towards the sides of the landfill and are made with compacted silty-clayey soils (where not available, admixed sand–bentonite liners are used successfully). As clayey soils possess low shearing strength ($\phi'$ decreases with increasing plasticity index), a reasonable compromise has to be pursued between hydraulic requirements and mechanical resistance). Anyway, the possibility of compound (rotational and translational) slope failures, favoured by the 'weak' clay layers, should be carefully investigated.

The stability of a landfill can be impaired by a bad performance of the leachate drainage system (giving origin to 'perched ground-water' supported by the aforementioned clayey layers), or by an uncontrolled biogas developing within the landfill itself. Instability phenomena can occur during landfilling and after completion of the landfill, according to complex mechanisms as 'slump-earthflows'.

Additional problems can arise for slope landfills, where wastes are often poorly compacted and drained. An outstanding example of catastrophic failure occurred in 1977 near Sarajevo (Yugoslavia), where a 60–75° steep slope landfill, following explosion of accumulated biogas, suddenly slumped; the slided mass turned to a rapid flow and was mobilized on a gentle slope, only 6° steep, to a distance of about 1 km (Gandolla *et al.*, 1979).

Concerning settlement, five different mechanisms are recognized to be responsible (Sowers, 1973):

1. mechanical consolidation (distortion, bending, crushing and reorientation of 'solid' particles), similar to consolidation processes occurring in organic soils, as peat;
2. ravelling (erosion or sifting of fine materials into the large voids between larger particles);
3. physicochemical changes (corrosion, oxidation and combustion).
4. biochemical decay (fermentation and decay, both anaerobic and aerobic);
5. interactions between physicochemical and biochemical processes.

Moreover, the mechanical settlement is conventionally divided into two parts (Sowers, 1973): Consolidation (or primary) settlement; secondary (viscous) settlement.

The former generally develops at high rate (in less than one month), with little pore pressure build-up; it can be roughly computed in one-dimensional (oedometric) conditions; according to Sower's data, the virgin compression index $C_c$ linearly increases with the initial void ratio $(e_0)$

$$C_c = (0.15-0.55) e_0 \qquad (2)$$

and with the organic content, the highest values being of the same order of the values measured on peats.

The secondary settlement develops at slow rate for the entire life of the landfill. The coefficient of secondary compression $C_\alpha$ increases with the initial vaoid ratio $(e_0)$

$$C_\alpha = (0.03-0.09) e_0 \qquad (3)$$

and tends to increase with the increase of conditions favourable to decomposition of wastes (Sowers, 1973).

Mechanical settlement of refuse occurs during the landfill construction. For a typical landfill with an average life of ten years, the summation of primary and secondary settlements may amount to about 30% of the landfill initial height (Oweis and Khera, 1986).

If no load is applied to the refuse fill, successive settlements are due to the other mechanisms: physicochemical changes, biochemical decay, interactions between them, and ravelling (NAVFAC, 1983; Oweis and Khera, 1986). The ravelling contribute cannot be rationally computed; however, this phenomenon seems to be favoured by recirculating leachate (Cheyney, 1983). As to biochemical and physicochemical mechanisms, a crude estimation for Italian urban refuse leads to a weight loss of about 10%, due to biogas production (Cancelli and Cossu, 1984).

## Stability and Settlement of the Soil–Landfill Ensemble

The global stability of the complex formed by the waste landfill and by the foundation soil is a typical problem of area landfills and also of some slope landfills. In many cases, waste embankments on flat ground can be an obliged design solution, on account of:

1. the lack of suitable depressions within a reasonable distance from the waste-producing zone;
2. the presence of ground-water table at low depth.

High, well-compacted ($\gamma = 1000$–$1100$ kg/m$^3$) refuse embankments cannot be built up rapidly on soft clay soils or other unconsolidated deposits. Owing to insufficient (undrained) bearing capacity of foundation soils, a general failure can occur, in form of rotational slip (at least in homogeneous and fairly isotropic geological bodies).

A crude estimation of critical height $H_c$ of wastes on soft clay foundation can be obtained by the simple bearing capacity formula

$$H_c = (2 + \pi)\, c_u s/\gamma \tag{4}$$

where $c_u$ is the undrained cohesion of the foundation soil, $\gamma$ is the unit weight of refuse and $s$ is a shape factor (the shear resistance of refuse can be conservatively neglected for this analysis).

An example of undrained base failure occurred in March, 1987, at the Finale Emilia landfill, situated between Modena and the Po River (Northern Italy). The landfill consisted of a 15 m high, lightly compacted ($\gamma \approx 700$ kg/m$^3$) refuse embankment; moreover, the landfill was bordered along two sides by a 2 m deep canal (Fig. 5). No preliminary investigation had been carried out before landfilling; the presence within the subsoil of a thick layer of poorly consolidated, alluvial, high plasticity (plasticity index PI = 55–70%), organic silty clays was assessed only after the failure.

The slided area was about 45 m wide (the canal was almost completely obstructed for the same length) and 40 m long. The slip surface, in form of circular arc, starting from the top of the landfill and outcoming at the toe of the canal bank, followed the weakest clay layer at a depth of 3–4 m below the original ground level (Fig. 5). The soft clay layer had a water content higher than the liquid limit, also passing 100%; by means of 4 static cone penetration tests (CPT), values of the cone resistance $q_c$ as low as $0.2–0.3 \, \text{MN/m}^2$ were measured, allowing to evaluate an undrained cohesion $c_u \approx 20 \, \text{kN/m}^2$.

Such a value corresponds to the cohesion that should be required for the equilibrium according to Equation (4), suggesting that the landfill mainly failed by undrained loading of the soft clayey foundation soil. The presence of the canal along the toe of the landfill can be regarded as a concurrent instability factor. If foundation soils are highly anisotropic, base failures can still occur, but in the form of compound slip surfaces. Two examples have been reported, respectively from Northeastern USA (Fig. 6) and from Southern Finland (Fig. 7).

In the former case, the foundation soils were not completely consolidated under the load of a 15-year-old landfill. Thus, failure occurred as a consequence of undrained loading due to the emplacement of new fill materials; flooding of the tidal area at the foot of the landfill and subsequent

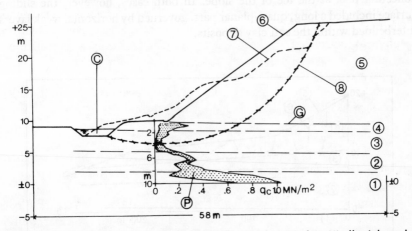

**Figure 5.** Base failure of Finale Emilia landfill (Northern Italy): (1) alluvial sands; (2) sandy silts and silts; (3) very soft, high plasticity, silty clays with some organic substances; (4) sandy and clayey silts; (5) lightly compacted refuse; (6) ground surface before failure; (7) ground surface after failure; (8) the most probable slip surface; (G) original ground level; (C) canal; (P) results of 4 static cone penetration tests: range of measured values of the cone resistance $q_c$ (data after Vandini, 1988).

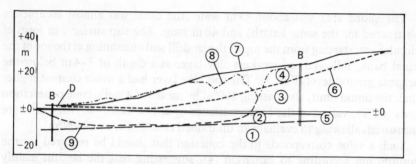

**Figure 6.** Base failure of a sanitary landfill in Northeastern USA coast: (1) stiff/dense Late Cretaceous deposits; (2) very soft tidal marsh deposits; (3) old refuse, deposited during 15 years; (4) recent refuse, deposited during 4–5 months; (5) original ground surface, depressed following subsoil consolidation; (6) ground surface before recent filling; (7) ground surface before failure; (8) ground surface after failure; (9) the most probable slip surface; (D) dike; (B) borings (after Dvirnoff and Munion, 1986).

have played a role too. In the latter case, the failure was primarily due to overestimation of the bearing capacity of the base ground; concurrent causes were an excessive thickness of the final covering and the location of a leachate collecting pool at the toe of the slope. In both cases, however, the sliding surface included a long, quasi-planar part, governed by horizontal weak layers interbedded within the soft clay deposits.

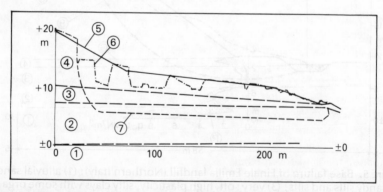

**Figure 7.** Base failure of Mankaa sanitary landfill near Helsinki: (1) moraine; (2) very soft clay, with silt lenses; (3) peat; (4) refuse; (5) ground surface before failure; (6) ground surface after failure; (7) the most probable slip surface. Note deformed horizontal/vertical scale ratio (interpreted after Tonteri and Lindroos, 1987).

It has to be remembered that, even if foundation failure does not occur, heavy waste landfills may induce settlements in the underlying soft clays, so damaging both surface water and leachate drainage systems. Also owing to this reason, most codes do not accept the emplacement of waste landfills on soft clay deposits.

## STABILITY EVALUATION FOR A SANITARY LANDFILL

An example of stability evaluation for the location, design and construction of a sanitary landfill in Northern Italy is presented. This landfill has been conceived for the needs of a group of municipalities all-around the township of Imola (Region of Emilia-Romagna). The morphology of this disposal plant is typical of slope landfills, presenting serious problems from the point of view of geotechnical stability.

### Geological and Geotechnical Site Conditions

The landfill area (Fig. 8) is located on a hillside in the Southern neighbours of Imola territory, close to the head of a small valley (named Rio della Rondinella), in turn tributary of Santerno River. The hillside, sloping to north-west with maximum values of 28–32°, somewhere showed gully erosion morphology and had been recognized as an area to be reclaimed (Monesi, 1981.

The lithological framework of the hill is formed by 'grey-blue, marly silty clays, locally slightly sandy', of marine Lower Pleistocene (SGI, 1969). These clays generally show laminated or (rarely) layered structure, with thin interbedded sandy seams. The layers dip almost uniformly downslope, with dip angle varying from 8 to 18° (Figs. 8 and 9). It is remarkable that these Pleistocene clays are underlaid by conformable Pliocene deposits, also formed by grey-blue, marly, silty clays with sandy seams, for an overall thickness of about 1400 m; the stratigraphic boundary between the two formations can be recognized only on paleontological criteria, while no substantial differences exist from a lithological point of view. As a consequence, ground-water circulation tends to develop only along the sandy interlayers. When designing the landfill, it was taken into consideration the possibility of ground-water inflow from the cut slopes, particularly from the South, favoured by the general dip to the North of the formation (see section I–I in Fig. 9).

Soil geotechnical properties have been determined by means of laboratory tests on undisturbed block samples; classification, oedometric compression and shear strength tests have been carried out.

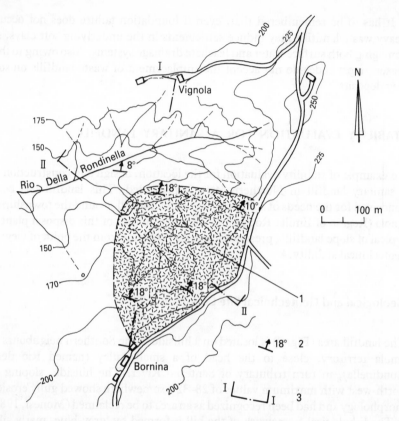

**Figure 8.** (1) Disposal area; (2) strike and dip of bedding; (3) trace of geological sections (simplified after Monesi, 1981).

The prevailing material can be defined as a grey, highly consistent, 'clay with silt' (containing clay and silt fractions in almost equivalent proportions and traces of fine sand). Plasticity analyses gave the following ranges of values: liquidity limit (LL) = 44–56%; plasticity index (PI) = 21–34%. Therefore, this formation is fairly homogeneous (at least, from the lithological point of view) and is formed of medium to high plasticity soils (groups CL and CH).

The natural water content resulted about 22.7% (average value): thus, close to the plasticity limit, in agreement with the high value of consistence. The low value of the standard deviation (2.2%) proves the good homogeneity of the formation.

The unconfined compression strength $q_u$ has been measured on two groups of cylindrical specimens, both having height/diameter ratio = 2 and

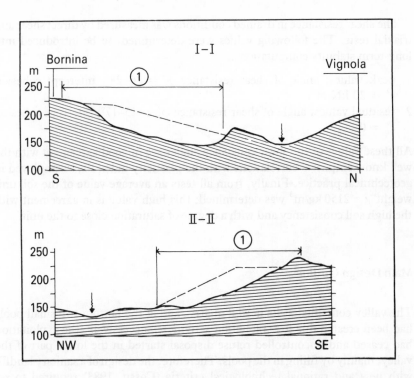

**Figure 9.** Geological sections through Rio della Rondinella valley, showing the general downslope dip of bedding and the location of disposal site area (1); dotted lines = final waste landfill profile.
Chapter 6.8/Legends/continued

diameter = 40 mm and 70 mm (respectively); average values of 1051 kN/m$^2$ for the former group and 930 kN/m$^2$ for the latter one were obtained. The highest values relate to intact soil material, while the lowest values, measured on larger specimens, were depending also on fissurative pattern; in this way, conservative value of the undrained cohesion $c_u = 450$ kN/m$^2$ was assumed for short-term stability analyses. In any case, such a very high value is consistent with the high overconsolidation degree of all Pliopleistocene deposits outcropping along the northeastern border of the Appenines (Esu and Martinetti, 1965; Cancelli *et al.*, 1984).

Oedometer compression tests assessed the high overconsolidation degree (overconsolidation ratio OCR higher than 40). Besides, a value of the coefficient of permeability $k \approx 3 \times 10^{-11}$ m/s (without considering fine sandy interlayers) could be evaluated. Therefore, the landfill substratum can be assumed as practically impervious and incompressible.

The shear resistance in drained conditions was measured by direct shear and triaxial tests. The following values were determined, to be introduced into long-term stability computations:

1. peak values: angle of shear resistance $\phi' = 26\text{-}28°$; intercept cohesion $c' = 20 \text{ kN/m}^2$;
2. residual values: angle of shear resistance $\phi'_r = 15\text{-}17°$; intercept cohesion $c'_r = 0$.

All these values of the shear resistance angles are in good agreement with the well know empirical correlations with the plasticity index PI, as employed in geotechnical practice. Finally, from all tests an average value of the soil unit weight $\gamma = 2150 \text{ kg/m}^3$ was determined: this high value is in agreement with the high soil consistency and with a degree of saturation close to the unit.

## Main Design Outlines

The valley contained at its head a quarry, and at the bottom two small pools had been created due to exploitation activities. Successively, the exploitation had ceased and uncontrolled refuse disposal started in the lower part of the valley, mainly by filling in the pools. Therefore, the design of a sanitary landfill with new and rational technological criteria (Cossu, 1983) required to re-profile and compact the old refuse at the bottom of the valley (Fig. 10). The old refuse was separated by the new one, by means of a 0.3 m thick compacted clay liner. At the same time, the upslope hillside was re-profiled, at slope inclination of about 34°, with both advantages of exploiting clay material for daily cover of refuse layers and obtaining an additional volume to be filled in with wastes.

The most significant geometrical characteristics of the new sanitary landfill are (Fig. 10):

1. landfill toe at elevation 150 m (above sea level);
2. re-profiled foundation gently sloping downwards (8°);
3. average inclination of cut-slope upwards (boundary with the in situ clay formation) equal to about 34°;
4. final upper surface of the landfill at elevation 220 m (the same of the municipal road at the SE border (see Fig. 8);
5. maximum height of the general excavation in the clay formation (and, consequently, maximum thickness of refuse fill) $H = 45$ m;
6. average inclination of the refuse fill slope $\beta = 23°$.

In more detail, the cut-slope is re-profiled in steps (each step being 5 m high,

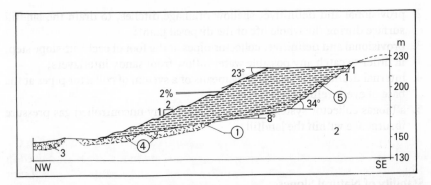

**Figure 10.** Schematical section of the refuse landfill, along section II–II of Figs 8 and 9: (1) old soil surface; (2) silty-clay substratum; (3) inert, demolition wastes; (4) old refuse, re-profiled and compacted; (5) new refuse, compacted in layers (simplified after Cossu, 1983).

2 m wide and 45° inclined) only after completion of the lower, compacted refuse layer; meanwhile, the refuse slope is also profiled in steps (each step being 26.5° inclined).

The waste is deposited and compacted in 2 m thick layers, interbedded by compacted clay layers (each 0.3 m thick). The layers are slightly inclined downwards (2%), in order to favour internal drainage of the landfill. According to recent experiences (Cheyney, 1983), compaction of wastes in 2 m thick layers allows to obtain an initial unit weight of about 600 $kg/m^3$, that increases to about 800 $kg/m^3$ under the effect of successive waste layers. As to the shear strength parameters, it should be possible to assume for compacted wastes $\phi' = 25$–$26°$ and $c'$ variable from 0 to 30 $kN/m^2$.

The very low permeability of the in situ clay formation is a warranty against pollution from the bottom of the landfill. Even if internal fissures and sandy seams are taken into account, the overall hydraulic conductivity of the clay mass (perpendicularly to the bedding) should not be higher than $10^{-10}$ m/s. Therefore, the foundation soil was assumed as practically 'impervious'; the only compacted clay liners are provided on the old refuse layers at the bottom of the valley and on the lateral cut slopes (owing to the presence of sandy seams and interlayers).

As far as the landfill drainage is concerned, the following design elements have to be remembered (Cossu, 1983):

1.  a continuous ditch, parallel to the crown of the disposal site, and a system of shallow ditches for the surface drainage of surrounding slopes;

2. provisional and definitive, shallow drainage ditches, to drain the landfill surface during the whole life of the disposal plant;
3. provisional and definitive, collector pipes at the foot of each cut-slope step, in order to catch any possible water inflow from sandy interlayers;
4. internal drainage of leachate, by means of a system of collector pipes at the base of every refuse layer;
5. a biogas collector system, in order to avoid any uncontrolled gas pressure insurgence within the landfill.

## Stability of Natural Slopes

Before the starting of re-profiling works for the new disposal plant, the slope was about 40 m high and the maximum inclination reached 28–32°. The natural hillslope was apparently stable and not affected by any mass movement, but only by some gully erosion. Besides, ground-water circulation and related water outflow (in form of small intermittent springs) were practically negligible, despite the gently downslope bedding (Fig. 9).

The long-term stability of the most inclined pre-existing slope (elevation 185–225 m) was verified, taking into account effective stresses and peak shear strength parameters. The slope profile was simplified, as represented in Fig. 11a; Spencer's method (Spencer, 1967) was adopted, under the assumption of homogeneous clay extending to infinite depth under the toe. According to this method, the pore pressure distribution along the slip surface was assumed to be represented by the pore pressure ratio ($r_u$):

$$r_u = u/\gamma h \tag{5}$$

in which $u$ = neutral pressure in a datum point along the slip surface; $\gamma$ = unit weight of the soil; $h$ = depth of considered point with reference to the ground surface.

The schematic section of Fig. 11a was verified for:

1. pore pressure ratio $r_u$ ranging from 0 to 0.5 (assumed to be constant along the whole slip surface; the so-introduced error is limited to a few percent units of the computed factor of safety $F$).
2. unit weight of the soil $\gamma = 2150$ kg/m$^3$;
3. shear resistance angle of the soil $\phi' = 28°$;
4. intercept cohesion of the soil $c' = 1$ t/m$^2$ and 2 t/m$^2$ (respectively, about 10 and 20 kN/m$^2$).

The results of computations are summarized in Fig. 11b and can be commented as in the following. At the starting of works for the new landfill, if

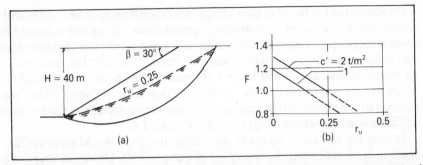

**Figure 11.** Stability analysis of natural slopes according to Spencer's method (Spencer, 1967): (a) simplified section of the slope, showing the piezometric surface under the hypothesis of $r_u = 0.25$ = constant along the slip surface; (b) effect of $c'$ and $r_u$ on the factor of safety $F$.

values of cohesion are assumed as determined by laboratory tests on undisturbed samples ($c' = 20$ kN/m$^2$), the natural slope was stable ($F$ higher than 1), provided the pore pressure ratio $r_u$ be lower than 0.25. On the other hand, ground-water flow was recognized as practically absent; thus, the assumed value $r_u = 0.25$ seems too conservative for the considered hillslope.

However, it has to be remembered that, also in natural slopes, hard fissured clays are subject to long-term, progressive reduction of cohesion, following a process of natural softening. If one assumes $c'$ to decrease to only 10 kN/m$^2$, failure could occur for $r_u = 0.15$: such a value cannot be excluded under particular unfavourable meteorologic and hydrogeologic conditions. Therefore, despite the apparent stability of considered natural slopes, the following specifications were introduced in the waste landfill design:

1. no definitive increase of the average slope angle (re-profiling every step at 45° being immediately followed by filling in with refuse, acting as partial counterweight at the toe of the cut slope;
2. realization of surface drainage systems, upslope and on both sides of the disposal area (so to inhibit any ground-water inflow).

## Stability of Cut Slopes

As said above, the first stage of landfilling operations requires a general excavation, to obtain a cut slope having an average slope inclination $\beta = 34°$ and a maximum height $H$ (Fig. 10).

On account of the fact that such a large excavation is going to be filled in with wastes, short-term stability analyses were carried out, taking into account the shear strength parameters as defined in undrained conditions (for saturated clay soils), employing an angle of shearing strength $\phi_u$ equal to zero and an undrained cohesion $c_u$ of 450 kN/m². By means of a simple Taylor's analysis for purely cohesive soils (Taylor, 1937), a critical height of about 115 m was computed; thus, the first stage general excavation should be more than sufficiently stable, with a factor of safety of about 2.4.

During the successive stages of landfilling, the slope should be re-profiled again in steps (each step having $H = 5$ m and $\beta = 45°$). However, every step should be re-profiled only immediately before emplacing the corresponding refuse layer: in this way, the local stability of every cut slope will be ensured.

In addition, provisional surface ditches and definitive collector pipes, running at the toe of every step, will provide the complete drainage of surface and subsurface water, during the entire life of the disposal plant (Cossu, 1983).

## Stability of the Compacted Waste Landfill

The final slope of the refuse embankment is designed at inclination of 26.5°, interrupted by horizontal beams every 15 m of elevation, so that the average inclination will decrease to about 23° (Fig. 10). The refuse is compacted in such a way, that the following parameters can be assumed for wastes involved into stability analyses: unit weight $\gamma = 800$ kg/m³; angle of shearing strength $\phi = 26°$; intercept cohesion $c' = 5$ kN/m² (this is a highly conservative value).

Also these long-term analyses were performed by Spencer's method (Spencer, 1967). According to the National Geotechnical Code (MLP, 1988), a minimum value was imposed for the factor of safety ($F \geq 1.3$): the corresponding value of the pore pressure ratio was computed ($r_u = 0.25$). This means that the long-term stability of the refuse mass is ensured, provided pore pressure increases are not allowed within the landfill. To this goal, drainage systems are designed for biogas, leachate, surface and subsurface water.

## Overall Stability of the Landfill

The long-term stability after landfill completion should be also intended as the stability of the ensemble formed by the waste landfill and by the underlying and surrounding clay formation.

Stability analyses were performed for the same section II–II, in which

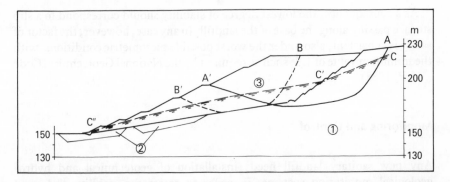

**Figure 12.** Section adopted for long-term, global stability analysis of completed landfill: (1) silty-clay substratum; (2) old refuse at the bottom of the landfill; (3) new refuse, compacted in 2 m thick layers, interbedded by 0.3 m thick compacted clay layers. A–A' and B–B': assumed, hypothetical, non-circular slip surfaces. C–C'–C'': assumed, maximum piezometric level.

two hypothetical slip surfaces are reported (Fig. 12). Both these sliding mechanisms are compound, depending on the high structural anisotropy of the hypothetically sliding mass, in turn due to:

1. the presence of sandy seams and interlayers within the clay formation;
2. the compacted clay liner at the boundary between old and new refuse;
3. the compaction of wastes in layers (2 m thick);
4. the presence of intermediate covering layers (0.3 m thick, every 2 m of compacted wastes and slightly inclined downslope), formed of compacted clay of medium to high plasticity and, for this reason, possible seat of sliding surfaces.

The well-known simplified Janbu's method, corrected in function of the sliding mass geometry and of the soil shear strength characteristics (Janbu *et al.*, 1956), was selected for analyses. Two different piezometric conditions were considered:

1. no pore pressure on the slip surface (admitting a fully satisfactory performance of all drainage systems).
2. hypothetical piezometric surface as indicated by C–C'–C'' in Fig. 12 (to be considered as the worst piezometric condition among those one can reasonably assume, if a bad performance of some drainage systems is admitted).

The results show values of the factor of safety between 2.05 and 2.21 with no pore pressure, and of 1.30–1.48 for bad performing drainage system.

As a consequence, the lowest degree of stability should correspond to a slip surface passing along the base of the landfill. In any case, however, the factor of safety is consistent, also under the worst possible piezometric conditions, with the minimum value of 1.3, such as required by the National Geotechnical Code (MLP, 1988).

### Monitoring and Control

Any new sanitary landfill needs installation of geotechnical and hydro-geological monitoring systems, in order to assess the stability and good performance of the waste landfill and to adopt whatever control measure be required for protecting the environment. To these aims, and bearing in mind the stability problems, the following physical quantities should be monitored:

1. the neutral pressures in the foundation soil and within the waste embankment;
2. the total vertical stresses applied by the waste embankment on the foundation soil, and also at intermediate levels within the landfill;
3. the horizontal movements induced by the earth and waste lateral pressures within the landfill and in the foundation soil;
4. the vertical movements (settlements) of the landfill base, in the foundation soils and within the landfill itself.

Therefore, the new sanitary landfill is instrumented with: electrical (no volume change) piezometric cells, within and under the landfill; total pressure cells, at the base of the landfill; vertical inclinometric columns through the entire landfill height and in the foundation soil, with fixed sensors at different elevations; horizontal inclinometric columns with fixed sensors, at the base of the landfill and at an intermediate level within the refuse mass.

### CONCLUSIONS

Sanitary landfills, even if thoroughly designed and realized, may present serious stability problems, arising from:

1. insufficient bearing capacity of subgrade soils, in particular of soft clays, peats and other compressible deposits (for area and also for some slope landfills);
2. instability of natural slopes in clay formations (for slope landfills);
3. poor stability conditions of cut slopes (for depression landfills into abandoned quarries and also for some slope landfills).

4. low shear resistance and excessive compressibility of refuse, due to the nature of wastes and to difficulties in putting in place and compacting them;
5. low shear resistance of intermediate compacted clay layers;
6. unsatisfactory performance of drainage systems (for surface and subsurface water, or for leachate, or for biogas), causing increase of pore pressures within the refuse mass;
7. possible, negative interactions between the refuse mass and the natural soils at the boundary;
8. natural softening of hard fissured clays, creep, progressive failure and other processes affecting the long-term behaviour of slope-forming materials, or of compacted refuse.

During recent years, significant advancements have been made towards analytical modelling of natural phenomena, also including slope instability, consolidation and settlements. Numerical methods (mainly the Finite Element Method) can be powerful tools for solving all geotechnical problems, provided reliable values of soil and refuse geotechnical parameters be available.

For sanitary landfills, the most serious difficulties derive from wastes, due to their extreme variability in nature, composition, deposition and long-term behaviour. For refuse, geotechnical parameters are generally assumed, and introduced into stability analyses, on the basis of a few published data and without any *ad hoc* investigation. New data, new experimental procedures to acquire them, and well documented case records, are needed in the future.

The reported case of the Imola sanitary landfill represents an example of landfilling on natural slopes that has been designed and is being realized in the respect of geotechnical stability criteria. The complex of field and laboratory, geological and geotechnical investigations allowed to reconstruct a reliable subsoil model and to solve the main stability problems, including:

1. the initial and long-term stability of natural slopes in the waste disposal area;
2. the short-term stability of cut slopes during the landfill construction;
3. the stability of compacted refuse layers;
4. the short-term and long-term stability of the whole complex, formed by the landfill and by its foundation soils.

## REFERENCES

Cancelli, A. (1986). 'Geological and geotechnical implications of hazardous waste disposal near the surface or at depth'. Proc., Seminar on Hazardous Waste Production, Control and Disposal, ISMES, Bergamo (Italy), 221–241.
Cancelli, A. and Cossu R. (1984). 'Problemi di stabilità negli scarichi controllati'. *Ingegneria Ambientale* 13, (11–12), 632–642.

506                                                                          Cancelli

Cancelli, A and Francani, V. (1984). 'Quarry reclamation in the Lombardy Plain, Italy'. *Bull. Int. Assoc. Eng. Geol.* **29**, 237–240.
Cancelli, A., D'Elia, B. and Sembenelli, P. (1987). 'Groundwater problems in embankments, dams and natural slopes', Gen. Rep., IX Eur. Conf. Soil Mech. Found. Eng., Dublin, 4. 1–30 (preprint).
Cancelli, A., Pellegrini, M. and Tonnetti G. (1984). 'Geological features of landslides along the Adriatic coast'. *Proc. 4th Int. Symp. on Landslides, Toronto* **2**, 7–12.
Cartier, G. and Baldit, R. (1983). 'Comportement géotechnique de décharges de résidus urbains', *Bull. Liaison Labor. Ponts et Chaussées* **128**, 55–64.
Chen, R.H. (1986). 'Slope stability analysis of a waste landfill'. Proc. Int. Symp. on Envir. Geotechnology (H.Y. Fang, ed.), Bethlehem, PA, 37–42.
Cheyney, A.C. (1983). 'Settlement of landfill'. Proc. Harwell Symp. on Landfill Completion, Harwell, UK.
Comitato Interministeriale (1984). 'Disposizioni per la prima applicazione dell'art. 4 del DPR 10 settembre 1982, n. 915, concernente lo smaltimento dei rifiuti', *Gazzetta Ufficiale della Repubblica Italiana* **253** (2).
Cossu, R. (1983). 'Progetto di scarico controllato per rifiuti solidi urbani in località Pediano di Imola'. Unpublished Report.
Dvirnoff, A.H. and Munion, D.W. (1986). 'Stability failure of a sanitary landfill'. Proc. Int. Symp. on Envir. Geotechnology (H.Y. Fang, ed.), Bethlehem, PA, 25–35.
EC (1987). 'Eurocode EC 7: Geotechnics, Design'. Ad-hoc Committee of the European Geotechnical Societies (N. Krebs Ovesen, chairman). Report prepared for the Commission of the European Communities, preliminary draft, 216 pp.
Esu, F. and Martinetti, S. (1965). 'Considerazioni sulle caratteristiche technice delle argille plio-pleistoceniche della fascia costiera adriatica tra Rimini e Vasto'. *Geotecnica* **12** (4), 164–185.
Fang, H.Y. (1983). 'Physical properties of compacted disposal materials'. Unpublished Report, Philadelphia.
Fang, H.Y., Slutter, R.G. and Koerner, R.M. (1977). 'Load bearing capacity of compacted waste disposal materials'. Proc. Spec. Sess. on Geotech. Eng. and Envir. Geotechnics, I, 265–278 (Z.C. Moh, ed.), IX Int. Conf. Soil Mech. Found. Eng., Tokyo.
Franzius, V. (1977). 'Der Sickerwasserabfluss aus Mulldeponien—Ein Matematisches Modell'. Wasserbau Mitteilungen, 16, Technische Universitat Darmstadt.
Gandolla, M., Grabner, E. and Leoni, R. (1979). 'Stabilitatsprobleme bei nicht verdichteten Deponien am Beispiel Sarajevo (Jugoslawien)', *ISWA Journal* 28–29.
Janbu, N., Bjerrum, L. and Kjærnsli, B. (1956). Veiledning ved losning av fundamenteringsoppgaver'. Norwegian Geotechnical Inst., Publ. 16, Oslo.
Ministero dei Lavori Pubblici (1988). D.M. 11.3.1988: 'Norme tecniche riguardanti le indagini sui terreni e sulle rocce, la stabilità dei pendii naturali e delle scarpate, i criteri generali e le prescrizioni per la progettazione, l'esecuzione e il collaudo delle opere di sostegno delle terre e delle opere di fondazione'. *Gassetta Ufficiale della Repubblica Italiana* **47**.
Monesi, A. (1981). 'Studio geologico ed idrogeologico dell'area situata in località Pediano in comune di Imola'. Unpublished Report.
Morgenstern, R.N. (1985). 'Geotechnical aspects of environmental control. State-of-the-Art Rep', Proc. XI Int. Conf. Soil Mech. Found. Eng., San Francisco, 1, 155–185.
NAVFAC (1983). 'Design Manual 7.3'. Dept. of the Navy, Alexandria, VA.

Oweis, I. and Khera, R. (1986). 'Criteria for geotechnical construction of sanitary landfills'. Proc. Int. Symp. on Envir. Geotechnology (H.Y. Fang, ed.), Bethlehem, PA, 205–222.

Oweis, I., Mills, W. and Leung, A. (1985). 'Slope stability of sanitary landfills'. ASCE Metropolitan Section Seminar, New York.

Sargunan, A., Mallikarjun, N. and Ranapratap, K. (1986). 'Geotechnical properties of refuse fills of Madras, India'. Proc. Int. Symp. on Envir. Geotechnology (H.Y. Fang, ed.), Bethlehem, PA, 197–204.

Sembenelli, P. and Ueshita, K. (1981). 'Environmental Geotechnics. State-of-the-Art Rep.', Proc. X Int. Conf. Soil Mech. Found. Eng., Stockholm, 4, 335–394.

SGI (1969). 'Servizio Geologico d'Italia. Carta Geologica d'Italia', Foglio 99, Faenza.

Sowers, G.F. (1973). 'Settlement of waste disposal fills'. Proc. VIII Int. Conf. Soil Mech. Found. Eng., Moskow, 2.2, 207–210.

Spencer, E. (1967). 'A method of analysis of the stability of embankments assuming parallel inter-slice forces', *Géotechnique* **XV** (1), 11–26.

STS Consultants Ltd (1985). 'Geotechnical slope stability analysis for perimeter slopes, DeKorte Park landfill'. Unpublished Report.

Taylor, D.W. (1937). 'Stability of earth slopes', *J. Boston Soc. Civil Engineers* **24**, 197–246.

Tonteri, K.J. and Lindroos, P.T. (1987). 'Stability problems of landfill site founded on peat bog'. Proc. X Eur. Conf. Soil Mech. Found. Eng., Dublin, 1, 337–342.

Vandini, R. (1988). 'Studio geologico di una frana avvenuta nella discarica di R.S.U. di Finale Emilia'. Curricular work, University of Modena (unpublished).

Wiemer, K. (1982). 'Qualitative und Quantitative Kriterion zur Bestimmung der Dichte von Abfallen in Geordneten Deponien'. PhD thesis, Technischen Universitat Berlin.

# 6.9 Occupational Safety at Landfills

VOLKMAR WILHELM

*Württemberger Gemeindeunfallversicherungsverband D-7000 Stuttgart 1, Panoramastrasse 11, West Germany*

## INTRODUCTION

The potential hazards posed by emissions from a landfill on the environment and local community usually gain much focus, while the occupational hazards on those actually working at the landfill often are overlooked. In contrast to factories, where measures to ensure the safety of personnel are of a high standard, the importance of safety at landfills has hardly been recognized. The measures necessary to ensure the safety of landfill personnel are only slowly being introduced.

The following provides a survey of the occupational hazards and adverse effects of employment at landfills as well as the measures necessary to be taken to deal with them.

## OCCUPATIONAL SAFETY: GENERAL REMARKS

The term 'Occupational Safety' here refers to all of those measures that need to be taken to protect personnel from the dangers and risks posed to them by their working conditions. They include, for example, measures to ensure safety from fire, explosion, injury by machines, health risks, as well as such questions as humane working procedures and working hours.

The implementation of occupational safety measures should be a self-evident responsibility, since protecting oneself and others from possible harm is one of the basic ethical and moral foundations of our society. On the other

509

hand, the implementation of occupational safety measures also makes economic sense, since unsafe working conditions and dangerous conduct at places of work can lead to serious economic losses. Clearly though, ethical and moral precepts and economic interests do not always have enough of a motivating force to impel people to act.

Experience has shown that only the passing of legislation with detailed provisions leads to the actual implementation of occupational safety measures in individual cases. In view of this fact, a number of national and international regulations have been introduced to ensure safety at work in factories. In West Germany there are, on the one hand, state regulations governing safety at work issued by the central government and, on the other hand, detailed regulations formulated by the state health insurance scheme. At the moment regulations are being drafted for landfills which are expected to come into effect by 1990. These new safety regulations will incorporate and codify current provisions for safety at landfills.

The numerous provisions that currently regulate occupational safety are, however, minimum requirements and do not by any means take into account every individual case, so that further, more extensive, measures may be necessary. For this reason, in order to make working places safe and free from health hazards, the dangers and health risks must first be determined. When they are known, the relevant safety measures must then be implemented. In this context, a clear hierarchy or sequence of safety measures can be made out. The primary task is to take those technical and organizational measures which will completely eliminate present hazards to person and health ('direct safety measures'). The complete elimination of hazards cannot be achieved, however, in many cases (e.g. it is not possible to prevent the forming of biogas in landfills) so that secondary steps must be taken to make sure people are not directly involved in hazardous situations. This can be achieved by putting either time or space between the individual and the potential risk ('indirect safety measures').

An example of an indirect safety measure would be a protective shield on a machine or tool. If, in spite of such measures, hazards remain, then a last step would be the use of 'directive safety measures', such as the wearing of personal protective clothing (over the shoes or the head, for example), or the use of respirators, furthermore, the measuring of the concentration of hazardous substances, the employment of qualified training staff, the placing of warning signs and insistence on the observation of regulations and instructions. These measures are, however, hardly effective since they are not automatic but require the cooperation of those involved. Frequently they are simply ignored, for example, when personnel enter gas-filled, underground shafts without respirators or measuring instruments.

## HAZARDS AND ADVERSE WORKING CONDITIONS AT LANDFILLS

The hazards and health risks posed to personnel by the working conditions at landfills can be of very different kinds and must be determined individually and characterized from case to case. Frequent hazards and impositions are, for example:

1. Contact with hazardous substances (e.g. biogas, acids, bases, silicon dust, infectious substances)
2. Accident risk through collision with heavy vehicles
3. Dangerous technical equipment (e.g. danger points on pumps, defective hydraulic pipes on scrapers, defective electrical and soldering equipment)
4. Vibration caused by heavy vehicles
5. Accidental contact with sharp or pointed objects
6. Impositions placed on personnel by dust, wet, cold, heat, smells and insects.

How these hazards and harsh working conditions at landfills can be eliminated or reduced will be discussed in the following sections by reference to several examples.

## HAZARDS AND ADVERSE EFFECTS OF BIOGAS

### Biogas Constituents

The term biogas is used to refer to the gaseous products of the microbial decomposition of organic waste deposited at landfills, as well as waste substances that have vaporized. As a rule, biogas consists of about 55–60% methane ($CH_4$) and 40–45% $CO_2$. Methane is a non-toxic, colourless, odourless, energy-rich gas, lighter than air. In concentrations of 5–15% volume in air, explosive mixtures form. It has a slightly narcotic effect; however, only in concentrations of more than 50% is it noticeably anaesthetic. Carbon dioxide is a colourless, caustic gas, with a slightly acidic taste and smell, 1.5 times heavier than air. High concentrations of $CO_2$ are toxic even when enough oxygen is available for breathing. Air containing 4–5% $CO_2$ leads to unconsciousness and concentrations of over 9% cause death within minutes.

In addition to the main constituents of methane and carbon dioxide, biogas can also contain a variety of trace substances. Table 1 gives an overview of the range of values of trace gases in biogas and the corresponding MAK values ('maximale Arbeitsplatzkonzentration' = 'maximum concentration at places

**Table 1.** Trace substances in biogas and their MAK values in mg/m$^3$ measured in airfree conditions.

| | Concentration in biogas | MAK value |
|---|---|---|
| Propane | 1.4–13 | 1800 |
| Butane | 0.3–23 | 2350 |
| Pentane | 0–12 | 2950 |
| Cyclohexane | 0.03–11 | 1050 |
| Hexane | 3–18 | 180 |
| Cyclohexane | 2–6 | 1015 |
| Heptane | 3–8 | 2000 |
| Octane | 0.05–75 | 2350 |
| Cumol | 0–32 | 245 |
| Ethylbenzol | 0.5–236 | 435 |
| Toluene | 0.2–615 | 750 |
| Xylol | 0–376 | 870 |
| Dichloromethane | 0–6 | 360 |
| Trichloromethane | 0–2 | 50 |
| Tetrachloroethane | 0–0.6 | 65 |
| 1,1,1-Trichloroethane | 0.5–4 | 1080 |
| Dichloroethylene | 0–294 | 40 |
| Trichloroethylene | 0–182 | 260 |
| Tetrachloroethene | 0.1–142 | 345 |
| Chlorobenzol | 0–0.2 | 230 |
| Hydrogen sulphide | 0–600 | 15 |
| Benzene | 0.03–7 | |
| Chloroethene | 0–264 | |

of work'). The MAK value list (TRGS 900) is compiled anew each year by the German Research Foundation.

The MAK values are the average permissible values for an eight-hour workshift. If these limits are observed then there is usually no risk to health. Excesses of short duration are permissible. The MAK limits do not apply to mixtures. A procedure for assessing the adverse effects of gas mixtures on the basis of a scientific analysis of the interaction of substances is not yet available. The system employed in the USA is not strictly scientific. It is based on the following procedure: if there is no positive evidence suggesting that the effects of the individual constituents of a gas mixture are independent of each other then it is assumed that their effects are cumulative. For each substance the ratio between its concentration in the air and its limiting value is calculated. If the sum of these ratios is greater than 1 then the limiting value for the mixture has been exceeded, if the sum is 1 or less than 1, then the limiting value has not been exceeded.

$$\frac{c_1}{MAK_1} + \frac{c_2}{MAK_2} + \frac{c_3}{MAK_3} + \ldots \leq 1$$

In spite of the criticisms laid against this method it can be extremely helpful, if the formula is not used to assess the interaction of substances but as a guideline for implementing practical safety measures.

Various trace gases are toxic (for example, dichloromethane and hydrogen sulphide), or they are potentially carcenogenic (for example, benzole and vinyl chloride). For substances which are potentially carcinogenic it is usually impossible to work out MAK values since cancer only becomes apparent after many years or even decades. Long-term exposure leads potentially to a large accumulation of mutations. Whether cell damage is repaired or not and, if so, to what extent, is at present unknown. Safe levels of exposure cannot at the moment be determined on the basis of epidemological statistics as neither adequate analytical data on the intake of substances over sufficient periods of time exist, nor sufficiently large numbers of exposed persons and control groups are available. In order to reduce the risk to health posed by carcinogenic substances at work, in West Germany, "Guideline Technical Concentrations" (TRK values) have been drawn up. The TRK value for benzole is 5 ppm and for vinyl chloride 3 ppm.

## Properties of Biogas and the Resulting Hazards

### Asphyxiation

Although biogas is, as a rule, somewhat lighter than air, nevertheless, escaping biogas is capable of displacing air and collecting in shafts, rooms or cavities. If the oxygen content of air decreases from an average 21% vol to below 17% vol then there is a threat of asphyxiation. With respect to biogas this means that when biogas stands in a ratio of 1 : 4 to air, asphyxiation is possible.

### Toxic Effects and Health Risks

The toxic properties of biogas derive mainly from the presence of carbon dioxide and hydrogen sulphide. Concentrations of more than 9% carbon dioxide can cause death within a matter of minutes. As biogas has a $CO_2$ content of between 10 and 50% it loses its toxicity when it is mixed with air to a ratio of 1 : 4. The amount of hydrogen sulphide present in biogas is usually below the toxic level of 800 ppm. However, higher values do at times occur, especially at landfills for hazardous wastes (up to 17 000 ppm). As long as the gas can escape over the whole landfill area and is sufficiently mixed with air, there is no immediate danger. In places where concentrated discharges of gas are possible, for example, from substantial fissures in the landfill cover or from vertical gas discharge shafts, the limits permitted at places of work may be exceeded.

There is no reliable data on the long-term effects on health of exposure to biogas. The difficulty here lies in evaluating the toxic properties of a gas mixture such as biogas. The physiological effects of the individual substances are known and corresponding limits have been laid down. When mixed, however, the toxic properties of the individual substances may decrease or increase. Toxicologists refuse to set limits for mixtures. Personnel employed at landfills frequently point to cases of colleagues dying of carcinomas and believe this to result from long-term exposure to biogas. Such claims have not yet been scientifically proved.

## Fire and Explosion Hazards

Methane, an important constituent of biogas, can, when mixed with air, cause fires, explosions and fulminations. Within the upper and lower limits of explosion (between 5 and 15% vol methane), methane can be detonated. Gas mixtures below and above the limits of explosion are not explosive but quantities $\geq$ 15% represent a fire hazard. When methane has been mixed with air to the usual ratio over the surface of the landfill then there is no immediate danger. Biogas can collect in shafts, trenches and cavities to form an explosive atmosphere, which can then be detonated by gas burn-offs, biogas compressors, open fires or lights or hot surfaces such as car or truck exhausts.

There have been many reports of explosions or fulminations at landfills, which have, however, rarely been considered in the relevant literature. In one case, explosive biogas was detonated at a distance of 150 m from a gas discharge shaft and then began to burn in the discharge shaft itself.

**Safety Measures**

In order to eliminate or reduce the risks created by biogas the following safety measures must be taken:

1. The waste delivered to landfills should, as far as possible, have had all hazardous substances removed; these should be collected, for example, at mobile or stationary collection points and then disposed of at special dumps or processing plants for dangerous materials. In this way substances which may vaporize and are responsible for the dangerous trace gases in biogas can be reduced.

2. Biogas must not have harmful effects on the health of the personnel at landfills. This can be achieved during the construction of a landfill by the installation of horizontal degassing equipment and by making sure that vertical vent pipes are properly sealed and discharged gas is prevented from collecting and concentrating. If these measures are not practicable then respirators using an independent air supply should be worn during work in the vicinity of escaping gas.

3. Buildings which have to be entered by personnel should be constructed in such a way that dangerous concentrations of biogas cannot form (for example, by observing the principle that all underground buildings should have the necessary technical ventilation devices installed).

4. Areas in which there may be dangerous concentrations of biogas (for example, shafts and enclosed leachate sumps) should only be entered by personnel after it has been ascertained by means of portable measuring instruments that the atmosphere does not contain dangerous concentrations of gas. When warning devices detect dangerous concentrations, then a safe atmosphere must be created by the deployment of ventilators. Ventilators may be regarded as sufficient if they fulfill the following technical requirements: (a) At the lowest point in shafts and underground gullies an air current of 600 $m^3$/h per $m^2$ of shaft cross-section is produced; (b) In other constructions such as pump stations the air should be renewed at a rate of six to eight times an hour.

5. Personnel must present themselves regularly for medical checks. This aspect is dealt with in greater detail in a following section.

## HEALTH HAZARDS FROM DUST AND INFECTIOUS SUBSTANCES

The substances deposited in landfills can present a great health risk. Infectious substances may be contained in, for example, treated sewage, refuse from doctors' surgeries or kitchen waste. Direct contact with these substances or the inhalation of bacteria and dust represent a potential health risk.

One of the most important risks to health is the inhalation of hazardous dust. Such dust may contain, for example, asbestos fibres or silicotic particles. Long-period exposure to even inert dust particles through deposition in the lungs, may be detrimental to the health of personnel. In order to reduce the health risk due to direct contact with infectious materials it is necessary that protective clothing be worn and that the necessary sanitary facilities for personnel are available. Working clothes contaminated by infectious substances should be stored separately from private outdoor clothes and should be cleaned at the landfill site so that the infectious materials do not contaminate the private homes of the personnel.

Hazards from dust can be eliminated or reduced by the following measures:

1. Filters fitted to cabs of landfill vehicles
2. Moistening of substances that tend particularly to give rise to dust (for example, the moulding sand from foundries before delivery to the landfill)
3. Prompt covering of substances likely to cause dust
4. Sprinkling dust with sprinkling vehicles
5. Wearing of dust masks (if the other measures fail).

Cab filters for vehicles are a new development, especially devised for the decontamination of disused dumps (Mascher, 1989). This basic idea is to seal the cab to such an extent that the air supply has to pass through a filter. The filter apparatus consists of separators for coarse and fine dust, a charcoal filter, a ventilator, and either a radiator or air-conditioner, while the cab itself has an excess pressure gauge and a warning device for sudden falls in pressure.

## HAZARDS PRESENT DURING REFUSE DELIVERY

Fatal accidents frequently occur during the delivery and covering of refuse at landfills. The main causes of this are the following:

1. Collection vehicles reversing without proper piloting
2. Unloading point and covering areas are not separated
3. Covering of refuse immediately after delivery
4. Lack of sanitary facilities or if available not open to use by people delivering refuse
5. Lack of conspicuousness of people who are not landfill personnel
6. People searching refuse for reusable objects.

For these reasons it is advisable to divide the landfill into separate areas for entry and exit, unloading, sorting, and covering. By separating these areas from each other it is possible to prevent delivery vehicles and covering vehicles from coming into contact with each other in a confined space, thus avoiding hazardous situations arising for both landfill and delivery personnel. A good solution is to have the refuse unloaded onto ramps at delivery and then to move it exclusively by means of landfill vehicles.

It should go without saying that sanitary facilities must be provided and open to use by all authorized personnel. All delivery personnel and landfill personnel engaged in refuse covering should be made easily conspicuous by the wearing of the appropriate clothing with warning marks. The rummaging in refuse after delivery should be prohibited.

## EFFECTS OF VIBRATION ON DRIVERS OF LANDFILL VEHICLES

Exposure to vibration at work can endanger the health and safety of employees. The measurement of vibration and its evaluation on the basis of international standards (ISO) and national guidelines enable the hazards to health and safety to be assessed, ISO 5379 (1986) and DIN-ISO 7096 (1984). A very broad and intensive study of the adverse effects of vibration on landfill drivers revealed that vibration can lead to the spinal conditions that derive from higher than average abrasion of the vertebrae (Köhne *et al.*, 1982). Intense vibration of the hands and arms deriving from adjustment levers can cause damage to bones and joints in the lower arms and lead to tissue and nerve damage in the hands.

It is not possible to prevent chassis vibration in landfill vehicles that have no independent suspension and work on rough surfaces. For this reason all technical means must be employed to reduce as far as possible the amount of vibration acting on the driver's body. This can be achieved by means of

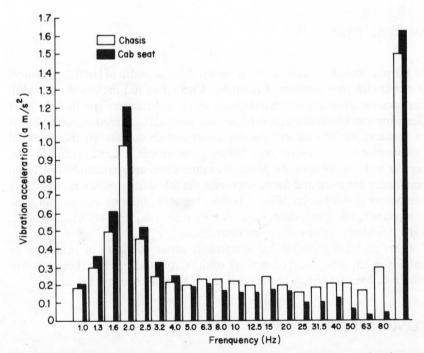

**Figure 1.** Frequency spectrum of vertical vibration acceleration on the driver's seat in a landfill vehicle in relation to a measuring point on the chassis, Köhne *et al.* (1982).

modifications in cab construction: vibration-proof seats and levers. Research has shown that cab seats frequently increase rather than decrease vibration acting on the driver's body (Fig. 1).

The reason for this is failure to harmonize the frequency-dependent suspension capability of the seats and the actual vibration of the vehicles. There is no increase in vibration if the transmitted rate of vibration acceleration does not exceed the value $a_{zw} = 1.25$ m/s$^2$. Seats which fulfill this requirement play a major role in reducing work-related stress factors.

## UTILIZATION OF TECHNICAL EQUIPMENT

Landfills employ a variety of different types of technical equipment. This includes covering devices and the tools for servicing and repairing them, landfill vehicles, instruments for measuring gas, and respiration devices. The safe utilization of these technical devices depends on their being in good condition and the personnel being properly trained to use them.

## MEDICAL CARE

In the previous sections it has been shown that the health of landfill personnel is put at risk by a number of hazards. Whether or not the health of landfill personnel is adversely affected depends on the substances deposited. In West Germany the law states that if hazardous substances exceed statutory limits or if particular jobs are carried out under certain conditions, then medical examinations are necessary both before employment is taken up and then at regular intervals afterwards. Medical examinations are required by law when respirators are worn and for contact with the following substances, which are frequently present at landfills: asbestos, benzene, fluorine and its inorganic compounds, silicogenic dust, sand-blasting materials, vinyl chloride, toluene, xylol, trichloroethylene and tetrachloroethene.

If an examining doctor has misgivings about the state of health of an employee, the latter must cease work until safety measures have been taken to eliminate harmful effects or the permissible limit is no longer exceeded.

## CONCLUDING REMARKS

Landfill personnel are exposed by the nature of their work to a variety of health risks. The most important hazards and methods of eliminating them have been discussed above. It has been shown that the problem of the adverse effects of biogas on the health of personnel has not been solved satisfactorily yet. Thus a toxicological assessment of biogas must become an urgent priority.

There are many regulations governing occupational safety; they do not, however, relate specifically to landfills. At the moment safety regulations are being formulated in West Germany which are expected to come into effect in 1990. These regulations will give detailed provisions for the safety measures required at landfills.

## REFERENCES

DIN-ISO 7096. (1984). 'Landfill vehicles — driver's seat — vibration transmission'.

ISO 5349. (1986). Mechanical vibration—Guidelines for the measurement and the assessment of human exposure to handtransmitted vibration. First edition 1986-05-15.

Köhne, G., Zerlett, G. and Duntze, H. (1982). 'Effects of vibration on landfill vehicles: development of suitable systems to decrease vibration'. Schriftenreihe HdA, Vol. 32, VDI-Verlag, Düsseldorf, 1982.

Mascher, W. (1989). 'Cab filters for landfill vehicles'. *In* Tiefbau-Berufsgenossenschaft, München, 2/1989.

# 7.   DESIGN

# 7.1    The Austrian Guidelines for Sanitary Landfills

PETER LECHNER

Technische Universität Wien, Karlsplatz 13, 1040 Wien, Austria

## INTRODUCTION

By order of the Ministry of Agriculture and the Ministry of Public Health the Institute for Water Quality and Waste Management of the Technical University of Vienna elaborated guidelines for designing and operating municipal solid waste landfills. This chapter deals in detail with the main and important subjects of these guidelines.

Austria produces over 2 million tons municipal solid waste a year. Approximately 80%, including the residues, of 15 composting plants and three incinerating facilities, are sent to landfills. In the opinion of the author, landfilling will always be the final element of solution after recycling, destroying and concentrating the residues.

These guidelines only deal with municipal solid waste. The co-disposal of industrial waste must be taken into consideration in every single case.

## SANITARY LANDFILL SITING

A landfill should serve a population of at least 50 000 inhabitants. To formulate hydrogeological requirements is of great importance because there is no guarantee that a lined site can provide 100% of containment for an indefinite period. Soil and ground-water conditions to depths that could potentially be affected by leachate must be explored by hydrogeologic cross-sections. In general, sanitary landfill sites must be located in areas where ground-water is not available neither today nor will be in the future or is either unsuitable for the utilization as drinking water or is of limited volume. The landfill, including the liners, must be located entirely above the seasonal high water table.

The landfill site shall include a 300 m distance around the entire filling area where uniform hydrogeological conditions are required. Bedrock or a thick clay layer must be encountered at a technically reachable depth. Austria has big carst areas. In these areas, no sanitary landfill sites shall be located. Also no sanitary landfill sites should be located in natural reserves areas. Sanitary landfill shall be sited at a minimum distance of 300 m from urban areas.

Failure of leachate collection systems may cause leachate to build up on top of the liner. This can lead to failure of the liner system and to leachate migration through subsurface soil into ground-water. To prevent the building up of leachate inside the sanitary landfill the leachate has to leave the landfill at the bottom by gravity. Therefore gravel pits are not suitable sites for sanitary landfills.

The above-grade (refuse mound) landfill-method should be preferred; therefore, many aspects of landfill design are dictated by refuse stability. Maximum slopes of 1–2.25 shall be the norm for a landfill height up to 30 m.

## SITE DESIGN

### Lining System

Site lining is required in all cases. Leachate can migrate through subsurface soil into ground and surface waters emitting pollutants which may threaten human health.

The proposed lining system must be constructed of two different materials that have appropriate chemical properties and sufficient strength and thickness to prevent failure. The system requires a primary flexible membrane liner of at least 2 mm of synthetic material, chemically resistant to waste and leachate. This primary flexible membrane liner must be in direct contact with a secondary compacted soil liner of low permeability ($k_f < 10^{-9}$ m/s). The low permeability soil component must be at least 60 cm of compacted material with an in place saturated hydraulic conductivity of $10^{-9}$ m/s or less. This soil liner should be made of three layers of 20 cm. The maximum grain size should be less than 36 mm (Fig. 1). As these two liners must be in direct contact no leachate collection system between these two liners is possible. The membrane liner should be protected by the drainage layer immediately. As an additional protective measure a geotextile can be placed above the plastic liner. The lining system must slope towards the leachate collection pipes at a minimum of 3% grade.

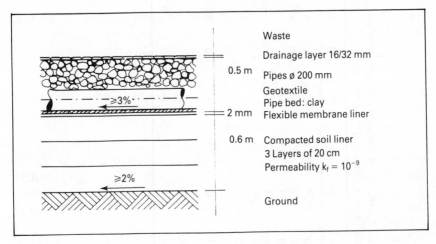

**Figure 1.** Lining and drainage system; Austrian guidelines for sanitary landfills.

## Drainage System

By means of the drainage system, the leachate, from any point of the landfill basis, has to be removed. The drainage system consists of a gravel layer and pipes. If flow through the pipes is restricted, the drainage requirements shall be met by the gravel layer alone. Very important is a careful pipe bedding since differential settling as a result of inconsistent slope can cause a build up of leachate locally. The pipe size shall be 200 mm in diameter, the minimum pipe slope shall be 2%. No bends should be in horizontal projection. To protect the pipes the drainage layers shall cover the pipes with 30 cm thick strata at least. Pipe outlets below a water surface will maintain anaerobic conditions in order to avoid chemical precipitation. Nevertheless, clogging or pipe deterioration caused by chemical attack cannot be excluded. Access should therefore be provided to all parts of the system. As a consequence the placing of manholes has to be done in a way that inspection and maintenance is possible. The distant between access points should be in maximum twice the operating distance of jet cleaning equipment. The periodically televisual direct inspection of the draining pipes is necessary. The leachate collection and removal system shall be established immediately above the lining system. The leachate collection and removal system should not be used as a gas-extraction system. As an additional protective measure against clogging a geotextile or 10–15 cm of a graded granular filter can be placed above the drainage layer.

**Table 1.** Water budget of the closed landfill of an above-grade landfill in the east of Austria ('pannonisches Klima') (Klaghofer, 1986). Vegetation = grasses; evaporation = +30% for above-grade landfill in this climate type; no surface run-off.

| Month | I | II | III | IV | V | VI | VII | VIII | IX | X | XI | XII | Year |
|---|---|---|---|---|---|---|---|---|---|---|---|---|---|
| $T$ °C | -4.3 | -1.5 | 5.2 | 7.5 | 14.6 | 18.1 | 19.9 | 19.5 | 17.4 | 10.4 | 4.8 | 2.2 | 9.5 |
| $N$ mm | 44 | 10 | 17 | 6 | 55 | 35 | 105 | 42 | 19 | 49 | 10 | 38 | 430 |
| $V$ mm | — | — | — | 120 | 129 | 147 | 136 | 149 | 89 | 33 | — | — | — |
| $k_c$ | — | — | — | 1 | 1 | 0.8 | 0.5 | 0.8 | 0.3 | 0.3 | — | — | — |
| $ET_c$ mm | — | — | — | 72 | 77 | 71 | 41 | 27 | 16 | 6 | — | — | — |
| $ET_c + 30\%$, $nWK = 55$ mm (topsoil: 50 cm sandy, gravelly soil) | | | | | | | | | | | | | |
| $S$ mm | 0 | 0 | 16 | 0 | 0 | 0 | 0 | 4 | 0 | 39 | 10 | 38 | 107 |
| $ET_c + 30\%$, $nWK = 160$ mm (topsoil: 50 cm waste compost) | | | | | | | | | | | | | |
| $S$ mm | 0 | 0 | 0 | 0 | 0 | 0 | 0 | 0 | 0 | 0 | 0 | 0 | 0 |

$T$, temperature.
$E_p$, evaporation for a plane area (class A).
$ET_c$, transpiration.
$N$, precipitation.
$k_c$, transpiration ratio.
$nWK$, water-retaining capacity.
$S$, infiltration ( = leachate generation).

## LEACHATE

The leachate generation should be predicted by using a water budget calculation of the closed landfill. Tables 1 and 2 show the estimated leachate generation for two different climatic types and an above-grade landfill in Austria.

For minimizing the leachate generation during the time of operation it is necessary to fill the landfill in small sections and to fill each landfill section to the final height in a short time. If this is not possible a temporary capping and revegetation of these sections should be made. To minimize the infiltration rate during the operation phase waste layers of 0.5 m height in maximum shall be formed and strongly compacted. The temporary and final slopes shall be carefully designed to inhibit infiltration within the site.

The leachate must be treated in a special leachate treatment facility or co-treated in a municipial waste water treatment plant. Leachate produced by landfills in the acid phase can be readily treated by a biological process. Leachate produced in the methanogenic phase requires an additional physico-chemical treatment. If there is no possibility to transport the leachate by the pipes to the treatment plant a storage basin shall be established. The capacity of this basin should correspond to the average one week precipitation height at the landfill area in operation. The leachate must flow to the storage basin by gravity. Lifting the leachate in front of the storage basin is prohibited. Also an underground storage basin is prohibited (Fig. 2).

## LANDFILL GAS

Gas extraction systems should be already used during landfill operation. The leachate drainage system shall not be used as a gas extraction system. Active gas extraction will be necessary if the final landfill cover system includes a barrier of compacted clay soil. The gas extraction pipes must be sufficiently stable and enable good dewatering facilities. Actively extracted landfill gas shall be treated by combustion (flares); utilization; deodorization by means of biofilter. For these treatment systems safety requirements are formulated.

Passive gas venting may be achieved if a thin final layer cap of sand or soil allows the gas to migrate through the completed landfill surface. In this case, the oxygen level is not so depressed that root growing can be completely inhibited. Difficult growing conditions for plants may appear only locally, particularly during the starting phase of revegetation. This form of 'open landfill' should be possible with the east Austrian climate (see Table 1). An open landfill will also be possible if there is no need to get the most out of the completed landfill site.

**Table 2.** Water budget of the final landfill cap of an above-grade landfill (hillfill) in the middle and west of Austria ('zentral-alpines Klima') (Klaghofer, 1986). Vegetation = grasses; evaporation = +15% for above-grade landfill in this climate type; 50% surface run-off in April–October.

| Month | I | II | III | IV | V | VI | VII | VIII | IX | X | XI | XII | Year |
|---|---|---|---|---|---|---|---|---|---|---|---|---|---|
| $T$ °C | −4.4 | −5.1 | −1.7 | 2.1 | 8.6 | 12.7 | 15.6 | 13.4 | 12.2 | 5.1 | 0.8 | 3.5 | 4.7 |
| $N$ mm | 219 | 6 | 100 | 27 | 57 | 191 | 82 | 187 | 55 | 53 | 56 | 99 | 1132 |
| $V$ mm | | | | | 80 | 55 | 52 | 54 | 45 | 30 | | | — |
| $k_c$ | | | | | 0.6 | 0.8 | 1 | 1 | 0.6 | 0.3 | | | — |
| $ET_c$ mm | | | | | 48 | 44 | 52 | 54 | 27 | 9 | | | — |
| $ET_c$ mm (+15%) | | | | | 55 | 51 | 59 | 62 | 31 | 10 | | | — |
| $AO$ mm (= 50% N) | 219 | 6 | 100 | 27 | 29 | 82 | 41 | 94 | 28 | 27 | 56 | 99 | — |
| **nWK = 55 mm (topsoil: 50 cm sandy, gravelly soil)** | | | | | | | | | | | | | |
| $S$ mm | 219 | 6 | 100 | 27 | 9 | 147 | 30 | 133 | 28 | 44 | 56 | 99 | 898 |
| **nWK = 160 mm (topsoil: 50 cm waste compost)** | | | | | | | | | | | | | |
| $S$ mm | 219 | 6 | 100 | 27 | 0 | 0 | 0 | 0 | 0 | 0 | 56 | 99 | 507 |

$T$, temperature.
$E_p$, evaporation for a plane area (class A).
$ET_c$, transpiration.
$N$, precipitation.
$k_c$, transpiration ratio.
nWK, water-retaining capacity.
$S$, infiltration ( = leachate generation).
$AO$, surface run-off.

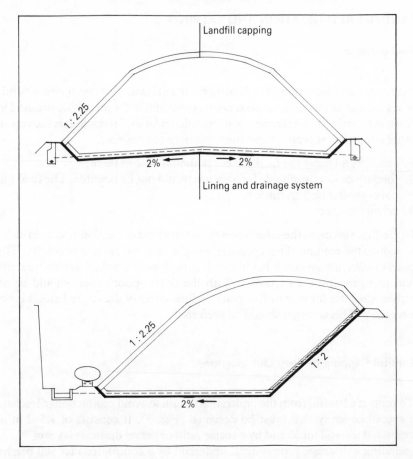

**Figure 2.** Preferred types of sanitary landfills; Austrian guidelines for sanitary landfills.

## LANDFILL OPERATION

The landfill should be filled in landfill sections in order to reach the final height as soon as possible. If this is not possible a temporary landfill capping and revegetation with compost shall be made. If daily cover is necessary only material with high permeability shall be used. Doing so leachate migration between lifts will not occur. The maximum area of a landfill section in operation should be 1 ha.

## LANDFILL REVEGETATION AND CAPPING

### Revegetation

Because of accidents caused by landfill gas the utilization of completed landfills as residential or industrial areas is not recommended. As there is no pressure in Austria to make such extreme use of completed landfill sites, as it is the case in other countries, several revegetation solutions are possible:

1. agricultural use—the final soil cover should be 2 m thick;
2. forestry use—a profitable forestry use would not be possible. The final soil cover should be 3 m thick.
3. ecological areas.

In the first two cases the installation of an active gas extraction system in order to reduce the volume of gas escaping into the final soil cover is necessary. The costs of covering an entire landfill with enough native soil for agricultural use and tree growing are excessive. With the third option, there should be no public after-use for at least five years. A final cover of the entire landfill is not necessary. This solution should be preferred.

### Landfill Capping to Keep Out Moisture

To protect a landfill from the infiltration of rain to avoid production of leachate a special cover system must be designed (Fig. 3). It consists of 1.5–2 m of vegetated top soil underlaid by a coarse sand or gravel drainage layer of 30 cm including a drainage pipe system, underlaid by a compacted clay soil barrier layer of 60 cm, underlaid by a drainage layer of 30 cm. An active gas extraction system must be established. To make the landfill cap resistant to physical forces (erosion) a dense vegetation should be established. The roots must be prevented from penetrating the clay cap. Therefore the top soil cover must be at least 1.5 m thick.

## GROUND-WATER MONITORING NETWORK

Before starting the landfill operation a monitoring network must be installed. The wells shall be located so that samples can be obtained showing the quality of upgradient ground-water (background wells) and downgradient ground-water adjacent to the landfill at various depths.

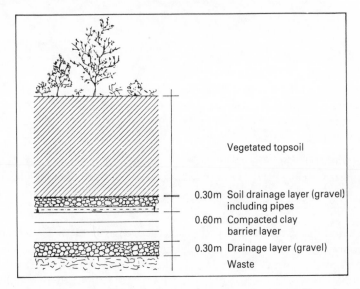

**Figure 3.** Landfill capping to keep out moisture; Austrian guidelines for sanitary landfills.

## REFERENCES

Klaghofer, E. (1986). 'Einfluβ von Klima und Topographie auf den Wasserhaushalt von Mülldeponien'. Anforderungen an Mülldeponien—Grundlagen zum Richtlinienentwurf, Abfallwirtschaft Band 7, S. 55–56, TU Wien.

Kremnitzer, P. (1986). 'Kunststoffdichtungsbahnen'. Anforderungen an Mülldeponien—Grundlagen zum Richlinienentwurf, Abfallwirtschaft Band 7, D. 149–165, TU Wien.

Lechner, P. and Pawlick, R. (1987). 'Richtlinienentwurf für Mülldeponien'. Auftrag der BM für Land- und Forstwirtschaft, Gesundheit, Umwelt und Familie.

Pregl, O. (1987). 'Geotechnische Anforderungen an Mülldeponien'. Anforderungen an Mülldeponien—Grundlagen zum Richtlinienentwurf, *Abfallwirtschaft* Band 7, S. 107–125, TU Wien.

Pregl, O. (1987). 'Geotechnische Probleme beim Entwurf und beim Bau von Mülldeponien'. Vortrag 2. C. Veder-Kolloquium, TU Graz.

# 7.2    Practice and Trends in Landfill in the UK

KEITH KNOX

*Knox Associates, 21 Ravensdale Drive, Wollaton,*
*Nottingham NG8 2SL, UK*

## INTRODUCTION

This chapter is in two parts. In the first part, current practice in UK landfilling is examined, particularly in relation to two main controversies over design concepts and the suitability of landfill for various wastes. In the second part, recent trends in both landfill practice and in the administrative process are discussed.

In examining current practice, a different picture may emerge to that which might be obtained from a reading of popular environmental magazines. Specifically, the following points will be stressed:

1. UK authorities do not encourage, grant planning permission for, or licence, the indiscriminate landfilling of all forms of industrial waste in all kinds of sites. They do encourage the very carefully controlled disposal of a restricted range of industrial wastes if they can be shown to be suitable for landfill, in a particular, defined context.
2. Of the two theoretical extremes in landfill design, there is not a general preference across the UK for dispersal sites rather than containment sites.
3. There is no nationwide 'British Concept' in landfill design, nor should there logically be one. A wide diversity of landfill designs is in use, and if sound technical and environmental assessment are to continue to be the overriding considerations, that diversity will continue.

In examining trends, a warning is sounded against allowing case by case application of logic, science and professional expertise to be displaced by a recipe-book approach and a mistaken belief that philosophy can be a remedy for bad practice and poor quality work.

533

## CURRENT PRACTICE IN THE UK

### The Physical and Administrative Background

For the design and assessment of any proposed landfill, the most important factors include rainfall, topography, hydrogeology and the nature/quantity of wastes to be landfilled. For the latter, the distribution of population and industry may be key influences. In each of these factors, the extent of variation across the UK is often surprising to outsiders, for a country of such relatively small size.

All geological periods from pre-Cambrian through to Quaternary are represented by appreciable areas of the country. In general, older rocks are represented in the north and west, giving way to younger rocks in the south and east. There are large areas of glacial and alluvial deposits in many places.

Annual rainfall varies from over 1500 mm in hilly areas in the west of the country (e.g. Wales, the Lake District, western Scotland, much of which lie above 300 m) to as little as 550 mm or less in some lowland eastern areas. Since actual evapotranspiration varies much less, tending to lie in the region of 300–600 mm over much of the country, there can be almost an order of magnitude difference in the volume of effective rainfall in different areas and, therefore, in the potential for production of leachate.

Approximately one-third of the UK's water supply comes from ground-water. However, the suitability of ground-water for use, and the extent of exploitation vary a great deal across the country. The major water supply aquifers are the Cretaceous Chalk, Lower Greensand and superficial deposits, mainly in the South East, the Permo-triassic sandstones in the Midlands and north west, and various other limestones. These, and some lesser acquifers, are abstracted on a large scale for potable supply by the Regional Water Authorities, of which 10 cover the whole of England and Wales and by private Water Companies. There are, in addition, many small abstractions, both inside and outside these major supply areas.

In many locations, even though not used for abstraction, hydrogeological considerations indicate the potential for rapid transmission of leachate to surface waters. On the other hand, there are areas where ground-water flow rates are slow, and where a significant unsaturated zone exists, or can be created, such than an attenuation layer could be designed and installed beneath potential landfill sites. In some cases the clay mineral content and the organic content of the underlying materials is such that the potential for adsorption and ion-exchange, as well as biodegradation, is enough to allow an unlined site to have little or no impact.

The anticipated rate of waste input can be an important factor in landfill design, particularly with respect to leachate control and the possible need for leachate disposal facilities. The concentration of population and industry in specific areas of the UK means that the quantities of waste available for a given site, or more importantly, the rate of waste input available within a given area may be very much greater in urbanized areas than rural areas.

If co-disposal of industrial wastes with domestic refuse is being planned, as opposed to joint disposal or mono disposal, the relative quantities of the two are important in order to ensure that domestic refuse decomposition processes dominate within the landfill and are able satisfactorily to attenuate the key components of the industrial wastes.

In addition to the variations in physical factors, the administration of waste disposal rests at a local level, leading to the potential for further variability in design and supervision across the country. The present arrangements have been in existence since 1976. Any waste management facility, be it landfill, incinerator, treatment plant or transfer station, must have a Site Licence, in order to operate lawfully. This licence is issued by the designated Waste Disposal Authority (WDA), which has the power and duty to impose detailed conditions on site preparation, waste inputs, operational methods, monitoring and restoration. In England, the WDA is the County Council, with the exception of special arrangements for some of the large metropolitan areas. There are 76 English WDAs. In Wales, WDA responsibility is at an even more local level, that of the 37 District Councils, although in practice the Welsh WDAs work through larger Regional Groups. In Scotland, the 56 District Councils are the WDAs, while in Northern Ireland the 26 District Councils are the WDAs.

Thus, over the UK as a whole, for a population of 57 million, there are 195 separate Waste Disposal Authorities with legal responsibility for surveying, planning and updating waste disposal activities in their area and for assessing, licensing, monitoring and enforcing Site Licences.

A direct and practical influence on landfill design is also exerted through the WDAs by the Regional Water Authorities (RWAs), of which there are ten covering England and Wales (River Purification Boards have a similar role in Scotland). RWAs are statutory consultees in any Site Licence application and it is normal practice for informal consultation and discussion to take place with them before formal application for a Site Licence. It is often the RWAs which determine the degree of water quality protection which is necessary, and the acceptability of a proposed landfill design. The RWAs act independently of each other, and some are further split into geographically separate divisions, with a considerable degree of autonomy. The scope for different combination of WDA/RWA attitudes and practices is therefore considerable.

Central government does not involve itself directly in planning or policing waste disposal, having taken the view that detailed control is best undertaken at a local level. However, it is involved in five influential ways.

Firstly, it publishes guidance in the form of a series of Waste Management Papers (WMPs), covering many classes of wastes, both industrial and domestic. Many of these contain Codes of Practice for particular classes of waste. They are heavily used by WDAs in preparing site licences and in assessing proposals for new landfills, or specific requests concerning specific wastes.

Secondly, the Secretary of State for the Environment is responsible for introducing and fine-tuning legislation, regulations and Codes of Practice, and also determines the outcome of any appeals made against Site Licences or conditions in them. The accumulation of appeal decisions (many of which are made public) has formed a significant body of 'case law' which often amplifies the guidance given in WMPs.

Thirdly, the Department of Environment has sponsored research into waste management since 1973, the results of which have proved extremely valuable in better design and operation of landfills.

Fourthly, a Waste Inspectorate was set up (now incorporated in Her Majesty's Inspectorate of Pollution, HMIP), with the power to visit any facility at any time and to publish its observations, with names, if it so decides. Although a non-statutory body (at the time of writing), the reports and recommendations presented have had a significant influence on the standards of operation of disposal facilities.

Finally, the Department of Employment is responsible for the Health and Safety at Work Act, 1974, through its Health and Safety Executive. This exerts an important influence on day to day working practices.

A survey of licensed landfill facilities in the UK undertaken in 1985, indicated an approximately pyramidal structure in their use for increasing hazardous materials, with the following rounded-off numbers:

≈ 5000 licensed landfills in total;
≈ 500 licensed to take difficult wastes;
≈ 50 licensed to take special wastes.

### The Debate on Disposal Philosophies

A persistent theme in much of the literature is that all landfills fall into two categories, 'containment' and 'dispersal', that each type is the physical embodiment of a distinct philosophy for the management of wastes, and that the two philosophies are fundamentally conflicting, the containment one often being broadly portrayed as a good and ethical philosophy and the other as bad.

This theme is based upon false premises: the two categories are rarely found in a pure form in practice, but represent two idealized extremes of landfill hydrogeology. While the phrases can be useful on occasion, in reality few landfills are found to conform to either extreme when subjected to detailed scrutiny. Nor is it necessarily the case that the two types reflect philosophical judgements.

Environmentally acceptable landfill need not to be dependent upon philosophy. It is dependent upon: good hydrogeological assessment and design; upon good quality engineering; upon good quality operation, technical control and monitoring; and upon good quality restoration and aftercare. In other words, it depends upon the proper application of entirely practical skills.

In considering whether a spectrum of landfill types might be suitable for a spectrum of waste types, rather than insisting on a single landfill type and upon the rejection of industrial wastes from them, it is instructive to examine the disposal concepts available for any problem material.

For any material which may be hazardous to man or damaging to this environment, there are three, and only three, distinct principles which can be incorporated into a course of action. They are:

1. Alter the nature of the material so that it is no longer hazardous;
2. Dilute the material or allow its release into the environment at a rate which gives rise to no hazard and does not cause unacceptable damage to the environment;
3. Isolate the material from humans and from the rest of the environment.

Properly operated co-disposal, as understood and accepted in the UK, attempts to minimize total reliance upon permanent isolation but to retain materials for long enough to maximize the extent of alteration and dilution. This approach greatly reduces the burden of perpetual surveillance necessary if permanent isolation is the major environmental defence in the disposal method.

There are present in a decomposing domestic refuse mass many mechanisms—physical, chemical and biological—which are similar to those used regularly in treatment plants and which can immobilize, or degrade, problem components of many residues. As a result, co-disposal can be practised in such a way that leachate quality is indistinguishable from that arising at a site receiving only domestic wastes. The principle of controlled co-disposal is to make use of this capacity, but only to the extent that:

1. inhibition of the normal refuse breakdown does not occur;
2. there is no effect on leachate quality which would make its disposal more difficult or its environmental impact more severe;
3. there are no unacceptable hazards to operators, visitors or to site neighbours;
4. restoration, after-care and after-use are not significantly hampered.

An important aspect of assessing any landfill proposal, whether for industrial, domestic or co-disposal purposes, is the degree of risk that something might go wrong, and the presence of additional lines of defence.

## Examples of UK Landfills

Having indicated in the preceding sections that there is a scope for a variety of landfill designs and practices, the resulting diversity will be illustrated by four examples which will be described briefly. They have very different characteristics of site design, waste inputs and methods of operation. However, all four are designed to keep environmental impact within acceptable limits in a manner which is derived from their own set of circumstances. The characteristics of the four sites are summarized jointly below.

Site 1. A site which would receive low inputs of domestic waste only, in a high rainfall rural area, was designed with the acceptance that leachate production was inevitable, and engineered to collect and remove leachate for on-site treatment and disposal.

Site 2. A site intended to receive high inputs of domestic waste in a low rainfall area, such that little leachate would be produced, was found to have a permeable base, with a hydraulic link to a nearby high quality stream. A semi-permeable attenuating blanket was therefore designed and installed in order to remove ammonia and COD from any leachate leaving the base of the site.

Site 3. A large co-disposal site, over seventy years old, with high waste input rates and low rainfall is located on a natural base which provides partial containment. The resulting build-up of leachate provides sufficient retention time for attenuation processes within the refuse mass to act upon components of the industrial wastes. The resultant leachate is indistinguishable from a purely domestic waste leachate and is treated biologically on site to a standard suitable for direct discharge to surface waters.

Site 4. A clay pit in a moderate rainfall area, intended to receive largely industrial solid and liquid wastes, has been engineered to reservoir standards to provide long-term isolation of wastes. Cellular infilling has been used to minimize rainfall ingress and to maximize the potential for liquid waste disposal. Leachate quality is considerable worse than from a co-disposal site.

Sites 3 and 4 epitomize the acceptance, in the UK, of landfilling as an outlet for a high proportion of the country's industrial wastes. Because this practice is the subject of much interest and criticism in other countries, the two sites will be described in greater detail than sites 1 and 2.

## Site 1; Bryn Posteg, Wales (Davies, 1985; Robinson, 1985)

This site, opened in 1982, was developed to provide facilities for domestic refuse disposal for 25 years in an upland area of Wales (altitude 350 m AOD). The area has an annual rainfall of about 1200 mm and monthly rainfall can be as high as 300 mm. The waste inputs were not expected to exceed approximately 50 t/day so that significant production of excess leachate was anticipated, in spite of the intention to infill in small cells.

The site is underlain partly by low permeability waste materials from a former lead mine and partly by a peat mire. These were assessed as being capable of being engineered to provide a considerable degree of protection for ground-water. However surface water quality in the area is high; it was judged to be necessary to minimize leachate losses by improving the site's containment characteristics and collecting leachate for treatment and disposal.

The base of the site was engineered to provide a 1 m thick clay seal. Additional works included cut-off systems to prevent surface run-on and shallow ground-water from entering the wastes. Leachate is routed to pumping chambers using drainage lines consisting of perforated pipe and rubber tyres, protected from the refuse by Terram fabric.

The collected leachate is pre-treated in an aerated lagoon on site, before being discharged for further treatment at a sewage works some 3 km away. The design and operation of the leachate treatment plant have been described in detail by Robinson (1985), and to date have worked very effectively.

The necessity to collect and treat the leachate is obviously a significant cost element for the landfill, and reliance on such a system has been criticized. However, the overall solution is environmentally acceptable and the overall costs of the site still make it the best option available to the local authority concerned, giving them guaranteed disposal for a long period.

## Site 2: Stangate, Kent (Robinson and Lucas, 1985; Robinson and Blackley, 1987)

This site is used for the disposal of domestic refuse with some commercial and trade waste. It is located in a rural area in a worked out limestone quarry. The waste input rate has been high since the site opened in 1981 (up to 1300 t/day), and the rainfall in the area is moderate ($\approx$ 750 mm/a). The potential existed to fill to final levels, cap and restore, before the absorptive capacity of the refuse was exceeded. With a low permeability cap, the long-term rate of leachate generation would be very low.

However, a high quality stream runs within 100 m of the site at a level below that of the quarry floor. A hydrogeological investigation showed that there was a hydraulic connection between the site and the stream. Calculations indicated that, of the key components expected in the leachate, chloride would be sufficiently attenuated by dilution alone, whereas ammonia and COD would require additional attenuation. Extra precautions were therefore needed to ensure that any leachate which might be produced would not cause pollution of the stream.

A semi-permeable attenuating blanket was constructed. This provided a 6 m unsaturated zone between the refuse and the ground-water table.

A waste product from the quarrying was used to construct this layer. It consists of a mixture of calcareous sands and silt and is probably an ideal material for this purpose, combining a fairly low permeability with high buffering and ion exchange capacities. Until now no contamination of the ground-water has been detectable, even though a saturated layer does exist in the base of the refuse.

## Site 3: Pitsea, Essex (Knox, 1983; Knox, 1985a)

This site has been built up since the early 1900s on $\approx$ 284 hectares of low-lying land, reclaimed several centuries ago from salt-marshes in the flood plain of the River Thames estuary. The site is the biggest co-disposal site in the UK and is used for domestic, commercial and industrial solid wastes, and for liquid industrial wastes.

The Cretaceous Chalk, a major potable water aquifer, lies at depth under the site. However, this is protected from potential contamination, by an extremely low permeability Tertiary clay, the London Clay. This ranges in thickness from 33 m to 68 m underneath the Pitsea landfill. Overlying the London Clay are $\approx$ 20 m of alluvial beds, including various clays, sands, silty sands and gravels, with brackish to saline ground-waters. An alluvial clay layer immediately underneath the site results in a partial degree of containment and hence partial saturation of the waste materials, which now have an average depth of $\approx$ 7 m. There is no subdivision into cells, and large-scale mixing and recirculation of leachate and liquid wastes within the refuse mass are encouraged.

Lateral migration of leachate is contained on two sides by a sea wall and on the two other sides it is intercepted in a perimeter ditch. From the ditch, leachate is recirculated back into the site via a large holding lagoon in the refuse and thence by various rain-guns and spray irrigation equipment.

The landfill had already been in use as a domestic waste site for several decades, before industrial waste disposal started on a significant scale in the 1960s. There was therefore already in place a large mass of partly decomposed, partly saturated refuse. In addition, since the start of industrial waste disposal, the rate of industrial waste input has nearly always been exceeded by the rate of domestic waste input, and the rate of liquid waste input always been exceeded by effective rainfall.

The inputs have been such that the landfill is dominated by domestic refuse decomposition processes. These processes, and the ditch/lagoon system provide environments in which many mechanisms can act to immobilize or degrade the components of a wide range of hazardous wastes.

Overall, the landfill may be considered as a reactor in which a range of inputs (domestic and industrial wastes and rainfall), whose quality, quantity and physical form varies widely and fluctuates with time, is converted into a single output (i.e. leachate).

Although this may still require management, which may mean a leachate treament plant, it is now a single, definable effluent. Its properties are very much more predictable and regular than those of the individual industrial inputs to the site, and the degree of hazard involved is dramatically reduced. Leachate quality at Pitsea is exactly the same as would arise at a methanogenic domestic waste landfill. The concept of the Pitsea landfill as a reactor is shown in Fig. 1 with approximate quantities shown to give an overall perspective.

Two points need to be stressed. The first is that it takes time to develop a landfill to the point where it is suitable for this type of co-disposal; it obviously

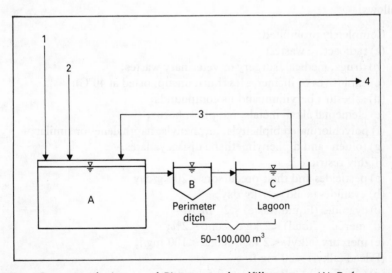

**Figure 1.** Conceptual picture of Pitsea as a landfill reactor. (A) Refuse mass ($\approx 10 \times 10^6$ t) and leachate volume ($1.5$–$2.0 \times 10^6$ m$^3$): anaerobic, low redox, biologically active, partly saturated; (B) perimeter ditch ($50$–$100\,000$ m$^3$ capacity): access of air, quiescent conditions; (C) lagoon ($50$–$100\,000$ m$^3$ capacity): access of air, quiescent conditions.

takes time to build up an in-place mass of refuse and, similarly, it takes time for methanogenic conditions to become fully established. Depending on local conditions, a lag of several years may be desirable between the start of domestic refuse disposal operations and the start of several liquid industrial waste disposal operations. Conversely, after domestic waste disposal has finished, the site may continue to be capable of receiving liquid industrial wastes for many years afterwards without there being any adverse effect on leachate quality. In planning co-disposal facilities this time lag needs to be considered carefully in advance.

The second point to be stressed is that although such a large site is very tolerant, it is possible, as with any reactor, to cause process upsets. The capacity of physical and chemical immobilization mechanisms can be exceeded, while biological processes may either be overloaded or inhibited. The overall quantities of industrial wastes deposited so far at Pitsea have had no detectable effect on refuse decomposition or on leachate quality for the site as a whole.

It is in the interests of the operators of such a site to ensure that waste inputs are controlled so that domestic refuse processes remain dominant. However, the Site Licence also imposes, externally, many restrictions on what may be deposited at Pitsea, and the methods of operation to be used. The principal licence restrictions on inputs reflect a political/technical balance and are as follows.

1. Completely prohibited:
    (a) radioactive wastes;
    (b) drugs, medical, surgery or veterinary wastes;
    (c) explosives or incinerables (burn unsupported at 40°C);
    (d) asbestos; beryllium and its compounds;
    (e) elemental alkali metals, red phosphorus;
    (f) polychlorinated biphenyls, terphenyls, napthalenes or similar;
    (g) toluene and diphenylmethane di-isocyanates.
2. Highly restricted:
    (a) pesticides and their precursors <20 kg/day;
    (b) cyanides (solid) <1 kg/day;
    (c) cyanides (liquid waste) <20 mg/l;
    (d) mercury (total) <2 kg/d; conc <2%;
    (e) mercury (alkyl) <2 kg/d; conc <100 mg/l;
    (f) lead (alkyl) <20 mg/l;
    (g) (As, Se, Sb) (elemental) <0.5 kg/day;
    (h) (As, Se, Sb) (non-elemental) <100 kg/week.

A prior Technical Control procedure is applied. Industrial wastes are only accepted for deposit at the site if they have first been assessed and declared suitable for disposal at Pitsea. This assessment will frequently involve the analysis of a sample of the waste at the operator's own laboratory. The analysis provides a specification for each load of waste and constitutes part of the contract between the operator and the customer. The technical control document also stipulates the area of the site where the waste must be deposited and any special safety and operational precaution which may be needed. Copies are sent both to the site management and to the customer.

The vehicle carrying the waste is only allowed past the site gatehouse if the driver is able to produce his copy of the technical control document, to match against a copy already held at the site. If the paperwork is found to be in order, the vehicle is then directed to an Industrial Waste Reception Area where a site chemist and marshals direct the final deposit of the waste.

Such a procedure is an important aspect of the management of difficult wastes, at any type of facility.

## Site 4: Choppington, Northumberland (Knox, 1985b)

This is one of the few (perhaps only) examples in the UK of an engineered secure landfill designed and used specifically for industrial residues, whereas it is what would appear to be, in many countries, the option preferred over a co-disposal site such as Pitsea.

This site was developed in a former clay extraction pit, covering 12 hectares, in the north-east of England, with the intention of making it suitable for the secure landfilling of a wide range of solid and liquid industrial wastes.

Before operations started, in early 1980, the site had been extensively engineered, following a thorough hydrogeological assessment, to provide a very high degree of containment. This involved constructing reservoir-standard water-retaining bunds to form the southern and western boundaries of the site.

The original pit was excavated from laminated clays, overlying boulder clays; these in turn overlie mudstones and siltstones of the Middle Coal Measures. The hydraulic conductivity of the boulder clay was found to be in the range $5.4 \times 10^{-10} - 9.2 \times 10^{-10}$ m/s. The Site Licence required that a minimum of 8 m of this clay be left between the site base and the rockhead; in practice under much of the site the actual thickness remaining is greater than this. In addition, the Site Licence requires completed cells to be capped with at least 0.9 m of the same clay.

The site is being filled in five cells. The system of operation is that a particular cell is filled with solid wastes and clay-capped before liquid wastes can be landfilled. Liquid wastes are then injected through a series of slotted injection pipes protruding through the cap. The use of this cell system allows clean surface water to be pumped off-site from unfilled areas and allows cells to be filled and capped without generating large volumes of leachate. This maximizes the quantity of storage capacity remaining for liquid wastes, which can be injected up to a control level set in the Site Licence.

The Site Licence restricts both the methods of operation and the materials which can be landfilled. Some of the more important restrictions on inputs are as follows.

1. Completely prohibited:
   (a) flammable wastes (capable of burning unsupported at 40°C or with a flashpoint less than 32°C);
   (b) berylliun oxides;
   (c) PCBs and analogues;
   (d) other halogenated organics;
   (e) isocyanates.
2. Highly restricted (to less than 100 ppm):
   (a) organic and inorganic peroxides;
   (b) arsenates and arsenites;
   (c) organolead compounds;
   (d) carbides and acetylides;
   (e) sodium and potassium cyanide;
   (f) other compounds which liberate toxic gases upon acidification.

Quantity limits are put on all other classes of hazardous waste, using the classification system of DOE. Solid wastes are primarily industrial in origin and the quantity of domestic waste going into the site has been very small.

It is likely that in most of the cells, by the time they are completed and injected with liquid wastes, roughly half of the leachate present will be derived from rainfall infiltrating during the infilling phase and half from injected liquids. This is quite a different picture from that at a co-disposal such as Pitsea where liquid wastes form only a small proportion of the leachate held within the site. The inputs to the Choppington site probably make it one of the sites with the highest proportion of the higher hazard materials landfilled anywhere in the UK, and there is little reason to expect domestic type decomposition processes to dominate.

## CURRENT TRENDS

### Gas Monitoring

Gas monitoring and control measures now receive much more attention by operators and regulators than previously. This trend has been accelerated by the publication, in early 1989, of a new Waste Management Paper specifically concerned with landfill gas. The risks of gas migration now receive more attention at the design stage, and measures to restrict or prevent migration are often part of new landfill designs. Site Licences now routinely require monitoring scheme to be undertaken to the satisfaction of the WDA and, if necessary, a control scheme to be installed. Gas extraction and utilization schemes are much more common now, and likely to become increasingly so. These trends must be welcomed: if damage to human health and property is considered, then landfill gas has proved much more of a danger than landfill leachate, as has inconsistent control of site vehicles.

## Site Hydrogeology and Landfill Practice

It is likely that fewer sites will be developed so as to allow significant losses of leachate, and more sites will be engineered to increase their retention of leachate, than used to be the case. Partly this is because the standards of hydrogeological assessment of new sites are now, generally, higher than they used to be; as a result some sites are found to be unsuitable for use without significant alteration, whereas a decade ago they might have been thought acceptable.

There is little sign that engineered attenuation, as opposed to engineered retention, is increasing. The results of current research projects, such as that at Stangate (Robinson and Lucas, 1985), will be very influential in this respect, but may take many years yet to reach the point where they are conclusive.

Although the principles of co-disposal are widely accepted, there is little sign that new sites for co-disposal of the type practised at Pitsea, will be created in future in the UK, unless there is a suitable change in the economic and political framework. Such sites need to be big, they require long-term regional planning, and they require guaranteed large inputs of domestic refuse. These requirements are hard to meet and are additional to the hydrogeological and environmental requirements.

## Site Licensing and Surveillance

There is a tendency towards uniformity and standardization of Site Licence conditions even on aspects where a site-specific approach would be more logical. The publication of a major government review of landfill practices, in 1986, has reinforced this trend towards a 'recipe-book' approach.

Site Licences now contain more conditions which require monitoring of aquatic domains, especially leachate within the site, and ground-water outside the site. This is, as far as it goes, a welcome trend since the amount of monitoring carried out at many UK landfills has been far from adequate, in either the public or private sector. This move towards more extensive monitoring is slow and the intensity of monitoring at sites across the country, and even within the boundaries of a particular WDA, is very uneven. There is currently no uniform pattern in the allocation of monitoring effort between WDA, RWA and site operators, nor any agreement as to what the allocation ought to be.

**Legislation and Policy**

EEC legislation has, as yet, had little direct effect on landfill design or practice. A Ground-Water Directive (80/68) has caused a great deal of interest and discussion but there is no clear picture of how it will, or should, influence future policy and hence projects. Nor is there a clear picture in the UK of its impact in other European countries.

There is an increasing amount of time spent discussing philosophies, concepts and policies, and less effort (apparently) directed at technical and environmental fundamentals. A common topic currently is Best Practicable Environmental Option (BPEO). While for some classes of waste, BPEO disposal may be clearly represented by a single disposal technique, there is no unified understanding of how the concept should affect landfill design and practice.

Within the UK, proposals for legislative changes have been published in early 1989. Among other things, they are likely to lead to a more regionalized approach to site licensing than has been the case so far.

**CONCLUSIONS**

UK conditions, both physical and administrative, lend themselves to a diversity of landfill designs, including the use of dispersal sites and the practice of co-disposal. Both can be environmentally acceptable options in the correct circumstances.

Diversity exists in current practices but may be threatened by a desire for uniformity of site design and site licensing and by reliance upon a recipe-book approach.

This tendency is based more upon a developing folklore and upon discussion of philosophies than it is upon sound investigative work, of which less is reported than used to be the case.

Greater attention is required to the quality and even handedness of landfill surveillance and enforcement.

Political and economic changes might be required to ensure that the appropriate locations and full potential of co-disposal are exploited.


## REFERENCES

Davies, J.N. (1985). 'A small high rainfall site with leachate treatment', Landfill Monitoring Symposium, Harwell, UK.

Knox, K. (1983). 'Treatability studies on leachate from a co-disposal landfill', *Env. Poll. (B)* 5, 157–174.

Knox, K. (1985a). 'Leachate treatment with nitrification of ammonia', *Wat. Res.* **19** (7), 895–904.

Knox, K. (1985b). 'Co-disposal sites: two case studies', Landfill Monitoring Symposium, Harwell, UK.

Robinson, H.D. (1985) 'Treatment of domestic waste leachate in a full-scale automated plant', Conference on New Directions and Research in Waste Treatment and Residuals Management, June. University of British Columbia, Canada.

Robinson, H.D. and Lucas, J.L. (1985). 'Attenuation of leachate in a designed, engineered and instrumented unsaturated zone beneath a domestic waste landfill', *Water Poll. Res. J. Canada* **20** (3), 76–91.

Robinson, H.D. (1987). 'Unsaturated zone aspects of landfills', October. ISWA Symposium: Processes, Technology and Environmental Impact of Sanitary Landfill. Cagliari, Sardinia, Italy.

Robinson, H.D. and Blakey, S.G. (1987). 'Attenuation of contaminants: how reliable a philosophy?' *Surveyor*, 5th February, pp. 12–15, UK.

## 7.3    Multi-barrier Concept in West Germany

KLAUS STIEF

*Umweltbundesamt, Bismarckplatz 1, 1000 Berlin 33, West Germany*

### INTRODUCTION

Landfilling of wastes is the ultimate permanent disposal at a site without having the intention of removing the waste in the future. The waste must be landfilled according to the Waste Avoidance and Waste Management Act (Abfallgesetz, 1986). The landfilling of waste must be carried out in such a way that there is no risk of polluting groundwater according to the Federal Water Act (Wasserhaushaltsgesetz). In practice, it is impossible to guarantee permanent pollution control at a landfill site by human-made or natural barriers as the site will be unevenly structured and therefore prediction of performance will not be 100% reliable. Therefore it would seem to be a realistic approach to make careful preparations for unforeseeable environmental impacts which may arise at a later date, even though the best technology available may have been used at the time of construction to prevent them. It will be necessary to obtain further knowledge concerning the landfills under construction, especially regarding the structure of the paths along which the pollutants will migrate from the site into the surrounding environment. Only then can one hope to be able to face less severe abandoned site problems in the future than those we have to cope with today. Encapsulation of waste in landfills is an efficient and necessary approach to protect the environment. However, the acquisition of as much knowledge as possible about the landfill and surrounding area must not be overlooked.

### THE MULTI-BARRIER CONCEPT

The basis of the German concept for planning, operation and after-care of

549

landfills is the multi-barrier concept (Stief, 1986). The following elements are considered to perform the role of barriers:

1. the site;
2. the bottom liner;
3. the landfill body (the waste);
4. the surface liner system (the cap);
5. the controlled post-closure use of the landfill area;
6. the long-term monitoring and control of the landfill behaviour.

From the environmental protection point of view, each of the barriers should, independently of the others, be permanently effective. However, the "human-made" barriers are likely to have a limited life span. It is therefore necessary to know the period in which their efficacy is most important and more so, what their probable life span will be.

It should be stressed that every barrier has to be selected and/or constructed according to the best available standards of technology (BATs). It is obvious that the BATs will be improved in time and that landfills ten years from now will be better, safer and more environmentally acceptable than they are today. However, we have to dispose of our waste now and our society will not be able to afford a 'ten-yearly' redisposal of the wastes disposed of in the past to adjust the buried barriers to the new BATs.

Considering the multi-barrier concept for each future landfill will help to keep the environmental burden within acceptable limits. These limits have to be pushed further and further according to the best available standards or to new knowledge regarding 'tolerable' impacts on the environment. The impacts from the landfills on the environment will diminish according to the threshold values set for tolerable emissions, and according to the quality and performance standards of the landfill barriers.

The multi-barrier concept might also help to gain better public acceptance of landfill sites for either municipal solid waste, treated chemical (hazardous) waste, demolition material or mineral waste (such as residues from incineration plants).

## THE LANDFILL SITE

The geological and hydrogeological settings of a landfill site will be the main barrier against movement of pollutants into the environment in the long-term. Attenuation processes in the unsaturated and saturated zone under the landfill bottom and downstream from the site are necessary to achieve acceptable groundwater quality within a defined distance downstream from the landfill site (Drescher, 1987).

Attenuation capacity should theoretically be sufficient to cope with all pollutants released from the landfill body into the underlying strata. The efficacy of the other barriers, i.e. the bottom liner and surface liner systems has to be neglected in evaluating the sufficiency or attenuation capacity.

Simulation of the attenuation should be an essential part of the landfill design prior to licencing of the landfill. However, in the case of simulation, only incomplete information will be available regarding the waste to be disposed and the leachate that will be generated. The mathematical model therefore involves a number of assumptions which may later need modification. Thus, the model must be improved during the operational and post-operational phase of the landfill (Schneider and Tietze, 1987).

However, if assurance can be obtained regarding the landfill body and leachate composition, it will be possible to leave a remedial action plan for effective hydraulic measures to future generations which will prove necessary if they should not accept the contamination we left.

## THE BOTTOM LINER SYSTEM

Soil and ground water should not be polluted deliberately; for this reason a landfill bottom liner system is required at every landfill site. From the author's point of view it should be at least a simple composite liner. More effective would be a twofold composite liner. However, construction technology is not yet available (Stief, 1987).

Since there is often severe public concern with regard to ground-water pollution, resulting in the need to detect and control leaks in liner systems, double liner systems will be installed at some sites. An important role in the bottom liner system is played by the leachate collection and removal layer from which the leachate is discharged to a leachate treatment plant. An accumulation of leachate causes an increased flow through leaks in membranes and/or through the low permeable clay layer.

Today, there are no requirements of any calculation of the breakthrough time of leachate through the bottom liner system. In addition, the absorption capacity of the clay layer for heavy metals is not calculated. The recommended values for the thickness of the various elements of the bottom liner system are more or less the result of good engineering judgement. However, it is strongly believed that the bottom liner will be effective for at least 30 years if a composite liner has been used. The period over which a bottom liner has to be effective should be much longer than the operation phase of the landfill and should cover a major part of the post-operation phase (during which, for example, a MSW landfill stabilizes).

Some years after closure of the site, the settlement of the landfill body in MSW landfills along with the underlying soil and leachate composition will have reached a steady state. By that time it will have become obvious what burden is going to be left for future generations. Then it may be possible to calculate more reliably the 'true' attenuation of leachate in soil and ground-water in case the bottom liner system and the cap system were ineffective. At least until such a time an effective bottom liner system is needed.

## THE SURFACE LINER SYSTEM (THE CAP)

The cap is to prevent leachate generation by minimizing rainwater infiltration into the landfill body and to control landfill gas migration into the air. The cap generally consists of the following elements:

1. landfill gas drainage layer;
2. low water permeable liner;
3. drainage layer;
4. soil cover;
5. vegetation.

If organic wastes have been landfilled, 'landfill gas' will be generated due to microbiological degradation. To protect public health and the environment, this gas has to be collected and removed. The landfill gas drain layer is for collection and removal of the landfill gas.

If one wants to avoid the so-called 'bath tube effect' (water head on the landfill bottom) within a landfill, one has to select a composite liner as a low water permeable liner in the surface liner system if a composite liner has been installed also as an element of the bottom liner system.

There are several good reasons for which leaks in the surface liner system must be detected. The main reason is that it will prove advantageous to apprehend immediately the likelihood of leachate generation. If leaks are detected it will be necessary to repair them as soon as possible. According to the available technology a double liner must be installed to reliably detect any leaks in the long term. A double liner has a leakage collection and removal layer (leachate drain) between the layers. The lower liner must be a composite liner in order to obtain sufficient leak detection sensitivity. The liner in the cap system is repairable thanks to the relatively low total thickness of the various layers placed over the upper liner layer. However, the layers above the upper liner layer must be thick enough to protect the liner against freezing as well as from damage by plant roots (Stief, 1987).

From results obtained through monitoring the performance of surface liner system, possible ways of improving design and construction of the soil cover may be gained. Moreover, the selection and planting of the appropriate vegetation can enhance the reaching of a water balance with a zero infiltration rate into the landfill body. Before these mechanisms are clarified, a liner should be installed in the cap system. In the author's opinion, a composite liner is necessary to prevent infiltration.

However, the moment of installation of a cap for landfills which must be considered as 'bioreactors' must be discussed. Capping too early will inhibit microbiological degradation of organic materials. Heavy settlement of the landfill body during the period of microbiological degradation of organic wastes and for some time after, is most probable and may cause damage to the surface liner. Therefore it would seem reasonable to allow the decomposition of organic waste until gas production and organic pollution of the leachate have ceased. If the landfill has already been capped, it will be necessary to infiltrate water into the landfill body to enhance degradation. During this period the bottom liner must of course be effective.

Landfills in which no biological processes take place due to the characteristics of the waste, may be capped after closure as soon as settlement has reached an acceptable steady state. No surprises are to be expected from unpredictable reactions within the landfill body.

The cap system might completely prevent the generation of leachate if the landfill body is at a sufficient distance above the ground-water table and as long as no surface water flows from the outside into the waste. If this performs perfectly there is no need to treat the leachate and migration of pollutants in the long-term need cause no concern.


## THE LANDFILL BODY

The generation of leachate and landfill gas, as well as settlement caused by the degradation of organic material depend mainly on the amount and composition of waste disposed of at the landfill site.

Despite all measures taken to encapsulate the landfill body it is necessary to assess the impact of leachate and gas emissions on the environment assuming that the bottom liner and cap systems were ineffective as there remains some uncertainty as to how long these will function adequately. The main purpose of this strategy is to be aware of the importance of the landfill body itself as a barrier against release of pollutants into the environment. The more immobile the pollutants contained in the landfill waste and the slower the remobilization of non-mobile pollutants by means of reactions within the landfill body, the smaller the environmental risk posed by the landfill. Moreover it is also true that the immobility of pollutants in the landfill increases with time (Göttner, 1987).

The smaller the quantity of organic and inorganic wastes co-disposed in a landfill, the lesser the unpredictable physical, chemical and biological reactions inside the landfill body will be and the more predictable the generation of leachate and gas. Thus, the possibility of foreseeing the amount of gas and leachate generation and therefore of calculating settlement, will render the landfill less of an environmental risk. Consequently, the main objectives of the German concept for landfilling waste are:

1. to avoid co-disposal of municipal solid waste and industrial waste;
2. to totally avoid landfilling of organic waste wherever possible; the results of leaching tests carried out on the waste should indicate less than approximately 3–5%, with a maximum of 10%, of organic material;
3. to landfill inorganic waste in monofills (only one type of waste from one waste generator) or in quasi monofills (comparable wastes with little or no reaction within the landfill body);
4. to avoid landfilling of water soluble waste, a maximum of 10% should be soluble according to the leaching procedure. Waste with higher content of solubles should be disposed of in salt caverns.

To achieve maximum stability of the landfill body the waste has to be highly compacted during the operational phase of the site. Only waste which can be compacted to a high density and therefore cause little settlement should be landfilled.

## CONTROLLED POST-CLOSURE USE OF THE LANDFILL AREA

Immobilized pollutants in the waste could become mobile once again if mobilizing agents were to infiltrate into the landfill body, e.g. oil, acids, etc. used on top of the closed landfill. In addition, the surface liner system might be damaged by impact from outside, e.g. by roots, animals, construction activity or even by vandalism. In order to contain the potential risk for human health and the environment represented by the landfill, it should be assured that the above-mentioned impacts do not occur. This can only be carried out through a controlled after-use of the closed landfill area. A closed landfill must not be abandoned. The owners of a landfill must assume responsibility for maintenance and repair of the surface liner system and for controlling of activities on top of the landfill. Moreover, they should insure that no plant roots will damage the liners, no construction work takes place on top of the landfill, which may damage the cap. Further, the owner should verify the performance of the leak detection system and cope with erosion of the soil cover.

Of course, it should be realized that after several decades or even centuries, concern about the potential environmental impact which could be caused by the 'forgotten' landfill, will no longer be so active. For this reason the long-term problem of controlling a closed landfill should be underlined. As a consequence, only waste which will not cause immediate human health problems of environmental impact, even if it should remain uncovered for some years, should be landfilled.

## DECLARATION OF LANDFILL BEHAVIOUR

It is not usually known at the moment of designing of the landfill what the main pollutants in the landfilled waste will be nor what quantities of these pollutants will be 'stored' in the landfill body at the time of closure. The potential risk can only be rather roughly assessed; therefore, the composition of the landfill body and of emissions must be continually measured. The results of monitoring should be compared with the assumptions made during design and licensing of the landfill and must then be evaluated and documented for further use.

A possible useful and necessary documentation is that of a regular 'Declaration of Landfill Behaviour' which should include:

1. amount and types of waste and their location within the landfill body;
2. mobility of pollutants in the waste landfilled;
3. amount and composition of leachate;
4. composition and production of landfill gas;
5. settlement;
6. density of the landfill body.

In addition the documentation of landfill behaviour should cover:

1. performance of the bottom liner system;
2. performance of the cap over the closed part of the landfill;
3. results of detailed investigation of the ground-water around the landfill site;
4. results of simulating the attenuation of pollutants by using improved data concerning leachate, and attenuation capacity of the ground.

By this Declaration of Landfill Behaviour, rendered at least once a year, it will prove possible to create data for the prediction of long-term landfill behaviour. To this regard, it will be necessary to measure leachate, gas, settlement and density of waste as well as the performance of other barriers on a daily, weekly or monthly basis (Bothmann, 1987; Tybus, 1987).

From the data collected, reliable information about the main pollutants necessary to assess risk and environmental impact will be acquired.

## SUMMARY

Applying the multi-barrier concept for design, operation and evaluation of landfill sites is the basic measure to ensure leaving acceptable landfills to future generations. During the phase of landfill design the attenuation of extreme leachate discharge into the surrounding area has to be simulated. The landfill bottom liner and surface liner systems, necessary for all types of landfill, should have a composite liner layer as sealing element. The use of a double liner system is recommended for capping to facilitate leak detection.

The landfill body should contain as little organic or soluble waste as possible. The waste should be highly compactable to reduce settlement. The post-closure use of a landfill should be controlled. The Declaration of Landfill Behaviour should be published annually and be based on regular measurements of emissions and barrier performance.

Landfills are, and will undoubtedly continue to be, potential sources of air, soil and water pollution. Landfilling of waste will inevitably result in consumption of land and consequently should be minimized.

By landfilling mainly inorganic and insoluble wastes with immobile main pollutants, selecting sites with high attenuation capacities and controlling the post-closure use of the landfill, we will achieve a slow release of only low concentrations of non-hazardous pollutants from the landfill into the environment.

## REFERENCES

Bothmann, P. (1987). 'Kontrollen an Deponien—Vorschlag für ein erforderliches Überwachungsprogramm' Stuttgarter Berichte zur Abfallwirtschaft, Band 24, Tagung "Zeitgemäße Deponietechnik', März 1987, ISBN 3-503-02055-1, Erich Schmidt Verlag.

Drescher, J. (1987). 'Schadstoffadsorptionspotential des Deponieuntergrundes—Ergebnisse des 9000 ha—Programmes des Landes Niedersachen'. Abfallwirtschaft in Forschung und Praxis, Band 19, Tagung 'Fortschritte der Deponietechnik', Mai 1987, Hrsg. Fehlau/Stief, ISBN 3-503-02683-5.

Dullmann, H. (1987). "Geotechnische und baubetriebliche Einflüsse auf die Dichtigkeit von Deponieabdichtungen aus Ton—Ergebnisse von Praxisuntersuchungen'. *Abfallwirtschaft in Forschung und Praxis*, Band 19, Tagung 'Fortschritte der Deponietechnik', Mai 1987, Hrsg. Fehlau/Stief, ISBN 3-503-02683-5.

Göttner, J. (1987). 'Möglichkeiten der weitergehenden Schadstoffimmobilisierung in Deponien für mineralische Abfälle'. *Abfallwirtschaft in Forschung und Praxis*, Band 19, Tagung 'Fortschritte der Deponietechnik', Mai 1987, Hrsg. Fehlau/Stief, ISBN 3-503-02683-5.

Schneider, W. and Tietze, K. (1987). 'Numerische Schadstofftransportmodelle als Beurteilungsinstrument für die Barrierewirkung des Deponieuntergrundes'.

*Abfallwirtschaft in Forschung und Praxis*, Band 19, Tagung 'Fortschritte der Deponietechnik', Mai 1987, Hrsg. Fehlau/Stief, ISBN 3-503-02683-5.

Stief, K. (1986). 'Das Multibarrierenkonzept als Grundlage von Planung, Bau, Betrieb und Nachsorge von Deponien'. *Müll und Abfall*, 18. Jg., Heft 1, Seite 15–20, Erich Schmidt Verlag.

Stief, K. (1987). 'Möglichkeiten und Vorteile der Kombinationsdichtung für die Deponiebasisabdichtung'. *Stuttgarter Berichte zur Abfallwirtschaft*, Band 24, Tagung 'Zeitgemäße Deponietechnik', März 1987, ISBN 3-503-02055-1, Erich Schmidt Verlag.

Stief, K. (1987). 'Praxisnahe und realisierbare Deponiestrategien'. Abfallwirtschaft in *Forschung und Praxis*, Band 19, Tagung 'Fortschritte der Deponietechnik', Mai 1987, Hrsg. Fehlau/Stief, ISBN 3-503-02683-5.

Tybus, M. (1987). 'Anleitung zur Auswertung und Bewertung von Betriebs—und Kontrolldaten', *Stuttgarter Berichte zur Abfallwirtschaft*, Band 24, Tagung 'Zeitgemäße Deponietechnik', März 1987, ISBN 3-503-02055-1, Erich Schmidt Verlag.

# 7.4 Landfill Design Concepts in the United States

JOHN PACEY

*EMCON Associates, 1921 Ringwood Avenue, San Jose, California 95131, USA*

## INTRODUCTION

The US government agency primarily responsible for the country's environmental policy is the Environmental Protection Agency (EPA). The EPA establishes minimum environmental standards (enforceable criteria) required 'to protect human health and the environment'. Such standards must be enforced throughout the 50 states that comprise the US; however, each state may develop and enforce more stringent standards. Environmental standards in the US, therefore, can vary widely among the states.

The EPA has established many standards for hazardous waste landfills, but required each state to provide its own standards for solid waste landfills. Until recently, solid waste landfill legislation was left primarily to each state's discretion. The EPA did establish minimum guideline recommendations in 1980 and required criteria for landfill gas control in September 1987, stipulating that methane be limited to no more than 5% at property boundaries and no more than 1.25% within occupied structures.

On the state level, solid waste regulations generally emphasize prevention of ground-water contamination through leachate control and containment. Regulations are based upon factors such as geography, hydrogeology, population and population centres, and waste types. In addition, each state has regulations that are specific to its physiography; for example, Florida's regulations address sinkholes, and California's regulations address fault zones.

This chapter examines solid waste landfill design within the framework of state and federal regulations. First, it defines the major considerations in landfill design. Secondly, it compares the existing standards in three states — Oregon, California, and Wisconsin — and discusses impending federal standards. Finally, it presents three case studies — the Kirby Canyon site in California, the Bacona Road site in Oregon, and the Libby site in Wisconsin — to illustrate typical landfill design concepts in the US.

## LANDFILL DESIGN CONSIDERATIONS

When designing a solid waste landfill, important considerations include (1) site location characteristics; (2) containment systems; (3) leachate control, collection, and removal; (4) gas migration control; and (5) final cover. If the landfill design properly addresses these considerations, an environmentally sound landfill can be constructed that safeguards air and water resources.

### Site Location

To determine whether a site is a suitable landfill location, its climate, hydrology, hydrogeology, soil characteristics, and other site-specific conditions must be evaluated. Existing vegetation and wildlife must also be investigated, since both state and federal regulations protect environmentally sensitive areas such as wetlands and the habitats of endangered species.

In developing the landfill design, hydrogeologic evaluations are particularly important. They are conducted during siting and follow-on design studies to determine local geology, in place soil properties, and surface- and groundwater characteristics and use. This information can be used to evaluate: (1) the potential for, and impact of, contaminant migration; (2) the need for special drainage or containment features; (3) the suitability of on-site soils for borrow materials (for soil liners and intermediate and final soil cover); and (4) temporary and final slopes.

### Containment System

Until recently, compacted clay soils were the materials most widely used to line municipal landfill sites. Geomembrane liners, however, have been used increasingly since the early 1970s for waste containment, because of their low permeability and long-term resistance to chemical attack. The hydraulic conductivity of a nearly flawless geomembrane is in the order of $1 \times 10^{-12}$ centimeters/second (cm/s), as compared with $1 \times 10^{-7}$ cm/s for a low-permeability, compacted soil. Geomembranes are not perfect, however; even with a careful construction quality assurance program, they may contain about two to five defects per hectare of liner. These defects may result in an effective permeability two to five orders of magnitude lower ($1 \times 10^{-7}$ to $1 \times 10^{-10}$ cm/s).

Another containment system currently considered at municipal landfills is the composite liner, consisting of a geomembrane underlain by compacted, low-permeability soil. By combining the attributes of both, a composite liner offers unique advantages over either the soil or geomembrane component alone. When intact, the geomembrane component maximizes the collection and removal of leachate by almost completely rejecting fluids. The soil component provides a back-up barrier if leachate leaks through defects in the geomembrane. This state-of-the-art liner is considered for municipal landfills where containment systems must prevent measurable leachate migration from the landfill.

## Leachate Collection and Removal

Leachate collection and removal systems (LCRSs) are standard design components in currently constructed landfills. A demonstrated technology, LCRSs have been proven through field use to be highly reliable, low-maintenance systems. The typical LCRS consists of the following components:

1. a drainage layer-designed to produce little or no head of liquid on the underlying liner, that enables rapid detection, collection, and removal of leachate;
2. a drainage pipe system of appropriate size and spacing to efficiently remove leachate;
3. a sump sized to collect the leachate discharged by the drainage pipes;
4. methods of measuring and recording fluid volumes in the sump.

Geosynthetics made from polymers are increasingly used for drainage or filters in the LCRs. A geosynthetic may be substituted for natural materials if it is: (1) chemically resistant to the waste and leachate; (2) resistant to loading stresses; and (3) compatible with the geomembrane liner (if used).

## Gas Migration Control

Bacterial degradation of organic waste deposited in a landfill produces various gases, principally methane and carbon dioxide. A landfill will undergo an initial, relatively short phase of aerobic (oxygen-requiring) decomposition, followed by an extended period of anaerobic (oxygen-free) degradation. During the aerobic phase, respiration-type processes consume available free oxygen and produce carbon dioxide.

The interest in gas production stems primarily from methane's combustibility when present in concentrations of 5–15% in air. Since oxygen in a landfill is rapidly depleted and is present in only small amounts when methane gas is being produced in quantity, the gas is not an explosion hazard in the landfill. As the gas migrates from the landfill, however, it mixes with air and passes through the combustion range: (1) within or near the cover soil; and (2) within a limited distance from the landfill. The project design, therefore, must consider development on the landfill and the potential for lateral gas migration.

The migration of gas from the landfill can be controlled by either passive or active systems. One example of a passive system is an impervious liner and vent system installed on excavation slopes before placing refuse fill. To assure that a passive barrier system is functioning effectively, gas monitoring probes should be installed. An example of an active system is a series of extraction wells installed near the landfill perimeter. In such a system, gas is drawn to the wells by a vacuum, which is introduced to each well through a lateral/header piping system powered by a motor blower unit. The collected gas discharges through flares that burn the gas, removing undesirable contaminants and the malodor.

## Final Cover

The final cover's primary function is to minimize infiltration of precipitation into a closed section of the landfill. Other functions include preventing contamination of surface water runoff, wind dispersion of wastes, and direct animal or human contact with waste.

Although EPA regulations currently apply only to hazardous waste landfills, the specified final cover components are applicable to municipal landfills. These components are:

1. a 0.6 m thick, vegetated soil layer (erosion control);
2. a 0.3 m thick granular drainage layer having a minimum permeability of $1 \times 10^{-2}$ cm/s;
3. a geomembrane, at least 0.5 mm thick, overlying a 0.6 m thick compacted clay barrier with a maximum permeability of $1 \times 10^{-7}$ cm/s.

For long-term performance and minimum maintenance, final cover must promote drainage, minimize erosion, prevent accumulation of gas pressures, and accommodate waste settling and subsidence. Final slopes are generally constructed no steeper than 33% (3 : 1) for stability and no flatter than 5% to prevent ponding of surface run-off and to facilitate drainage.

## SOLID WASTE LANDFILL REGULATIONS

Solid waste landfills have been primarily regulated by the states, although the EPA has developed design and operation guidelines. However, due to growing concern over the disposal of household and 'small quantity generators' hazardous waste in solid waste landfills, the federal role is becoming more active. ('Small quantity generators' are those that produce less than 100 kg of hazardous waste per month; typical wastes include used lead-acid batteries, spent solvents, and waste oil.)

Federal environmental legislation was amended in 1984 to require the EPA to assess the hazardous waste threat at solid waste landfills and develop standards necessary to 'protect human health and the environment'. EPA currently is developing minimum standards, to be issued by March 1988, that will apply to solid waste landfills throughout the country.

### State Requirements

A comparison of regulations in the states of California, Oregon, and Wisconsin illustrates the states' approaches to landfill design. Each state's solid waste regulations generally are based upon factors such as geography, hydrogeology, population and population centers, and waste types. In addition, each state has regulations that are specific to its physiography.

### California

California site location standards include geologic considerations: no new landfill or expansion may be sited on a Holocene fault (a fault that has been active in the last 11 000 years), the landfill must be adequately separated from ground water, and there must be protection from a 100-year flood.

Where site characteristics are inadequate to protect surface and ground-waters, California requires a single clay liner or native soils with a permeability of $1 \times 10^{-6}$ cm/s or less. Cut-off walls and grout curtains may also be employed. Where liners are used, the state also requires a leachate collection and removal system that must be capable of collecting and removing twice the maximum anticipated daily leachate volume.

If an appropriate state or local agency believes a landfill gas nuisance or hazard may exist, the site owner must monitor the presence and movement of gas and, possibly, construct a gas control system. The monitoring programme must meet the agency's specifications and continue after the landfill closes, unless authorized otherwise by the agency.

Final cover regulations require that the final cover be designed to prevent surface water infiltration (to minimize leachate), ponding, and erosion. Final cover must have a minimum thickness of 0.6 m, a maximum permeability of $1 \times 10^{-6}$ cm/s, and a 15 cm thick topsoil cover to support native vegetation.

## Oregon

Oregon has neither site location nor liner requirements, but does require the owner or operator to minimize and monitor leachate. The state may require leachate collection and treatment; if leachate control is required, the state determines design and/or performance standards for each site on a case by case basis.

Oregon does specify methane gas monitoring controls: the landfill gas concentration must be below 25% of the lower explosive limit (LEL) in structures and less than the LEL at the site's boundaries. Further, malodorous decomposition gases must be controlled to prevent a public nuisance.

Final cover must be a minimum of 1 meter of compacted soil, with a minimum 2% grade on top (to restrict ponding and surface water infiltration) and a maximum 30% grade for side slopes (to control erosion). Native grasses must be used to vegetate the cover.

## Wisconsin

Wisconsin has the most specific location requirements of the three states examined here. Landfills are prohibited in areas such as floodplains, wetlands, and habitats of endangered species; and near water supply wells, lakes, navigable rivers and streams, highways, and public parks. Wisconsin also has a general rule that the landfill can have no detrimental effects to waters located beyond its boundaries.

While not stipulated by law, the state has developed design criteria for liners and for leachate detection, collection, removal, and treatment systems. These criteria indicate that liners should be compacted soil having a minimum thickness of 1.5 m and a maximum permeability of $1 \times 10^{-7}$ cm/s.

Wisconsin landfills must employ effective means to prevent the migration of explosive gases from within the limits of the waste fill. At no time shall the concentration of explosive gases in any facility structures (excluding gas control or recovery systems) exceed 25% of the LEL for those gases. Similarly, the concentration of explosive gases in the soil at or beyond the property boundary may not exceed the LEL. To monitor the effectiveness of gas control systems, the state may require that gas monitoring devices be installed and sampling and analysis programmes be implemented.

The final cover must meet specific design criteria. It must consist of a minimum 0.6 m of compacted, fine-grained soil and 15 cm of topsoil; have top slopes of at least 2% and side slopes less than 33%; and be planted with native vegetation. Surface water must be diverted from the site as further protection against leachate formation.

## Proposed New Federal Requirements

The US Congress amended environmental legislation in 1984 and increased the federal role in solid waste regulation. These amendments — the Hazardous and Solid Waste Amendments of 1984 — direct the EPA to revise the criteria for solid waste facilities that may receive household hazardous waste or hazardous waste from small quantity generators. The intent of the 1984 amendments is to assure that solid waste landfills are designed, constructed, and operated to minimize potential ground-water contamination. At a minimum, the 1984 amendments require the EPA to develop enforceable standards for (1) ground-water monitoring adequate to detect contamination; (2) the location of new and existing facilities; and (3) corrective action. The standards 'shall be those necessary to protect human health and the environment'.

The EPA has drafted such standards for municipal solid waste landfills. Requirements proposed in the EPAs 12 March 1987 draft included location standards and liner and LCRS requirements. The proposed standards will undoubtedly change as the EPA receives public comments and further evaluates the requirements necessary to 'protect human health and the environment'. However, the draft proposal does indicate the EPA's direction toward criteria that are much more specific and stringent than the earlier requirements and guidelines. States with less stringent standards than those adopted by the EPA will be required to adopt the EPA's standards. States where existing standards are at least as stringent as the EPA's probably will not be affected.

## CASE HISTORIES

Three case histories demonstrate typical landfill designs in the United States: the Kirby Canyon site in California; the Bacona Road site in Oregon; and the

Libby site in Wisconsin. In each case, the landfill design addressed client needs, state regulations, and site-specific conditions.

## Kirby Canyon Landfill, California

The Kirby Canyon Landfill, located in a series of canyons in foothills east of San Jose, California, was opened in 1986 to meet San Jose's urgent need for additional landfill capacity. With a refuse capacity of over 28 million cubic meters, the landfill's estimated service life is 52 years. Only one canyon is currently used for landfilling; other areas will be developed during later phases.

### Location Characteristics

The Kirby Canyon site has many desirable features, including: (1) location adjacent to a major highway; (2) remoteness from residential areas; (3) large refuse disposal capacity; and (4) favorable hydrogeology. The climate, too, is favorable, with mean annual precipitation of about 48 cm and net annual evaporation of about 68 cm; a water balance indicates that no leachate will be generated (EMCON, 1983).

Hydrogeologic conditions at the site were investigated through an extensive subsurface exploration programme, which included exploratory borings, backhoe test pits, seismic refraction surveys, vertical electrical soundings, and magnetic profiles. This investigation indicated that the site is located within a massive bed of serpentine bedrock that contains little ground water (Van Heuit and Leach, 1984). In addition, ground water within Kirby Canyon is hydraulically isolated from adjacent watersheds. The investigation concluded that, if leachate were generated, the site's hydrogeology would help minimize the potential for leachate migration and ground-water contamination.

### Design

The favourable natural features of Kirby Canyon are enhanced by engineered design features. The design relies on the low permeability of the unweathered serpentine bedrock to contain leachate that might migrate to the landfill base. To prevent lateral leachate migration, a barrier consisting of a relatively impervious clay berm and chemical grout curtain has been constructed at the toe of the current development (Figs. 1a, b). These design features are primarily contingency measures, however, since no leachate is expected to be generated.

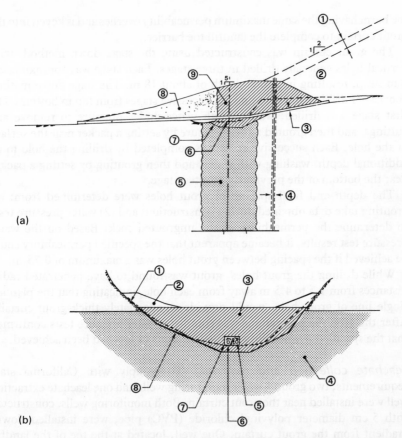

**Figure 1.** Kirby Canyon Landfill chemical grout curtain and toe berm section. **(a)**: (1) future intermediate soil cover; (2) future refuse fill; (3) temporary storm drain; (4) ground-water extraction well; (5) chemical grout curtain; (6) concrete anti-seep collar; (7) approximate base of excavation; (8) compacted clay fill ($k \le 10^{-6}$ cm/s). **(b)**: (1) original ground surface; (2) buttress fill (typical); (3) clay fill berm ($k \le 10^{-6}$ cm/s); (4) chemical grout curtain; (5) temporary storm drain; (6) 15 cm PVC leachate collection pipe; (7) concrete anti-seep collar; (8) approximate base of initial excavation. Excavation stepped for keying of fill; (9) compacted clay fill ($k \le 10^{-6}$ cm/s).

*Containment.* The aminoplast resin (urea formaldehyde) grout curtain is the subsurface barrier portion of a leachate control system designed to prevent leachate from migrating down the canyon. Previous testing established that: (1) the unweathered bedrock at depth beneath the site canyon has the required permeability of $1 \times 10^{-6}$ cm/s or less; and (2) the fine fractures in the surficial bedrock zone could be sealed by injecting chemical grout into the formation. A toe berm having the same maximum permeability overlies and is keyed into the grout curtain to complete the landfill toe barrier.

The grout curtain was constructed using the stage down method, with vertical holes generally drilled in three stages. Each stage was approximately 6 m deep, resulting in a total depth of about 18 m. The stage down method consists of drilling and grouting in successive stages from top to bottom. The first stage was drilled, washed with water under pressure to remove any cuttings, and then grouted under pressure by setting a packer near the surface of the hole. Each successive stage was completed by drilling the hole to an additional depth, washing with water, and then grouting by setting a packer near the bottom of the previously grouted stage.

The depth and final spacing of grout holes were determined from: (1) grouting take data obtained during construction; and (2) water pressure tests to determine the permeability of the ungrouted rock. Based on the water pressure test results, it became apparent that the specified permeability could be achieved if the spacing between grout holes was a maximum of 0.75 m.

While drilling the grout holes, grout was found to have penetrated radial distances from 1.5 to 4.5 m away from each hole, indicating that the planned single line of grout holes would provide an adequately thick grout curtain. After the grouting programme was complete, water pressure tests confirmed that the specified permeability of $1 \times 10^{-6}$ cm/s or less had been achieved.

*Leachate collection and removal.*   To comply with California state requirements, two ground-water monitoring wells and one leachate extraction well were installed near the grout curtain. Both monitoring wells, constructed with 5 cm diameter polyvinyl chloride (PVC) pipe, were installed down-gradient from the grout curtain. One well, located at the toe of the landfill cell, was drilled in weathered serpentinite from the ground surface to a depth of 15 m, and in fresh serpentinite to the final depth of 18 m. The second monitoring well, located at the mouth of the main site canyon, was drilled in sandy clay from the ground surface to a depth of 1.2 m and in weathered serpentinite to the final depth of about 16 m.

A leachate extraction well was installed about 6.5 m upgradient from the grout curtain centre (Fig. 1a). It extends through about 6 m of clay fill in the toe berm and then through serpentinite to the final depth of about 28.5 m. To complete the well, the drillers installed 15 cm diameter PVC casing, screened in the 7.5–25.5 m interval.

*Gas management.*   Gas migration control is provided by the naturally low permeability of the site bedrock and also by separation between the landfill and the site boundary. The refuse limit is set back a minimum of 61 m from the westerly and easterly site boundaries. If any structures are built within 125 m of the landfill, monitoring probes will be installed at the site boundary nearest the structure. The probes will be monitored monthly (during landfill operation), and if methane is detected, the need for a control system will be evaluated. After the landfill is completed, gas monitoring will be reviewed and adjusted based on previous monitoring results.

*Final cover.* Filling to final grades is designed to create a surface configuration compatible with adjacent natural terrain. To minimize erosion, however, final slopes will be no steeper than 3 : 1 (horizontal to vertical), and they will be no flatter than 3% to ensure enough slope for storm-water run-off. The slopes will be constructed with 6 m wide benches at 15 m vertical intervals.

A 1.2 m thick final soil cover will minimize water infiltration and provide for vegetation. Consisting of 0.3 m of intermediate soil cover, 0.3 m of clayey soil, and 0.6 m of vegetated soil, it will be planted with native California grasses and plant species.

## Proposed Bacona Road Landfill, Oregon

Bacona Road, located in northwestern Oregon, is a proposed landfill site designed to accept waste from the Portland, Oregon metropolitan area. As currently designed, the active landfill area will be 138 hectares, with 72 million cubic meters total volume available; the in-place refuse volume will be about 56 million cubic meters (CH2M-Hill, 1987). If only unprocessed refuse is landfilled, the site will have a design life of about 47 years; if a combination of unprocessed refuse and incinerator ash is landfilled, the design life will increase to about 60 years.

### Location Characteristics

In a draft feasibility study conducted for the Oregon Department of Environmental Quality (DEQ), CH2M-Hill (1987) described the Bacona Road site's characteristics, including geology, hydrogeology and climate. The following discussion draws from that study.

The Bacona Road site is located about 56 km west of Portland at an elevation of about 490 m above mean sea level (MSL). Due to the proximity of the Pacific Ocean and the Coast and Cascade mountain ranges, the site receives abundant precipitation which falls primarily during winter. Annual average rainfall is about 127 cm; for snowfall, the annual average totals about 90 cm.

Four geologic units underlie the site: (1) tertiary marine sedimentary rocks—primarily sandstone and siltstone; (2) Columbia River Basalt—typically broken, weathered, and eroded; (3) suspected landslide debris; and (4) alluvium. Landslide movement has affected much of the basalt exposed on site.

Depth to ground-water at the Bacona Road site ranges from 1.5 to 3.7 m. Data indicate that the site is a local ground-water discharge area and that all geologic units underlying the site are saturated. The hydraulic interconnections between the units appear to be complex and are little understood.

Based on the site characteristics noted above, the following design constraints were identified:

1. evidence of landsliding indicates potential slope instability;
2. depth to ground water is less than 6 meters. Excavations below 3 meters, therefore, could encounter ground water and require dewatering before liners are placed;
3. the near-surface alluvial soil could potentially be used as soil lining material. In addition, most of the site soils are suitable for use as daily, intermediate, and final cover.

## Design

In contrast to the other landfill sites discussed in this paper, the proposed Bacona Road Landfill includes complex design considerations. This is due to DEQ's regulatory requirements, developed primarily to address conditions (such as high precipitation and shallow ground water) specific to the Bacona Road site. DEQ has indicated that the landfill will be required to have a (1) a double liner; (2) a LCRS with leachate treatment; and (3) a leak detection system between the liners.

*Containment.* To prevent leachate from entering ground water beneath the landfill, the lining system will consist of two layers of a 1.5 mm thick high-density polyethylene (HDPE) geomembrane, each underlain by a layer of compacted clay (Fig. 2). Approximately 0.6 meter of sand will be placed over the upper liner to provide a leachate collection layer and to protect the liner from damage.

*Leachate Collection and Removal.* The liners will slope at least 2% to allow leachate to flow to perforated collection pipes placed above the liner. These pipes will be connected to larger, buried leachate lines, which will convey the leachate to a leachate pumping station. From this station, the leachate will be pumped to a pretreatment facility.

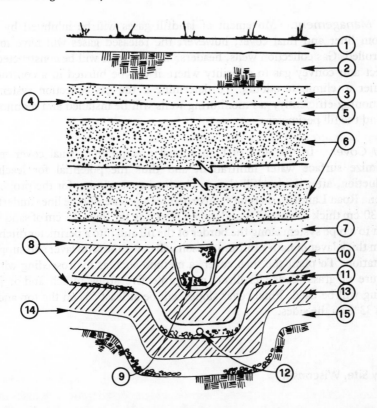

**Figure 2.** Bacona Road site landfill siting study—cover and liner detail (not to scale); from CH2M Hill (1987): (1) 15 cm of topsoil; (2) 45 cm of native soil; (3) 15 cm of sand; (4) geomembrane cap; (5) 30 cm of silt; (6) waste and daily cover; (7) 30 cm of sand or gravel; (8) geomembrane liners; (9) primary leachate collection pipe (15 cm perforated pipe) in drain rock wrapped in geotextile; (10) 45 cm of clay; (11) 15 cm of pea gravel; (12) secondary leachate collection pipe (10 cm perforated pipe); (13) 60 cm of clay; (14) geotextile; (15) 20 cm of drain rock.

The recommended leachate treatment and disposal system is a covered, activated sludge, biological pretreatment facility at the landfill site. The pretreated leachate would flow to an approximately 49 million litre, lined lagoon for flow equalization and storage. During the rainy season (November through April), the treated leachate would be pumped from the storage lagoon to a nearby public sewage treatment plant for further treatment and discharge to the Tualatin River.

*Gas Management.*   Movement of landfill gases will be inhibited by the bottom liner and final cover; however, the released gases will have to be controlled. Gas collection wells, headers, and eductors will be constructed to collect and convey gas to a facility where it can be burned in a controlled manner to eliminate odours. To determine if the gas collection system is functioning efficiently, gas monitoring wells will be installed and monitored around the fill perimeter.

*Final Cover.*   In this area of high annual precipitation, final cover must minimize surface water infiltration, and thus the potential for leachate production, after the landfill closes. The final cover design for the proposed Bacona Road Landfill, therefore, incorporates a geomembrane liner underlain by a 30 cm thick native soil foundation. Overlying the liner, 15 cm of sand will drain to a pipe system along the border of the fill and into a perimeter ditch. A 0.6 m thick layer of native soil and topsoil will cover the sand layer and support vegetation. To plant the final cover, a method such as hydroseeding with a mixture of grass seeds, fertilizer, and protective straw mulch and/or jute netting will be used. Final cover slopes will be at least 5% on the top and at most 33% on the sides.

## Libby Site, Wisconsin

The proposed Libby site is located south of Madison, Wisconsin near the western shore of Lake Waubesa. Designed as an area fill operation, the proposed fill area will occupy 24 hectares of a 71 hectare parcel. The proposed design capacity (refuse and daily cover) is about 2.48 million cubic meters. At an average refuse loading rate of about 750 tons/day, the site life will be about 7.5 years.

### Location Characteristics

Residuals Management Technology (RMT) (1986) investigated the proposed site from 1981 to 1983 to determine its suitability as a landfill location. The following descriptions are drawn largely from this investigation.

The Libby site is located in an area of ground moraine deposits that create a topography of gently rolling hills. Site elevations range between about 260 and 275 m above MSL. The climate is temperate with hot, humid summers and cold, dry winters. Annual average precipitation is about 79 cm with about 75% occurring from March through September and 25% occuring from October to February. Actual evapotranspiration is about 50 cm.

The upper bedrock unit in this area is Cambrian sandstone, which lies about 60 m below ground surface. Overlying unconsolidated glacial deposits at the site have been identified as:

1. a 0.3–3 m thick surface layer of silty clay;
2. an underlying 0–6 m thick layer of silty sand till that extends throughout most of the site (vertical laboratory permeability tests on compacted till samples indicated the permeability ranges from $2.6 \times 10^{-5}$ cm/s for samples at natural moisture to $1.2 \times 10^{-7}$ cm/s for samples at optimum moisture);
3. a $0-+12$ m thick sandy outwash deposit below the till;
4. a 0.6–3.5 meter thick lacustrine clay encountered within the till and outwash deposits.

In reviewing the site investigations, the Wisconsin Department of Natural Resources (DNR) summarized the advantages of the Libby site (RMT 1986):

1. great depth to bedrock;
2. good quality clays;
3. the silt and clay content of the sandy till soils;
4. proximity to sewers and a wastewater treatment facility for treatment of leachate.

The DNR has expressed concern about the shallow ground water, located at depths ranging from 2.4 to 15.8 meters below ground surface. Because the DNR will require a minimum 3 meter separation between the water table and the bottom of a clay liner, portions of the proposed site could require backfilling.

Eight private wells are located within 365 meter of the proposed site; therefore, the site does not meet one of Wisconsin's location criteria. Based on the area hydrogeology, however, promoters of the site have requested an exemption from this criterion.

## Design

The design concepts for the Libby site were based upon the hydrogeologic and soils information developed in the site investigation; information on ecological resources and land use; and concerns expressed by citizens at public meetings. Primarily to protect ground-water, the design incorporates several containment features:

1. a compacted clay liner;
2. a full leachate collection system beneath all refuse;
3. a minimum 3 m separation between the bottom of the clay liner and the ground-water table.

*Containment.* The proposed landfill liner will consist of 1.5 m of compacted clay on the facility's base and side-walls. On-site clay materials used in its construction will contain a minimum clay size fraction of 25% and have a maximum permeability of $1 \times 10^{-7}$ cm/s, a minimum liquid limit of 30%, a minimum plasticity index of 15%, and a minimum compaction of 90% Modified Proctor. Base grades will maintain a 4% minimum slope to allow leachate to flow to the leachate collection system.

*Leachate Collection and Removal.* Before filling with refuse, a 0.3 m thick granular blanket will be placed over the clay liner to increase leachate collection efficiency. The collection system itself will consist of perforated 15 cm diameter PVC pipes.

Leachate collected in the pipes will flow to a header pipe and exit the site in a clay encased transfer pipe that gravity-drains to external manholes or a lift station. The leachate will flow by gravity from the manhole or be pumped from the station to an existing manhole in a double-piped transfer pipe. At each location where the transport line connects with the existing sewer line, a 37 850 l holding tank will be installed to provide storage in case discharge to the sewer line must be halted. Collected leachate will be transferred to the nearby waste-water treatment plant during the site life and for 20 years after closure.

*Gas Management.* Gas control will consist of a passive gas venting system composed of: (1) surface gas vents and trenches; and (2) a sidewall gas venting system. The surface vents will include a total of 32 vents installed over the site. Spaced approximately 91 m apart, they will be connected with 1.5 m deep gravel trenches. Gas vents will also be installed near the break in slope of the final grades around the entire site.

Constructed as part of each phase development, the sidewall gas venting system will consist of a 0.6 m thick granular fill layer on the sidewall of the entire site. At approximately 76 m spacing, 28 15 cm diameter, perforated PVC vent pipes will be installed in the layer. This system provides better control for gas generated at various depths that may not be able to move up through the fill because of compacted refuse, cover layers, and/or moisture pockets.

At least quarterly, methane and oxygen will be monitored in eight wells installed at the property line. In addition, the site and vicinity will be routinely visually inspected for vegetation that may be stressed from potential gas generation and migration. If an active control system is warranted, the sidewall vents could become part of such a system by connecting them to a header and blower system. Valves could be installed at necessary intervals to control gas migration in a particular area.

Before the landfill is completed, a study will be made to assess the potential for gas migration. If the study indicates that additional gas control is needed, the appropriate section of the sidewall gas collection system will be activated or a separate system will be installed outside the site to prevent gas migration.

*Final cover.* Final cover will consist of 0.6 m of compacted clay capped with 15 cm of topsoil. Sufficient soil for berms and cover material will be obtained from: (1) excavating landfill base grades; (2) on-site borrow areas; and (3) an off-site borrow area. Designed to return the site to recreational use and/or open space, the site's final topography will blend into the surrounding topography as closely as possible. To minimize ponding and erosion, final grades will be 5% minimum on the top with side slopes of 4 : 1.

## CONCLUSION

The preceding case studies provide an overview of current solid waste management design practices in the United States. It is clear that the overall picture presented is one containing both similarities and contrasts.

Similarities in the design approach from one state to another lie primarily in intent: the desire to protect water quality, air quality, and human health. Secondarily, they are similar in that they deal with a relatively fixed set of natural physical parameters: local geology, hydrogeology, precipitation, site location, and so forth.

The dissimilarities in solid waste landfill design approaches are primarily due to human considerations, rather than nature's dictates. These factors include, but are not limited to, state and local regulations, available technology, suitable economics, and equipment.

It is the complex variability and interplay of these factors that combine to make the proper design of solid waste landfills so challenging. The final goal is the protection of human health and the environment; it is up to the engineer to use knowledge and experience to select the tools and techniques that allow this goal to be achieved.

## REFERENCES

CH2M Hill (1987). 'Draft Feasibility Study Report, Portland Metropolitan Area, Landfill Siting Project, Bacona Road Site'. Prepared for the Oregon Department of Environmental Quality.

EMCON Associates (1983). 'Hydrogeologic Investigation and Conceptual Landfill Design Study, Kirby Canyon Class II-2 Disposal Site, San Jose, California'.

EMCON Associates (1986). 'Ground-water Monitoring and Leachate Extraction Well Installations', Kirby Canyon Landfill. May 1986.

EMCON Associates (1986). 'Construction Report — Cell 1 Toe Berm, Kirby Canyon Sanitary Landfill, San Jose, California', 10 June 1986.

Environmental Reporter (1984). 'California Solid Waste Management Regulations: California Administrative Code, Title 14, Natural Resources, Division 7 — Solid Waste Management Board, Chapters 1 through 5 and 9 (as amended through February 3, 1983)', 4 May 1984. The Bureau of National Affairs, Washington DC.

Environmental Reporter (1986). 'Oregon Solid Waste Management Regulations: Oregon Administrative Rules, Chapter 340 — Department of Environmental Quality, Division 61 — Solid Waste Management in General', 28 February 1986. The Bureau of National Affairs, Washington DC.

Environmental Reporter (1986). 'Wisconsin Solid Waste Management Regulations: Wisconsin Administrative Code, Department of Natural Resources Chapter 180 — Solid Waste Management', 3 October 1986. The Bureau of National Affairs, Washington DC.

Residuals Management Technology (1986). 'Appendix B — Department of Natural Resources Review of Initial Site Report (3 December 1981) and Department of Natural Resources Review of Feasibility Study (8 October 1985)'. Madison Landfills, Inc., Libby Site Feasibility Report.

Residuals Management Technology (1986). 'Appendix O — Water Balance'. Madison Landfills, Inc., Libby Site Feasibility Report.

US Environmental Protection Agency (1987), Working Draft, Subtitle D Criteria, 12 March 1987.

Van Heuit, R.E. and Leach, R.J. (1984). 'Design, Construction, and Operating Provisions, Kirby Canyon Sanitary Landfill, San Jose, California'. EMCON Associates, San Jose, California.

# Index

577